Geography

2nd Edition

by Jerry T. Mitchell

Geography For Dummies® 2nd Edition

Published by: **John Wiley & Sons, Inc.,** 111 River Street, Hoboken, NJ 07030-5774, www.wiley.com

Copyright © 2022 by John Wiley & Sons, Inc., Hoboken, New Jersey

Published simultaneously in Canada

No part of this publication may be reproduced, stored in a retrieval system or transmitted in any form or by any means, electronic, mechanical, photocopying, recording, scanning or otherwise, except as permitted under Sections 107 or 108 of the 1976 United States Copyright Act, without the prior written permission of the Publisher. Requests to the Publisher for permission should be addressed to the Permissions Department, John Wiley & Sons, Inc., 111 River Street, Hoboken, NJ 07030, (201) 748-6011, fax (201) 748-6008, or online at http://www.wiley.com/go/permissions.

Trademarks: Wiley, For Dummies, the Dummies Man logo, Dummies.com, Making Everything Easier, and related trade dress are trademarks or registered trademarks of John Wiley & Sons, Inc. and may not be used without written permission. All other trademarks are the property of their respective owners. John Wiley & Sons, Inc. is not associated with any product or vendor mentioned in this book.

LIMIT OF LIABILITY/DISCLAIMER OF WARRANTY: WHILE THE PUBLISHER AND AUTHORS HAVE USED THEIR BEST EFFORTS IN PREPARING THIS WORK, THEY MAKE NO REPRESENTATIONS OR WARRANTIES WITH RESPECT TO THE ACCURACY OR COMPLETENESS OF THE CONTENTS OF THIS WORK AND SPECIFICALLY DISCLAIM ALL WARRANTIES, INCLUDING WITHOUT LIMITATION ANY IMPLIED WARRANTIES OF MERCHANTABILITY OR FITNESS FOR A PARTICULAR PURPOSE. NO WARRANTY MAY BE CREATED OR EXTENDED BY SALES REPRESENTATIVES, WRITTEN SALES MATERIALS OR PROMOTIONAL STATEMENTS FOR THIS WORK. THE FACT THAT AN ORGANIZATION, WEBSITE, OR PRODUCT IS REFERRED TO IN THIS WORK AS A CITATION AND/OR POTENTIAL SOURCE OF FURTHER INFORMATION DOES NOT MEAN THAT THE PUBLISHER AND AUTHORS ENDORSE THE INFORMATION OR SERVICES THE ORGANIZATION, WEBSITE, OR PRODUCT MAY PROVIDE OR RECOMMENDATIONS IT MAY MAKE. THIS WORK IS SOLD WITH THE UNDERSTANDING THAT THE PUBLISHER IS NOT ENGAGED IN RENDERING PROFESSIONAL SERVICES. THE ADVICE AND STRATEGIES CONTAINED HEREIN MAY NOT BE SUITABLE FOR YOUR SITUATION. YOU SHOULD CONSULT WITH A SPECIALIST WHERE APPROPRIATE. FURTHER, READERS SHOULD BE AWARE THAT WEBSITES LISTED IN THIS WORK MAY HAVE CHANGED OR DISAPPEARED BETWEEN WHEN THIS WORK WAS WRITTEN AND WHEN IT IS READ. NEITHER THE PUBLISHER NOR AUTHORS SHALL BE LIABLE FOR ANY LOSS OF PROFIT OR ANY OTHER COMMERCIAL DAMAGES, INCLUDING BUT NOT LIMITED TO SPECIAL, INCIDENTAL, CONSEQUENTIAL, OR OTHER DAMAGES.

For general information on our other products and services, please contact our Customer Care Department within the U.S. at 877-762-2974, outside the U.S. at 317-572-3993, or fax 317-572-4002. For technical support, please visit www.wiley.com/techsupport.

Wiley publishes in a variety of print and electronic formats and by print-on-demand. Some material included with standard print versions of this book may not be included in e-books or in print-on-demand. If this book refers to media such as a CD or DVD that is not included in the version you purchased, you may download this material at http://booksupport.wiley.com. For more information about Wiley products, visit www.wiley.com.

Library of Congress Control Number: 2022931558

ISBN: 978-1-119-86712-8 (pbk); ISBN 978-1-119-86713-5 (ebk); ISBN 978-1-119-86714-2 (ebk)

Contents at a Glance

Introduction 1

Part 1: Getting Grounded: The Geographic Basics 5
- **CHAPTER 1:** Geography: The Why of Where and Why You Should Care............ 7
- **CHAPTER 2:** Thinking Like a Geographer 19
- **CHAPTER 3:** Lining Up Locations 29
- **CHAPTER 4:** Truthiness in Mapping 41
- **CHAPTER 5:** Telling a Spatial Story 59

Part 2: Let's Get Physical: Land, Water, and Air 79
- **CHAPTER 6:** Shape-shifting Earth 81
- **CHAPTER 7:** A Nip and a Tuck: Giving Earth a Facelift......................... 103
- **CHAPTER 8:** Making a Splash on Earth 119
- **CHAPTER 9:** Warming Up and Chilling Out: Why Climates Happen.............. 137
- **CHAPTER 10:** Connecting Climates and Vegetation 159

Part 3: Peopling the Planet 179
- **CHAPTER 11:** Nobody Here But A Few Billion Friends 181
- **CHAPTER 12:** Shift Happens: Migration... 203
- **CHAPTER 13:** Culture: The Way We Live 219
- **CHAPTER 14:** Good Fences Make Good Neighbors............................... 239

Part 4: Putting the Planet to Use 261
- **CHAPTER 15:** Takin' Care of Business 263
- **CHAPTER 16:** Earth's Resources: Always Hungry for More 283
- **CHAPTER 17:** Downtown to the 'Burbs: Urban Geography 301
- **CHAPTER 18:** Only One Home: Impacts on the Environment 323

Part 5: The Part of Tens 343
- **CHAPTER 19:** Ten Organizations for Geography in Action..................... 345
- **CHAPTER 20:** Ten Interesting Career Paths for Geographers 351
- **CHAPTER 21:** Ten Things You Can Forget 357
- **CHAPTER 22:** Ten Great Places for Online Geography 365

Index 371

Table of Contents

INTRODUCTION 1
    About This Book. 2
    Foolish Assumptions. 2
    Icons Used in This Book 3
    Beyond the Book 4
    Where to Go from Here 4

PART 1: GETTING GROUNDED: THE GEOGRAPHIC BASICS 5

CHAPTER 1: Geography: The Why of Where and Why You Should Care 7
    Geography: Making Sense of it All 8
        From ancient roots 8
        . . . To modern discipline 9
    Exposing Misconceptions: More Than Maps and Trivia. 11
    The Geographic Advantage 11
        What is the capital city of Indonesia? 12
        Why is Jakarta the capital of Indonesia? 12
    Getting to the Essentials. 13
        Where things are in the world: The world in spatial terms 14
        What locations are like: Places and regions 15
        Why things are the way they are: Physical systems 16
        Giving that human touch: Human systems 16
        Interacting with the world around us: Environment and society ... 17
        Putting geography to use: Uses of geography 18

CHAPTER 2: Thinking Like a Geographer 19
    Changing the Way You Think — Geographically. 20
    Case Study #1: Where Something is Located 21
        A fraction of its former self 22
        Where lions hang out 23
        What gives with grasslands? 23
        Extinction made easy 24
        Fewer lions? So what? 25
        Summing up. 26
    Case Study #2: Where Something Should be Located. 26
        Summing up. 27
        Looking ahead 28

CHAPTER 3: **Lining Up Locations** 29
    Welcome to Gridville 29
    Feeling Kind of Square 30
    Telling Someone Where to Go 31
        Relative location 31
        Absolute location 32
        The best location to use 32
    The Global Grid: Hip, Hip, Hipparchus! 32
        Avoiding gridlock 33
        The naming game 34
    Getting Lined Up 36
        Latitude 36
        Longitude 38
        Graticule 39
        Minutes and seconds that don't tick away 40

CHAPTER 4: **Truthiness in Mapping** 41
    Seeing the Light: Map Projections 42
    Realizing Exactly How Flat Maps Lie 44
        Singapore, please. And step on it! 45
        Wading through lies in search of the truth 47
    Isn't there a truthful map anywhere? 48
        The one and only honest map: The globe! 49
        Honesty is the best policy, except 49
        Telling the truth, but telling it skewed 50
    Different Strokes for Different Folks: A World of Projections 51
        All in the (map) family 52
        Five noteworthy liars 53
    Mapping a Cartographic Controversy! 57

CHAPTER 5: **Telling a Spatial Story** 59
    Why We Need Tal(l) Dogs 60
    Taking It to Scale 61
        Going the distance 61
        Comparing Earth at different scales 62
    Showing the Ups and Downs: Topography 65
        Spot heights 65
        Contour lines 66
        Shading or Color 66
    Using Symbols to Tell the Story 67
        Point symbols 68
        Line symbols 68
        Area symbols 70

New Ways of Seeing: How Technology has Changed
　　　How we Make and Use Maps..................................72
　　　Geographic Information Systems............................72
　　　　　Global Positioning Systems...........................74
　　　　　Remote sensing......................................75
　　　Making Maps Yourself!...................................77

PART 2: LET'S GET PHYSICAL: LAND, WATER, AND AIR ... 79

CHAPTER 6: Shape-shifting Earth 81
　　　Starting at the Bottom: Inside Earth.....................82
　　　Moving Continents: Big Pieces of a Big Puzzle............84
　　　　　Where have you gone, Gondwanaland?...................85
　　　　　Alfred Wegener, mover and shaker....................85
　　　　　Puzzle solved!.....................................86
　　　Getting Down to Theory..................................86
　　　Making Mountains Out of Molehills.......................88
　　　　　Folding the crust..................................88
　　　　　Whose "fault" is it?...............................90
　　　　　Plate tectonics: A four-letter word!................91
　　　Experiencing Earthquakes: Shake, Rattle and Roll!........92
　　　　　Splitsville in California..........................92
　　　　　People at risk.....................................93
　　　　　How earthquakes kill and maim......................94
　　　　　A matter of magnitude..............................96
　　　Subducting Plates: Volcano Makers.......................98
　　　　　"The Ring of Fire".................................98
　　　　　Subduction: Another four-letter word?...............99
　　　Categorizing Tectonic Processes........................102

CHAPTER 7: A Nip and a Tuck: Giving Earth a Facelift 103
　　　Getting Carried Away...................................104
　　　　　Weathering Earth..................................105
　　　　　Wasting away.....................................108
　　　Changing the Landscape.................................109
　　　　　Staying grounded: Gravity transfer.................109
　　　　　Going with the flow: Water........................110
　　　　　The chill factor: Glaciers........................115
　　　　　Making a deposit: Wind............................117

CHAPTER 8: Making a Splash on Earth 119
　　　Taking the Plunge: Global Water Supply.................120
　　　　　Those ice caps are really cool!...................121
　　　　　Getting out: Oceans, seas, gulfs, and bays.........122
　　　　　Coming inland: Lakes..............................123

Shaping Our World: Oceans...124
    Going where the action is: The continental shelves...124
    Claiming ocean ownership...126
    Getting a rise out of oceans...128
Getting Fresh with Water...129
    The stages of the water cycle...131
    Run-off: Going with the flow...132
    Infiltration: Out of sight, not out of mind...132
    Good to the very last drop...134

CHAPTER 9: Warming Up and Chilling Out: Why Climates Happen...137

Getting a Grip on Climate...138
Playing the Angles...139
    Making hot and cold...139
    Making rain and snow...140
Tilt-a-World: The Reasons for the Seasons...141
    Special lines of latitude...141
    Defining the seasons...142
    Special lines of latitude revisited...144
Hot or Cold? Adjust Your Altitude...145
    Warming the atmosphere...145
    Weighty matter...145
    Seeing (and feeling) is believing...145
    The lapse rate...146
    Windward slope, leeward slope...147
Gaining Heat, Losing Heat...148
    Afternoon versus evening...148
    Summer versus winter...149
Oh, How the Wind Blows...150
Going with the Flow: Ocean Currents...150
    Warm currents, cold currents...151
    Going against the norm: El Niño and La Niña...152
Living Under Pressure...154
    Pressure belts...155
    Monsoons...155

CHAPTER 10: Connecting Climates and Vegetation...159

Giving Class to Climates...160
Mixing Sun and Rain: Humid Tropical Climates...162
    Tropical rainforest...163
    Tropical monsoon...165
    Savanna (tropical wet and dry)...165

    Going to Extremes: Dry Climates 167
        Desert 168
        Semi-desert (steppe)..169
    Enjoying the In-between: Humid Mesothermal Climates 172
        Humid subtropical..172
        Mediterranean..172
        Marine west coast 173
    Cooling Off: Humid Microthermal Climates 174
        Humid continental..175
        Subarctic..175
    Dropping Below Freezing: Polar Climates 176
        Tundra...176
        Ice cap 178

PART 3: PEOPLING THE PLANET 179

CHAPTER 11: Nobody Here But A Few Billion Friends 181
    Going by the Numbers 182
        Opportunity for livelihood 183
        Urban growth...185
    Going Ballistic: Population Growth 185
    Checking Behind the Curve: Population Change 187
        Dealing with births and deaths: Natural increase 187
        Increasing for a reason: The demographic transition model 189
        Making connections 193
    Considering "Overpopulation"..196
        Neo-Malthusians 197
        Cornucopians..199

CHAPTER 12: Shift Happens: Migration...................................203
    Populating the Planet 204
        Bridging the oceans...204
        Voyaging afar 205
        Making colonial connections 206
        Forcing involuntary migration 207
    Choosing to Migrate 207
        Coming to America 208
        Migrating at home...210
        Relocating within America 211
    Giving a Good Impression 214
        Playing the mental game 214
        Getting an image adjustment 216
        Putting your best image forward 216

CHAPTER 13: Culture: The Way We Live 219
    Being Different Thousands of Times Over 220
        Counting cultural diversity 221
        Isolating people 222
        Adapting to new surroundings 222
    Spreading the Word on Culture 223
        Relocating one's culture 224
        Coming down with culture 224
        Doing what the big boys do 226
    Calling a Halt: Barrier Effects 226
        Getting physical 227
        Socializing effects 229
    Getting Religion: How It Moves and Grows 230
        Putting diffusion to work 231
        Getting effects into action 232
        Creating local character 232
    Getting in a Word about Language 234
        Diffusing languages 235
        Checking the physical effects 236
        Playing the landscape naming game 236
    Creating a Single Global Culture 237
        Promoting cultural divergence 238
        Promoting cultural convergence 238

CHAPTER 14: Good Fences Make Good Neighbors 239
    Drawing and Re-Drawing the Boundaries of the World 240
    Typecasting Boundary Lines 241
        Ethnic boundaries 242
        Natural (physical) boundaries 243
        Geometric boundaries 244
    Living with the Consequences 245
        Ethnic intrigues 245
        Positional disputes 248
        Functional disputes 250
        Resource disputes 251
        Land-locked states 252
        Questions of size and shape 254
    Drawing Electoral District Boundaries 256
        Gerrymandering: Rigging the outcome 257
        Meeting the letter and spirit of the law 259

PART 4: PUTTING THE PLANET TO USE 261

CHAPTER 15: **Takin' Care of Business** 263

Categorizing Economic Activity 264
- Primary activities 264
- Secondary activities..265
- Tertiary activities 265
- Quaternary activities.......................................266
- Activity distribution around the world 266

Putting Economic Systems into Place..............................267
- Subsistence economies 267
- Commercial economies 268

Understanding Location Factors...................................269
- Proximity to raw material(s)................................272
- Proximity to market(s) 272
- Cost of labor...274
- Accessibility 275
- Cost of land/rent 279
- Taxes 279
- Climate 280

CHAPTER 16: **Earth's Resources: Always Hungry for More** 283

Defining Resources and Assessing Their Importance 284
- The central role of culture 284
- Culture change, resource change 285
- Resources and power 286
- Resources and wealth.......................................287

Differing Life Spans: Which Resources Are Here Today
or Gone Tomorrow 288
- Non-renewable resources 289
- Renewable resources 294
- Perennial resources 295

Trading-off Resources: The Consequences of Resource Use 298

CHAPTER 17: **Downtown to the 'Burbs: Urban Geography** 301

Studying the Urban Scene 302
Getting a Global Perspective 303
Getting Started: Urban Hearths 305
Finding Sites for Cities 306
- Confluence 307
- Protected harbor...307
- Head of navigation 308
- Defensive sites..308

Table of Contents **xi**

Getting Big: Urban Growth............................309
    Rural-to-urban migration........................309
    Changing means of transportation....................310
    Automobile ownership..........................311
    Low-cost fuel..............................311
    Home mortgage deductibility......................312
Looking Inside the City..............................312
    The central business district (CBD)..................313
    Residential areas............................314
Leaving Downtown, Living Downtown.....................317
    Moving out of downtown........................317
    Moving back downtown........................319
Facing up to Environmental Issues.......................320

CHAPTER 18: Only One Home: Impacts on the Environment....323

Grasping the Basics — Environmentally Speaking...............325
Contributing Factors: Pollution on the Move.................326
    Making an impact...........................326
    Spreading the mess..........................326
    Focusing on food chains........................329
Going Global: Environmental Issues Affecting Us All..............332
    Deforestation.............................333
    Biodiversity loss............................333
    Soil degradation............................333
    Ocean acidification..........................334
    Overfishing..............................335
    Acid precipitation...........................335
    Climate change............................339
Taking on the Challenges of Tomorrow.....................342

PART 5: THE PART OF TENS..........................343

CHAPTER 19: Ten Organizations for Geography in Action....345

American Association of Geographers (AAG).................345
American Geographical Society (AGS).....................346
National Aeronautics and Space Administration (NASA)............346
National Council for Geographic Education (NCGE)..............347
National Geographic Society (NGS)......................347
National Oceanic and Atmospheric Administration (NOAA).........348
Population Reference Bureau (PRB).....................348
Royal Geographical Society (RGS)......................349
United States Census Bureau.........................349
United States Geological Survey (USGS)...................350

CHAPTER 20: Ten Interesting Career Paths for Geographers 351
    Area Specialist 351
    Educator 352
    Environmental Manager...352
    GIS Technician 353
    Health Services Planner 353
    Location Analyst..353
    Market Analyst...354
    Remote Sensing Analyst..354
    Transportation Planner 355
    Urban Planner 355

CHAPTER 21: Ten Things You Can Forget 357
    The Bermuda Triangle 357
    Cold Canadian Air 358
    "Coming Out of Nowhere".......................................359
    "The Continent"...359
    The Democratic Republic of 360
    The Flat Earth Society 360
    Land of the Midnight Sun..361
    "The Rain in Spain Stays Mainly on the Plain" 361
    The Seven Seas 362
    Tropical Paradise...362

CHAPTER 22: Ten Great Places for Online Geography 365
    Any County/Local GIS Department 365
    Geocaching..366
    Geoguessr...366
    Geoinquiries...367
    Google Earth...367
    Google Lit Trips 368
    Perry-Castañeda Library Map Collection 368
    World Bank Open Data...369
    World Factbook 369
    Your State's Geographic Alliance................................369

INDEX..371

Introduction

Has there been a better teacher of geography than COVID-19? I don't mean that as a joke. That deadly serious disease showed how interconnected our world has become. Within a matter of months, people worldwide were exposed to something that they could not see that up-ended everything else that they could.

I was in Sweden in March 2020 when the world was shut down. I was traveling with colleagues to several universities to explore partnerships, discuss joint research opportunities, and so on. Being in charge of 14 other people is a challenge by itself, but managing them after being forced home by one's government as airlines cancel flights is quite another. Gothenburg to Charlotte via Frankfort was the original plan. Then Frankfort fell out. Agents suggested routes home via Brazil (!) and then Canada. Finally, an option opened through Brussels, but a stop in Washington D.C. was necessary as the government declared only a few airports could accept international travelers. How we organize travel networks across space — and what can happen when links within them fail — is of crucial geographic importance, as you can see.

Later we learned of the intricacy of spatial business connections as supply chains were stressed, and then political scale (who is responsible for public health decisions? The state or federal government?), and then spatial ethics as well-off countries secured vaccine supplies in numbers far greater than their poorer neighbors. COVID-19 laid bare how understanding *where*, *why there*, and *why should I care* — the essence of geographic thinking — is paramount in living on 21st century Earth.

Even at its worst — such as in the middle of a global pandemic — Earth is a very fascinating place. The purpose of geography is to convey the wonderment of it all and to explain how the world works. In this book I emphasize the interactions between the various things that characterize Earth's physical and human features and how they relate to everyday life.

Hopefully, the chapters that follow will instill in you some measure of the excitement I have for understanding our home, and whet your appetite for more.

About This Book

Introductory books on geography generally come in two varieties. This one takes a *topical* approach to the subject. That means the chapters focus on topics of interest to geography, such as maps, climate, population, and culture. I wanted this book to focus on the key concepts of geography and introduce you to a wide range of geographic information. Basically, I thought those goals could best be achieved by taking a topical approach.

The alternative was to take a *regional* approach to geography, which is like a world tour. You know what I mean, right? Chapter 4: Europe. Chapter 5: Africa. And so forth. In all candor, I didn't think I could give you a decent world tour in the allotted pages. Besides, books like that are already on the market, so why reinvent the wheel?

More importantly, I wanted *Geography For Dummies* to emphasize geography rather than the world *per se*. That may cause you to say, "Wait a minute! Isn't geography all about the world?" The answer is yes, but in a larger sense, geography is about a whole lot more. Specifically, it's about concepts and processes and connections between things, plus maps and tools and perspectives that combine individual "world facts" and give you big pictures that are so much more meaningful than their myriad components.

Parenthetically, there's a curious thing about those geography-as-world-tour books. They all seem to start by telling you geography is so much more than facts about the world, and then spend 350 pages telling you facts about the world.

Foolish Assumptions

I'm going to assume that you are an average person who is curious about the world but who just happens to have a limited background in geography. And I firmly believe "average" means intelligent, so nothing is out of bounds because of the gray stuff between your ears. Instead, in my view, you are completely capable of digesting the stuff of geography. You may be 14, or 44, or 84. It doesn't matter. As far as I am concerned, you're ready for prime-time geography. Please understand I'm not talking wimpy stuff like "What's the capital of Nevada?" No way. I'm talking big league stuff like how you can have a rainforest on one side of a mountain range and a desert on the other; or how to choose a good location for a wind farm; or how ocean currents help to determine the geography of climates.

I'm also going to assume that, generally speaking, you know your way around the world. Thus, when you see terms like Pacific Ocean, Nile River, Europe, or Japan, some kind of mental map pops up inside your head and allows you to "see" where they are located. On the other hand, when you meet up with terms like Burkina Faso or Myanmar, you may need some outside help. For that reason, it will be helpful to have an atlas or online reference handy.

Finally, if this book were a beer, then I'm assuming you went to your bookstore to pick up some Geography Lite. That is, you want the real thing, but figure you don't need all the calories. One of my goals is to make this book a painless — and indeed a pleasurable — experience. A lite-hearted read, if you will, that also communicates some serious geography and leaves you with a well-rounded exposure to the subject. If that sounds about right, then I invite you to keep reading.

Icons Used in This Book

From time to time you will encounter icons in the margin of the text. The purpose of these icons is to alert you to the presence of something that is comparatively noteworthy amidst the passing prose. That may be something I regard as particularly important, or something you may wish to take your time to think about, or something you may wish to skip. In any event, here are the icons and their meanings.

This icon identifies a major concept or helps to make sense of something. Likewise, you will sometimes come across a sentence or phrase that captures the essence of a principle or the theme of a chapter or of the entire book. Those kinds of tidbits are especially worth remembering and are identified by this icon. If you take away from this book only the information flagged with this icon, you'll have the basics of geography in your pocket.

Like many subjects, geography contains some specialized and perhaps arcane vocabulary terms that cause normal, well-adjusted people like you to scratch their heads. I could bypass this geo-jargon altogether, but then you really wouldn't be discovering more about geography, would you? In addition, geography involves elements of math, science, technology, ecology, modeling, and other technical stuff. Some will show up in this book because they are relevant to a well-rounded geographic education even at this introductory level. I do appreciate, however, that some people may find these a bit too complicated, so this icon alerts you to the presence of such technical stuff. You can skip it if you wish.

Some aspects of geography are a little involved, so it's always nice to encounter information that helps you simplify a process or make things easier to comprehend. Those are the kinds of items this icon pinpoints.

Beyond the Book

In addition to the abundance of information and guidance related to geography that we provide in this book, you get access to even more help and information online at *Dummies.com*. Check out this book's online Cheat Sheet. Just go to *www.dummies.com* and search for "Geography For Dummies Cheat Sheet."

Where to Go from Here

I recommend you read this book from start to finish as you would a novel. To some extent, geographic knowledge is cumulative. That is, there are basic concepts and information that provide a foundation for understanding other concepts and information.

Accordingly, the parts and chapters of this book follow a certain logical progression. In short, I do believe the content of this book will make more sense to you if you read this volume from start to finish.

However, if you wish, you can dive into chapters at random — each chapter is set up to be self-contained. The choice is yours!

1 Getting Grounded: The Geographic Basics

IN THIS PART . . .

Each and every academic discipline has its own particular and peculiar subject matter. Geography is no exception, but my, how things have changed!

For the longest period, geography was concerned primarily with mapping the world and acquiring facts about places. It has since become a much more analytical pursuit. Thus, the time-honored imperative to know where things are located is complemented by an equally strong (if not stronger) desire to know why they occur where they do. Geography is an applied discipline, seeking to identify the best, efficient, and fair locations for all sorts of human activities. Importantly, geography also works to understand places of difference and inequity with an eye toward making life better there — for people and the environment.

In this part, you will discover the key concepts and methods of contemporary geography as well as the principal tools and techniques of the trade. Among other things you will see how exciting technologies are giving geographers unprecedented perspectives on where and why.

IN THIS CHAPTER

» Contemplating a complex planet

» Unearthing myths

» Tracing the ancient roots of geography to the modern discipline

» Finding a new way to look at geography

» Going over some basic concepts

Chapter 1
Geography: The Why of Where and Why You Should Care

"We should cross *here*."

Staring at the broad expanse of the Galana River in southern Kenya, my response to our guide was to tilt my head to the side and say incredulously, "Are you sure?" After all, I could look further upstream and see that the river was narrower and there were some rocks we could use to hop across. Why on Earth should we cross at the widest and deepest part? I don't mind getting wet, but a chest-deep slog just didn't make sense to me.

"Well, we could cross up there," our guide said while pointing toward the rocks, "but that's also where the crocodiles hang out. We will do better down here to walk in a group, splashing as we go to mimic a large elephant." And so we did, and no one in our group became a croc's lunch.

What we discover from this anecdote is that what makes sense in one place — say, something as simple as crossing a river on a set of rocks — is a really bad decision somewhere else. *Place matters.* In this case, it really was a life-or-death situation.

That short story should also make it quite plain that you live on a very interesting planet. Earth is a world of never-ending variety — mountains and plains, deserts and forests, oceans and croc-infested rivers. If, as Shakespeare once wrote, "All the world's a stage," then one could hardly imagine a greater range of sets and scenery than exists on planet Earth.

You are an actor on that stage, and you are not alone. The entire cast numbers nearly 8 billion, and they are as diverse as their Earthly stage. They practice dozens of religions, speak many hundreds of languages, and display thousands of cultures. They live in scattered farmhouses, large cities, and every size settlement in between. They practice every kind of livelihood imaginable and, in innumerable ways great and small, have interacted with and changed the natural environment forever.

So "interesting planet" and "never-ending variety" turn out to be code for "complex." Truly, this is a complex world in which no two areas are exactly alike. On the one hand, this complexity makes for a very fascinating planet. But on the other hand, the prospect of learning all about this complexity can be overwhelming, or at least sometimes seems to be. Fortunately, one subject seeks to make sense of it all and, usually, does a pretty good job: Geography.

Geography: Making Sense of it All

People are fascinated by the world in which they live. They want to know what it's like and why it is the way it is. Most importantly, they want to understand their place in it. Geography satisfies this curiosity and provides practical knowledge and skills that people find useful in their personal and professional lives. This is nothing new.

From ancient roots . . .

REMEMBER

Geography comes from two ancient Greek words: *ge,* meaning "Earth," and *graphe,* meaning "to describe." So, when the ancient Greeks practiced geography, they described Earth. Stated less literally, they noted the location of things, recorded the characteristics of areas near and far, and used that information in matters of trade, commerce, communication, and administration.

Disputed paternity

A Greek named Eratosthenes (died about 192 B.C.) is sometimes called the "Father of Geography" since he coined the word "geography." The Greeks themselves called Homer the "Father of Geography" because his epic poem, *Odyssey*, written about a thousand years before Eratosthenes was born, is the oldest account of the fringe of the Greek world. In addition to these gentlemen, at least two other men have been named "Father of Geography," all of which suggests a very interesting paternity suit. But I digress. That the story goes back to the days of the Greeks tells us that geography is a very old subject. This is not to say that others, say Arabs or the Chinese, were not also thinking about how to describe Earth. People of every age and culture have sought to know and understand their immediate surroundings and the world beyond. They stood at the edges of seas and imagined distant shores. They wondered what lies on the other side of a mountain or beyond the horizon. Ultimately, of course, they acted upon those speculations. They explored. They left old lands and occupied new lands. And as a result, millennia later, explorers such as Columbus, Magellan, and others found humans almost everywhere they went.

Links to exploration

REMEMBER

Geographers from ancient Greece through the 19th century were largely devoted to exploring the world, gathering information about newly found (to them!) lands, and indicating their locations as accurately as possible on maps. Sometimes the great explorers and thinkers got it right, and sometimes they did not (see the sidebar called "Measuring the Earth"). But in any event, geography and exploration became intertwined; so, "doing geography" became closely associated with making maps, studying maps, and memorizing the locations of things (see Chapters 3 through 5 for information on locating things and creating and reading maps).

. . . To modern discipline

During the past two centuries, and especially during the past several decades, geography has blossomed and diversified. Old approaches that focused on location and description have been complemented by new approaches that emphasize analysis, explanation, and significance. On top of that, satellites, computers, and other technologies now allow geographers to record and analyze information about Earth to an extent and degree of sophistication that were unimaginable just a few years ago.

MEASURING THE EARTH

TECHNICAL STUFF

In the third century B.C., the Greek scholar Eratosthenes made a remarkably accurate measurement of Earth's circumference. At Syene (near Aswan, Egypt), the sun illuminated the bottom of a well only one day every year. Eratosthenes inferred correctly this could only happen if the sun were directly overhead the well — that is, 90° above the horizon. By comparing that sun angle with another one measured in Alexandria, Egypt, on the same day the sun was directly overhead at Syene, Eratosthenes deduced that the distance between the two locations was one-fiftieth (1/50th) of Earth's circumference. Thus, if he could measure the distance from Syene to Alexandria and multiply that number times 50, the answer would be the distance around the entire Earth.

There are diverse accounts of the method of measurement. Some say Eratosthenes had his assistants count camel strides (yes, camel strides) that they measured in *stade*, the Greek unit of measurement. In any event, he came up with a distance of 500 miles between Syene and Alexandria. That meant Earth was about [500 x 50 =] 25,000 miles around ("about" because the relationship between stade and miles is not exactly known). The actual circumference is 24,901 miles at the equator, so Eratosthenes was very close.

Interesting fact: The circumference is 41 miles less pole to pole; more on that in Chapter 4!

About a century-and-a-half later, another Greek named Posidonius calculated Earth's circumference and due to differences in the lengths of Roman versus Greek stadia, others reported his measurement as 18,000 miles. Posidonius' measurement became the generally accepted distance mainly thanks to Strabo, the great Roman chronicler, who simply did not believe that Earth could be as big as Eratosthenes said it was. About 18 A.D. Strabo wrote his *Geography*, which became the most influential treatise on the subject for more than a millennium. *Geography* credited the calculations of Posidonius and rejected those of Eratosthenes. And that leads to an interesting bit of speculation. Columbus was familiar with *Geography*, so he was aware of the official calculation of Earth's circumference — 18,000 miles. Had he known the true circumference was 25,000 miles, like Eratosthenes said, Columbus would have known that China was thousands of miles farther to the west than Strabo suggested. And if he had known the true distance to China, would Columbus ever have set sail?

As a result, modern geographers are into all kinds of stuff. Some specialize in patterns of climate and climate change. Others investigate the distribution of diseases, or the location of health care facilities. Still others specialize in urban and regional planning, or resource conservation, or issues of social justice and equality, or patterns of crime, or optimal locations for businesses — the list goes on and on. Certainly, the ancient *ge* and *graphe* still apply, but geography is much more than it used to be.

Exposing Misconceptions: More Than Maps and Trivia

Geography is a widely misunderstood subject. Many people believe it's only about making maps, studying maps, and memorizing locations. One reason is that polls and pundits occasionally decry the "geographic ignorance" of Americans, which usually means the average person doesn't know where important things are located. Presumably, therefore, if you memorize the world map, then you "know geography." Another reason is that on many TV quiz shows, contestants are occasionally asked "geography questions." Almost always, the answer is a fact that can be understood by studying a map and/or memorizing the locations of things or events.

Knowledge of the location of things is important and useful. Everything happens somewhere; and if you know the where, then the event has meaning that it otherwise would not. So map memorization is cool, but you need to keep it in perspective. Memorizing locations is to geography what memorizing dates is to history, or what memorizing the multiplication table is to mathematics. Namely, it's a foundation — a base — upon which you can build and develop deeper understandings.

TIP

Have you ever asked an English professor if they know the 26 letters of the alphabet? Of course not! It's silly. But care to guess how many times I have been asked to rattle off a list of state or country capitals? It's equally as silly. Just as letters build words, and then words build sentences, and then those sentences become ideas to share, so too are places. For a geographer, places are like our alphabet, a starting point to explain the complexity of Earth. The bottom line is: There is more to geographic awareness than *where* something is. As other geographers have stated, we need to think about where, why there, and why we should care.

The Geographic Advantage

Geographers still make maps and study them, and certainly, geography still consists of subject matter that cries out to be memorized. But map memorization and descriptive studies take a back seat to analysis, explanation, and significance. Geographers have a unique lens by which they try to understand Earth, and this approach comes with several advantages.

Geographer Susan Hanson described the Geographic Advantage as a focus on the

» relationships between people and the environment

» importance of spatial variability

» processes operating at multiple and interlocking geographic scales

» integration of spatial and temporal analysis

What this means is that geographers, more than other scholars, look at how people interact with the natural world, appreciate how interactions vary from place to place and from the local to the global, and link those processes and changes over time.

REMEMBER

A favorite definition of mine for geography encapsulates much of this and comes from Chinese–American geographer Yi-Fu Tuan. He stated quite simply and elegantly that "Geography is the study of Earth as the home of people." That says it all, doesn't it? If geography is just the study of Earth, well then isn't also geology or oceanography? If geography is just the study of the home of people, we then isn't also anthropology? It is the combination of the two together, understanding physical and social systems jointly operating in this one space — Earth — that makes all the difference.

What is the capital city of Indonesia?

To highlight the difference between old, descriptive geography (what far too many studied as "geography" in school) and analytical geography, first consider this question: What is the capital city of Indonesia? Do you know? The question is classic "old geography," and the answer is Jakarta. Right?

Why is Jakarta the capital of Indonesia?

REMEMBER

Now consider this question: Why is Jakarta the capital city of Indonesia? That's right, "Why?" This question involves analysis and explanation. The capital of Indonesia could be any number of cities. Indeed, several cities have been over its history with Jakarta finally assuming the role in 1949. But there's a catch to this question now. Indonesia has a plan in place to create a *new* capital city farther east in East Kalimantan on the island of Borneo. This is serious business as a country doesn't just decide to move its capital every day. But this has been done before (just ask Brazil or Nigeria). So why are Indonesians considering moving theirs? Here are a couple of reasons:

>> **An unpleasant setting:** Jakarta is densely populated, has overburdened infrastructure, is sinking due to over withdrawal of groundwater, and floods frequently. A new location would allow growth and it would be generally free from environmental hazards such as volcanoes and earthquakes.

>> **In the middle of it all:** Jakarta is on the western edge of the country. East Kalimantan is in the middle. Having the capital in the center of the country is important because Indonesia is flung across thousands of miles. A central location does more to integrate the citizenry and a central location also maximizes access to the seat of power.

REMEMBER

To sum up, I asked two questions: "What is the capital of Indonesia?" and "Why is Jakarta the capital of Indonesia?" Nothing is wrong with either question. But I trust you agree that the second is the more profound of the two. It calls for a deeper, more analytical brand of thinking and it leaves you with a more penetrating perspective on the geography of Indonesia and the significance of a number of factors. It can also lead us to other questions such as "What would it take for Indonesia to consider moving its capital? And if it did, what geographic conditions would be necessary in deciding on a new location?" Chapter 2 expands on how to "think" geographically.

Getting to the Essentials

To get you accustomed to thinking geographically, this volume makes use of unifying concepts that will help you to understand the breadth and structure of the discipline. But what are these unifying concepts? Yogi Berra once supposedly ordered a pizza pie and was asked if he wanted it cut into four slices or eight. He opted for four and explained, "I don't think I can eat eight." Whether or not the story is true, a pizza pie is a pizza pie, no matter how you slice it up. The same is true of geography. In a manner of speaking, it's a very big pizza pie. Over the years, geographers have devised different ways to cut it up in order to help people like you grasp its breadth and content. If you are a school teacher, you may have heard of the Five Themes of Geography or maybe even the Four Traditions. As I said, there have been many attempts to do this!

The "geography pizza slices" I'm going to introduce you to are The Six Essential Elements. They were developed by several professional geography organizations as part of the (United States) National Geography Standards, which describe in detail "what a geographically informed person knows and understands." The National Geography Standards were written with the advice and input of professionals who specialize in diverse aspects of geography and, accordingly, represent a broad consensus of its scope and structure. Specifically, therefore, I have chosen The Six Essential Elements, which are:

- » The world in spatial terms
- » Places and regions
- » Physical systems
- » Human systems
- » Environment and society
- » Uses of geography

These may sound somewhat imposing, but rest assured, they refer to simple concepts that you encounter in your everyday life. Indeed, you are already familiar with each of them, though perhaps not by their formal titles. I can prove it to you.

Where things are in the world: The world in spatial terms

You probably have a preferred grocery store, clothing store, and restaurant, plus a map in your head that tells you where they are and how to get to them. What's more, you could probably conjure up a route to visit all three in a single excursion and draw me a sketch map of the itinerary. If so, then you are already familiar with the world in spatial terms.

REMEMBER

Spatial refers to the location and distribution of things and how they interrelate. Accordingly, the world in spatial terms responds to geography's most fundamental question: Where? Getting a handle on this element involves:

- » Knowing how to use and read maps and atlases, whether paper or digital, and identify how they can lie to you (yes, you read that correctly).
- » Acquiring a general understanding of the tools and techniques that geographers use to accurately locate things.
- » Being able to indicate the location of something using the system of latitude and longitude, or plain language.
- » Seeing relationships that explain the locations of things.
- » Recalling from memory the location of things on Earth's surface.

These are basic skills to build on. On top of that, you'll never have to worry if somebody tells you to "Get lost!"

Chapter 2, which shows you how to think like a geographer, is very much about understanding the world in spatial terms. Chapters 3, 4, and 5 are devoted to location and maps, and, therefore, focus rather directly on this element. In addition, several other chapters will contain at least one map. Thus, you will encounter the world in spatial terms again and again throughout this book.

What locations are like: Places and regions

What'll it be for your next vacation? The mountains? The shore? Chances are you have mulled over questions like these that concern different areas with different characteristics. If so, then you are already familiar with places and regions.

Place: What a location looks like

Place responds to another important geographical question: "What is it like?" *Place* refers to the human and physical features that characterize different parts of Earth and that are responsible for making one location look different from the next. The terminology may puzzle you, because in everyday speech, people commonly use location and place interchangeably. In geography, however, these two terms have separate and distinct meanings. *Location* tells you *where. Place* tells you *what it's like*. In other words, places are locations to which humans have assigned meaning.

Take, for example, the proliferation of streets in the United States named after Martin Luther King, Jr. Are they locations? Sure, they are. They have specific addresses along them and they occupy space in hundreds of cities. But I think that we can also agree that they are imbued with much meaning and create unique places. Further, where they are located also says a lot about the people, history, politics, and so on in the neighborhoods where we find them.

Region: A bunch of locations with something in common

A *region* is an area of Earth, large or small, that has one or more things in common. So when you say "I'm going to the mountains" or "I'm heading for the shore," you refer to an area — a region — that has a certain set of characteristics over a broad area.

Regions make it easier to comprehend our Earthly home. After all, Earth consists of gazillions of locations, each of which has its own particular and peculiar characteristics. Knowing every last one of them would be impossible. But we can simplify the challenge by grouping together contiguous locations that have one or more things in common — Gobi Desert, Islamic realm, tropical rainforest, Chinatown, the Great Lakes, suburbia — each of these is a region. Some are big and some are small. Some refer to physical characteristics. Some refer to human characteristics. Some do both. But each facilitates the task of understanding the world.

TECHNICAL STUFF

Lest you think that this categorization idea is unique to geography, consider history or geology. History uses *eras* to group together like periods of time. Geology uses the term *epochs*. Each of these also has unique geographic characteristics. If I use the historic era *The Renaissance* as an example, your mind races not only to a specific time, but probably also a place: Italy.

Features that characterize different locations on Earth and, therefore, epitomize the concept of place, will be the focus of several chapters. These include landforms (Chapters 6 and 7), climates (Chapter 10), population (Chapter 11), culture (Chapter 13), economic activity (Chapter 15), and urbanization (Chapter 17). Each of these characteristics, of course, pertains not only to particular locations, but also to large areas as well. Thus, they also serve to characterize and define regions.

Why things are the way they are: Physical systems

I bet you have a favorite time of year, a favorite season. You probably also have a least-favorite season. No doubt you can tell me why you like some seasons more than others, and you can probably sprinkle your rationale with personal anecdotes about good times and bad. If that sounds about right, then you are already familiar with physical systems.

Atmosphere, land, and water are the principal components of the physical world. Geography seeks to understand how these phenomena vary from one location to the next and why. Geographers aren't content to know what the world looks like. They also want to know how it works. Why do islands like Aruba (near South America) and Socotra (off the Arabian peninsula), thousands of miles apart, have similar climates but differing land features? That involves understanding the natural processes that shape and modify Earth's surface (see Chapters 6 and 7), cause particular climates to occur in particular places (see Chapters 9 and 10), or result in some parts of Earth having too little water and others too much (see Chapter 8).

Giving that human touch: Human systems

Have you ever visited a locale that has many more or many fewer people than where you live? Have you ever moved a long distance? Have you ever visited a foreign country? Have you ever noticed that most of your shoes and clothing are made in a foreign country? Have you ever attempted to talk to someone, only to discover that person does not speak your language? Have you ever tried a very different cuisine (such as polishing off a mopane worm in South Africa as I have done)? If so, then you are already familiar with human systems.

Human beings characterize Earth's surface. That is, not only do humans live here, but by constructing cities, making farms, laying railways, and building other things, humans are an actual part of Earth's surface. Culture, trade, commerce, and government largely determine the specific ways in which people are part of Earth. And because these institutions are so diverse, so, too, are the human characteristics that are part of Earth's surface. Key aspects of human geography will be dealt with in separate chapters. They include population characteristics (see Chapter 11), movement and migration (see Chapter 12), cultural attributes (see Chapter 13), division of Earth into political units (see Chapter 14), economic activity (see Chapter 15), and urbanization (see Chapter 17).

Interacting with the world around us: Environment and society

Do you remember a farm or piece of countryside that is now a shopping center or a housing development? Have you ever experienced air pollution or water pollution? Have you ever had to cope with a severe storm, flood, or earthquake? If so, then you are already familiar with environment and society.

Human beings and the natural environment interact in many ways. For example, people play a very important role in shaping and modifying the natural world. Some results of this interaction may be visually pleasing, such as the skyline of Paris, or the terraced rice paddies of Southeast Asia, or the English countryside. But other results may be troubling, such as pollution and global deforestation, or the landscape devastation and human health problems stemming from Chinese uranium mining in Africa. References to human impact on the environment will appear in several chapters, particularly the ones dealing with water (see Chapter 8), natural resources (see Chapter 16), and urbanization (see Chapter 17). Most importantly, an entire chapter will be devoted to matters of environmental quality (see Chapter 18).

And while people impact the environment, environmental phenomena impact people. Climate affects agriculture and other human activity (see Chapters 9 and 10). Landforms and related processes and hazards affect life and property — think how Hurricane Katrina in 2005 upended the entire Mississippi coastline (see Chapters 6 and 7). The geography of water impacts settlement and commerce (see Chapter 8). In a nutshell, relationships between environment and society are pervasive, profound, and often political — and for those reasons will manifest themselves in several chapters.

Putting geography to use: Uses of geography

Have you ever used a mapping website to plan a trip? Have you ever visited a historical site and looked at maps and exhibits that help you understand the past? Have you ever attended a meeting or read an article concerning a proposal that would change the physical character of your neighborhood? If so, then you are already familiar with the uses of geography.

You can use geography to understand the past, interpret the present, or plan for the future. That is, you can use geography to understand the extent of former empires, to understand why your city looks the way it does, or to choose the location of a new factory. Geography is, therefore, a very useful and powerful tool.

IN THIS CHAPTER

» Thinking geographically

» Taking a look at two case studies

Chapter 2
Thinking Like a Geographer

REMEMBER

Geography is as much a way of thinking about the world as it is a body of information and concepts. Therefore, if you want to become good at geography, you must learn to think geographically. Remember when you were in the third grade and the teacher said, "Let's all put on our thinking caps"? Cute line, wasn't it? Well, I'm asking you to put on your thinking cap — your geography thinking cap, that is.

Thinking geographically is a process that involves a discreet set of skills. Therefore, this chapter is very different from the rest because it's not, on the whole or in part, about the content of geography. Certainly, you will encounter a fair amount of information about a particular part of the world. If you remembered it, great, but that's not the point. Instead, the goal is for you to learn how to think geographically and see that doing so facilitates a deeper understanding of the human and natural phenomena that geographers study.

Changing the Way You Think — Geographically

In Chapter 1, the content of geography was likened to a pizza pie, and The Six Essential Elements were presented as a way to "cut it up." The same National Geography Standards that give us those Elements also present a series of related skills that together constitute the process of thinking geographically. They include:

» **Asking Geographic Questions:** Thinking geographically typically begins with the questions "Where?" and "Why?" Sticking with pizza, one might want to know where all of the pizza shops in town are located and why they are there. Conversely, a person going into the pizza business may want to know where a good location would be to open a new pizza shop, and why.

» **Acquiring Geographic Information:** *Geographic information* is information about locations and their characteristics. If you want to know where all the local pizza shops are and why, then a first step may be to consult the internet (where you'll probably even find a map!). You may also visit the sites and acquire information about their characteristics. Similarly, someone going into the pizza business may do the same thing in order to learn the locations and characteristics of the sites that competitors have previously chosen.

» **Organizing Geographic Information:** After geographic information has been collected, it needs to be organized in ways that facilitate interpretation and analysis. This may be achieved by grouping together relevant notes, or by constructing tables, diagrams, maps, or other graphics. Thus, the person who wants to understand the geography of pizza shops might produce a map of them based on information previously acquired. The person who is considering going into the pizza business may do the same.

» **Analyzing Geographic Information:** Acquiring and organizing geographic information paves the way for analyzing geographic information. This is when the most heavy-duty thinking occurs. Analysis involves making comparisons, seeking relationships, and looking for connections between geographic information. What factors explain the locations of existing pizza shops? What factors make for a great location for a future pizza shop? Analyzing geographic information is kind of like playing a mystery game in which you use the information you previously acquired and organized to solve a puzzle.

>> **Answering Geographic Questions:** The process of thinking geographically culminates in the presentation of conclusions and generalizations based on the information that has been acquired, organized, and analyzed. It may reveal, for example, that pizza shops tend to be located in places that are readily accessible to a large number of people or that have lots of passers-by. Those conclusions may, of course, prove very useful to the person who wants to open a new pizza shop and is looking for the best possible location.

REMEMBER

Thinking geographically entails two lines of thought that are similar as well as different. They are alike in that both involve the bulleted points listed previously. The difference is that one approach focuses on where things *are* located, while the other ponders where things *should be* located. A geographer friend of mine named Phil likes to call these the international versus local rationales, respectively, for studying geography. To highlight this difference, the discussion above repeatedly referred to two people. One was trying to understand where pizza shops *are* located, and the other who was trying to determine where a pizza shop *should be* located. The following case studies help reinforce these perspectives. Each poses a geographic question (one international and one local, if you will) and challenges you to analyze geographic information before you arrive at an answer. That is, each has you thinking geographically. In doing so, you begin to acquire and develop important conceptual skills that constitute major mileposts in becoming a true geographer.

Case Study #1: Where Something is Located

Where are African lions located and why? Obviously, they live in Africa, and I know this to be true after nearly stepping on some — seriously — while in Kenya.

I was quietly walking up to some cape buffalo for a photograph and thought I could sneak up on them by being hidden behind a tree. As I approached the tree, suddenly three lions sprang out from the grass underneath it. I can only guess that the lions were enjoying the shade or hoping to surprise one of those same cape buffalo. Spooked, the lions ran off to the west. Spooked, I pondered a change of underwear.

So, we know that lions are in Africa but not in all parts. Why? That geographic question is central to our first case study.

I'd love to be able to pack you off to Africa (with your own change of underwear) and to have you acquire relevant geographic information, but that's not very practical. Instead, I simply refer you to Figure 2-1, which presents geographic information that has been acquired and organized in a map. So where are African lions located? What's the message of the map?

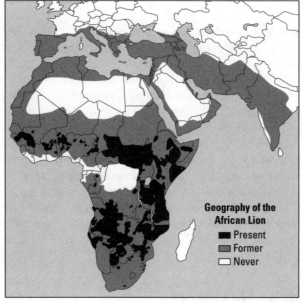

FIGURE 2-1: A map of the historical geography of the African lion.

(© John Wiley & Sons Inc.)

The answer is that African lions are much less widespread than they used to be. The map tells you this by using two kinds of shadings, the meanings of which are shown in the map legend. One shade shows areas where lions are found at present. Another depicts where lions formerly roamed. The last, blank space indicates areas where, so far as anyone can tell, lions have never lived.

A fraction of its former self

Today, African lions in the wild live only in the handful of patches shown on the map, mainly the ones in southern and eastern Africa. But the map also tells us there was a time when the lion's homeland consisted of a vast and contiguous hunk of Africa that stretched all the way from the Mediterranean coast in the north to the southernmost tip of the continent. Look at the map and visually compare the amount of territory that is lion country today versus the amount of former territory. Clearly, present-day lion country is a fraction of its former size and one estimate suggests that there are fewer than 23,000 left in Africa (a 40 percent loss in just the last two decades!).

What in the world — or rather, what in Africa — happened to cause such a reduction in the size of lion country? Why did it happen? And what is the significance? I do not really expect you to have the answers at your fingertips. But take a few moments again, and this time see if you can't come up with some possible reasons as to why lions live where the map says that they do, and why lion country has decreased so substantially.

Where lions hang out

First of all, where do lions live? No, I'm not asking you for a street address; but rather, in what kind of environment do lions tend to hang out? Here are a few choices of where your average well-adjusted lion might live:

- In a forest
- In a desert
- In a grassland
- Anywhere it darn well pleases

Although the last choice has considerable merit, the best response is "in a grassland." Lions generally live in grasslands (you can read more about where they don't live in the sidebar titled "Animal Geography, Hollywood Style"). You may have known the answer because just about everybody has seen a TV wildlife documentary, which, in graphic detail, shows lions killing their next meal and then eating it. But just in case, next time you see one of those programs, concentrate on the physical setting instead of the kill. That's right, skip the build-up . . . . the eyeing of the herd . . . . the stalk . . . the chase . . . . the cute little impala meeting its untimely end. Instead, focus your attention on the surrounding countryside, and what you are bound to see is that this life and death drama is playing out on what is essentially an extensive grassland.

What gives with grasslands?

But what gives with grasslands? Or rather, why do lions choose to inhabit grasslands? Here are a few choices as to why lions live in grasslands:

- Green is their favorite color.
- That's where those cute little impalas live.

>> They run into few trees.

>> The rents are low.

Although each choice could be correct, the best response is "that's where those cute little impalas live." Lions love impalas.

Indeed, they truly love them to death. Like all wild animals, lions tend to live in places where they can find relatively abundant food to their liking. So, lions hang out where impalas, zebras, wildebeests, and other animals are on the menu. Lions, of course, are *carnivores* — meat-eaters. And nearly all the animals on the menu are *herbivores* — grass-eaters. Lions prefer to live in a grassland because, as far as they are concerned, it's one big meat market.

Extinction made easy

Time to stop beating around the bush — and around the grassland, for that matter. The main message of the map is that lion country is a small fraction of its former size. And although the animal itself is not on the brink of extinction, things would appear to be headed in that direction. So what happened?

ANIMAL GEOGRAPHY, HOLLYWOOD STYLE

Movies may be responsible for more environmental *mis*information than any other source. Thus, in the world according to Hollywood, animals have a maddening tendency to show up in locations where they have no business being. Sometimes the errors are rather obscure. For example, in the nativity scene at the start of the 1959 movie *Ben-Hur*, a Holstein calf prances by the manger. Holsteins are those dairy cattle with the black and white splotches. The problem is the Holsteins come from Schleswig-Holstein, the part of Germany that borders Denmark. Two thousand years ago, there would not have been a Holstein anywhere near Bethlehem. Like I said, sometimes the errors are rather obscure. Then again, sometimes the errors are downright outrageous, and, in that regard, nothing beats Hollywood's treatment of the African lion. Check out just about any of the old *Tarzan* movies, *George of the Jungle,* or a host of other flicks set in a rainforest. Almost inevitably, one or more lions show up. The problem, of course, is that a lion has a whole lot less business being in a rainforest than does a Holstein in Bethlehem. Lions do not live in rainforests. Could they be near one? Sure, as habitats do rub up against each other. But an African lion really isn't the King of the Jungle and the reason is simple. A lion has virtually nothing to eat in a rainforest — except maybe Tarzan.

Perhaps it would be better if I personalized the question. Let's say you really have it in for the king of beasts and want to get rid them. I'm talking extinction. What is a safe, easy, and effective way to go about it? You have a handful of options:

» Shoot every last one of them
» Teach impalas self-defense
» Destroy their habitat
» Pack them off to Australia

Although each response has some possibilities, the best choice is "destroy their habitat." And that is indeed the main reason for the reduction of lion country from its former dimensions to its present ones and is also the reason why the lion is located where it is now.

A natural habitat can change for natural reasons or for unnatural reasons. As regards to the former, climate change is a major possibility. Natural grasslands are the result of a specific set of climatic characteristics. So if those climatic factors change, you would expect grasslands to change, too. Now, ample evidence exists of climate change in Africa. But the nature and extent of it is insufficient to explain the wholesale disappearance of grasslands over the wide area indicated on the map. So, climate is not the primary culprit. Instead, the fault lies elsewhere and mainly takes the form of human beings. Those humans are building roads, converting land to agriculture, and building settlements. Some people also kill lions as sporting trophies or in retaliation for killing livestock.

Fewer lions? So what?

What, if anything, is the significance of the map and the story behind it? Is there any relevance? I believe so.

The pressure on natural habitats continues (and not only in Africa). Unless something is done to halt the tide, the great grasslands will continue to diminish and so, too, will the lions. Governments in affected areas are increasingly committed to heritage conservation and view protection of natural habitats and wildlife as part of that process. Thus, the average lion in the wild today lives in a national park or national game preserve. But pressure is being put on governments to open the parks to grazing and other activities that constitute "multiple use." Local officials must make choices that concern balancing the desire for conservation with the needs of citizens.

The situation is relevant to other lands, including the United States. Lions don't live in the wild in the U.S., but other animals do. And in many cases, their stories mimic the lion's. That is, they are much less widespread than they used to be. National parks and preserves have helped stem their decline and some species have been successfully re-introduced to some areas. But human population growth, coupled with pressure for land development and multiple uses, make the future uncertain. In the U.S., as in Africa, choices must be made. Looking at the locations of animals and their habitats and thinking geographically about them help clarify the issues and processes that are involved and encourage informed decision-making.

Summing up

The answer to our geographical question (Where are African lions located and why?) is that lions are located in the parts of Africa shown on Figure 2-1 mainly because of habitat reduction that is human in origin. After posing the question, we analyzed geographic information that led to the answer, after which we pondered the implications of our findings to wildlife conservation elsewhere in the world. All in all, the focus was on thinking geographically so as to understand where things (African lions) *are* located.

Case Study #2: Where Something Should be Located

Where should a gas station be located, and why? Those questions are central to our second case study.

Thinking geographically about where something *should be* located has many important and useful applications. For example, consider the occupational endeavors called *planning*. That includes urban planning, regional planning, and transportation planning, to name just three. All are intimately concerned with the question of where things *should be* located. The business world also provides lots of useful applications. Choosing a good location is often an important determinant of whether an enterprise succeeds or fails. The questions posed previously call for a business decision based on the process of thinking geographically.

In this case study, assume that you want to go into the gas station business. Therefore, your relevant geographic question is "Where should my gas station be located?"

Similar to the first case study, I'd love to have you go around town and acquire pertinent geographic information. That would include finding prospective sites for your gas station, and identifying the factors that appear to be contributing to the success of existing gas stations that clearly are doing a lot of business. The latter is important because it helps you choose the prospective site that offers the best chance for success. But that's a bit much to ask. So once again, assume that the footwork has been done, that relevant geographic information has been acquired, and that it has been organized in ways that include a map (which happens to be Figure 2-2).

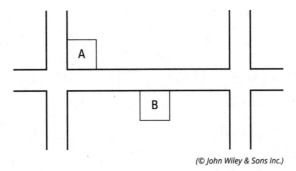

FIGURE 2-2: Potential sites for a gas station.

(© John Wiley & Sons Inc.)

The map shows two land parcels that are indicated by "A" and "B." Assume each has an identical size, an affordable price, a busy thoroughfare alongside, and that other prospective sites for your gas station have been eliminated from consideration. Your final choice with be either "A" or "B." Is one location clearly preferable?

Analysis of the geographical information indicates the two properties have one key difference: Property A is located on a corner lot, while Property B is in the middle of a block. Is that difference significant? Think about the location of every gas station you have ever seen. Is it on a corner or in the middle of a block? It's almost always on a corner, isn't it? And the main reason is that, on a daily basis, more cars (potential customers) pass by a corner lot as opposed to a middle-of-the-block lot because the corner adjoins two roadways rather than one. In addition, corner lots are somewhat easier to enter and exit. Accordingly, the answer to your geographic question (Where should my gas station be located?) is lot A.

Summing up

In the process of choosing a location for your gas station, you have been thinking geographically once again. Only this time, however, you began by considering where something (a gas station) *should be* located. You then proceeded to acquire and organize (map) pertinent geographic information, analyze it, and answer the question.

Looking ahead

TIP

At the beginning of this case study, I mentioned that thinking geographically about where something *should be* located has important applications in the fields of planning, business, and industry. Indeed, virtually every tool, concept, and content area of geography has useful applications. To reinforce this point, and to help you recognize the practical value of geography, some of the chapters that follow include a specific example with a sidebar whose title begins with "Applied Geography."

> **IN THIS CHAPTER**
>
> » Getting it right with a grid
>
> » Pointing someone in the right direction
>
> » Discovering a common theme: Degrees, minutes, and seconds
>
> » Locating a street address and global address

Chapter 3
Lining Up Locations

Back in 1992, a cargo ship lost a container and launched almost 29,000 bath toys into the northern Pacific Ocean. Plastic beavers, frogs, turtles, and ducks, all bobbing along, wave after wave. Dubbed "the Friendly Floatees," these plastic toys — cute as they were — added to a growing ocean pollution problem. But they had one useful benefit, and that was to help model ocean currents.

As the toys washed ashore or were seen at sea, a very useful piece of information was collected: their location. To relay that information to others, humans have devised a number of ways to express location on Earth's surface. In this case, the ducks and their friends were identified using *latitude* and *longitude*, a set of invisible lines and coordinates that describe where things are.

But how does this imaginary system work? Let's visit Gridville to find out.

Welcome to Gridville

Welcome to Gridville, the cute little burg shown in Figure 3-1. You and I are going to pay this town a quick visit because it looks like a great locale to review basic concepts of location (latitude and longitude), the topic of this chapter. I say "review" because if you are like most people, then you probably learned about these things during elementary or middle school, but may have forgotten some or most of it later on.

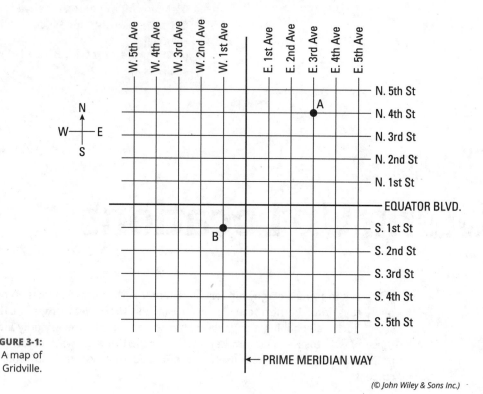

FIGURE 3-1:
A map of Gridville.

(© John Wiley & Sons Inc.)

Knowledge of latitude and longitude gives you basic location and orientation skills regarding our planet Earth. It also affords the opportunity to learn all sorts of little tidbits, which, in addition to impressing your friends, can greatly enhance your understanding of geography.

Feeling Kind of Square

To get started, look at Figure 3-1 and familiarize yourself with Gridville. In particular, note the following:

>> The roads are aligned with the cardinal directions — that is, they run north-south or east-west. The result is a *grid* pattern of north-south roads that

intersect east-west roads at right angles. So, getting right with Gridville means getting used to a city that is all right angles and nothing but right angles. Thus, I'll understand if this town leaves you feeling a little square.

- » North is toward the top of the map; south is toward the bottom; east is toward the right; and west is toward the left. This is a near-universal rule in map-making, but you should always carefully examine the map you are looking at and confirm which way is which.

- » Gridville has a principal east-west road named Equator Boulevard, and a principal north-south road named Prime Meridian Way. The two roads cross in the middle of Gridville.

- » Every other road in Gridville has a name that refers to its location relative to those two roads. Thus, streets are numbered consecutively north and south of Equator Boulevard. Avenues are numbered consecutively east and west of Prime Meridian Way.

- » A big dot and a letter mark two intersections. I'll refer to these shortly.

Telling Someone Where to Go

Because geography involves locations and directions, it affords ample opportunity to tell someone where to go. Suppose you live in Gridville and are standing on the sidewalk at Point A, the corner of North 4th Street and East 3rd Avenue. A stranger from out of town comes up to you and asks for directions to Gridville Hospital — can you help her?

Of course, you can. You know the hospital is located at Point B on the map. And you can convey that information to the stranger by stating either the hospital's relative location or absolute location.

Relative location

In the first instance, you can tell the stranger how to get to the hospital from Point A. For example (pointing west along North 4th Street), "Go that way four blocks, turn left, and walk five more blocks." This is called *relative location* because the information you gave is relative to Point A. Give those directions verbatim to the stranger at any other intersection in Gridville, and the result is a lost stranger.

Absolute location

As an alternative, you can convey the location of the hospital with respect to its *grid coordinates* — that is, its location within the grid system. For example, "Go to the corner of South 1st Street and West 1st Avenue." This is called *absolute location* because theoretically, those directions work anywhere in Gridville, not just at Point A.

The best location to use

Both relative location and absolute location have the potential of getting the stranger to the desired destination. And chances are you have used both types of location to direct someone to a destination in your town, neighborhood, or environs.

But in a global context, absolute location is far superior to relative location. When you think about it, the task of directing somebody to a location halfway around the world by means of relative location (for example, "Go that way 11,238 miles and turn right") is rather mind-boggling. And even if you could do it, that information would only work at the one location where that information was given. It would be far better if every place on Earth had an absolute location such as that hospital in Gridville. Of course, that would be contingent on the existence of a global grid that basically mimics what we've seen in Gridville. Fortunately, such a grid exists.

The Global Grid: Hip, Hip, Hipparchus!

Like Gridville, the world as a whole possesses a grid whose coordinates may be used to identify the absolute location of things. Indeed, that is why a Greek named Hipparchus invented the global grid more than 2,000 years ago. Though not necessarily the first with this idea — many credit the earlier ideas of Eratosthenes — he did leave us with what most people use today. (You can read more about Eratosthenes in Chapter 1).

As chief librarian at the great library in Alexandria, Egypt, Hipparchus compiled information about lands and cities all over the expanding Greek world. He saw the value of accurately locating objects on a map, but in those days that was easier said than done. Maps were notoriously inaccurate, due in good measure to lack of a systematic means of stating the location of things. So, Hipparchus set out to rectify the situation, and came up with the global grid that is still in use today (see Figure 3-2).

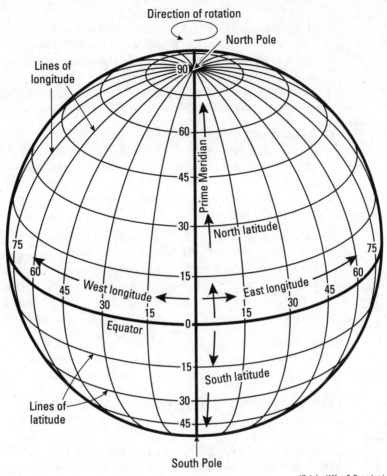

FIGURE 3-2:
The basics of the global grid.

(© John Wiley & Sons Inc.)

Avoiding gridlock

TIP

Proper use of a grid coordinate system to state the absolute locations of things depends on a handful of prerequisites. Think of these as ways of avoiding gridlock:

» **Familiarity breeds success.** Knowledge of the naming and numbering of grid components is essential. If, for example, that stranger were not familiar with Gridville's grid, then telling her the hospital is at "the intersection of South 1st Street and West 1st Avenue" would have made no sense whatsoever. The same is true with respect to the global grid. That is, knowing how the lines are named and numbered is essential if you are to use the grid successfully.

- **Unique components.** Each road in Gridville and each line on the global grid must have a unique name. In Gridville, for example, there must be only one road named South 1st Street, and only one named East 1st Avenue. If multiples exist, then more than one site could satisfy "the intersection of South 1st Street and East 1st Avenue." And that would rather defeat the concept of absolute location, whether in Gridville or around the globe.

- **No double-crossing allowed.** Don't take that as a threat or accusation. What I mean is two roads in Gridville may cross each other only once. The same goes for two lines on the global grid. If they have multiple junctions then, such as the last point, there would be two or more intersections of, say, South 1st Street and East 1st Avenue. And again, that would defeat the concept of absolute location.

- **Full names, please.** You must use the full name of each road in Gridville and each line on the global grid. Again, the absolute location of the hospital is the intersection of South 1st Street and West 1st Avenue. Now suppose you had told that stranger, "The hospital's at the corner of 1st Street and 1st Avenue." Well, if you look carefully at the map of Gridville, you find four locations where a 1st Street crosses a 1st Avenue. Obviously, the potential for location confusion here defeats the purpose of absolute location. The remedy is to use the full name of each grid component.

The naming game

REMEMBER

While the Gridville grid consists of real roads, the global grid consists of imaginary lines of latitude and longitude (see Figure 3-2). *Latitude* lines go across the map — latitude comes from the Latin *latitudo*, meaning breadth, or the measure of the side-to-side dimension of a solid. *Longitude* lines run from top to bottom — longitude comes from the Latin *longitudo*, meaning length. This makes sense because when viewed on a globe, lines of longitude are generally lengthier than lines of latitude.

Similar to the roads in Gridville, the global grid contains a principal line of latitude (*the equator*) and a principal line of longitude (*the prime meridian*). All other lines of latitude and longitude are named and numbered respectively from these starting lines. It makes sense, therefore, that if you want to make like Hipparchus and draw a grid on a globe, then these are the first two lines you would draw. But where would you put them, and why?

The equator

Because Earth is sphere-like, no compelling locale cries out and says, "Use me to locate the equator!" So where to put it? Old Hipparchus might simply have said,

"It's Greek to me!" and placed it anywhere. Instead, he wrestled with the challenge and came up with an ingenious solution.

He knew that Earth is sphere-like and that it rotates around an imaginary line called the *axis*. Look on a globe and you find two fixed points, halfway around Earth from each other, where the axis intersects Earth's surface: the *North Pole* and the *South Pole*. So, Hipparchus drew a line that ran all the way around the globe and was always an equal distance (hence, equator) from the two Poles. The result is a latitudinal "starting line" from which all others could be placed on the globe.

The prime meridian

The longitudinal "starting line" is called the *prime meridian*, which signifies its importance as the line from which all other lines of longitude are numbered. Locating this line proved more problematic than locating the equator. Quite simply, no logical equivalent of the equator exists with respect to longitude. Thus, while the equator came into general use as the latitudinal starting line, mapmakers were perfectly free to draw the longitudinal starting line anywhere they pleased. And that is what they did.

Typically, mapmakers drew the prime meridian right through their country's capital city. By the late 1800s, lack of a universal prime meridian had become a real pain in the compass. International trade and commerce were growing. Countries were claiming territory that would become colonial empires. But one country's world maps did not agree with another's, and the international climate made it increasingly advisable that they do so. I have a map hanging in my living room to prove the point! There are different longitude coordinates at the top of the map compared to those at the bottom. Both were given to aid a map reader more used to a coordinate system beginning with a different meridian.

As a result of this growing confusion, in 1884 the International Meridian Conference was convened in Washington, D.C. to promote the adoption of a common prime meridian. Out of that was born an agreement to adopt the British system of longitude as the world standard. Thus, the global grid's prime meridian passes right through the Royal Greenwich Observatory (which is in the London suburb of Greenwich) as well as parts of western Europe and Africa, and the Atlantic Ocean. The British system was chosen largely because in 1884 Britain was the world's major military and economic power, and also had a fine tradition of mapmaking.

Getting Lined Up

With the starting lines in place, one can now contemplate putting all of the other lines of latitude and longitude on a globe. In doing that, Hipparchus used the notion that 360 degrees (°) are in a circle (But why 360? See the sidebar "Why is Earth 360° round?"). Accordingly, he drew lines of latitude such that each and every one is separated by one degree of arc from the next. He then did the same with longitude. This is why lines of latitude and longitude are referred to as degrees.

Latitude

The system of latitude lines has the following characteristics:

» Lines of latitude run across the map (east-west) and are called *parallels* because each line of latitude is parallel to every other line of latitude.

» The equator (Latitude 0°) divides the world into the *Northern Hemisphere* and the *Southern Hemisphere*.

WHY IS EARTH 360° ROUND?

The ancient Sumerians believed there were 360 days in a year. Like other civilizations way back when, the Sumerians equated their gods with celestial objects. Not surprisingly, the sun god was especially important. Because it took the Earth 360 days to travel around the sun (or so they believed), the Sumerians figured the number 360 had extra-special significance. As a result, they developed a system of mathematics based on multiples of 6 and 60. Nowadays, we would call it *base-6 mathematics* or (get ready for this) a *sexagesimal* system. In any event, 6 times 60 equals 360.

The ancient Egyptians adopted the ancient Sumerians' numerical ideas, and eventually discovered the error concerning the length of the year. But by then, however, the number 360 had achieved such acceptance and status that the Egyptians decided not to mess with it. Accordingly, they kept the 360-day year but, being fun-loving people, added an annual 5-day holiday.

The ancient Greeks, like the ancient Egyptians, were adept at adopting things and ideas from civilizations more ancient than they. So, when Hipparchus, in about 140 B.C., began fiddling with the notion of dividing a circle (and Earth) into degrees, he chose the number 360.

- » Starting from the equator, each successive line (degree) of latitude is numbered consecutively both to the north and to the south as far as the North Pole (Latitude 90° North) and South Pole (Latitude 90° South).

- » Except for the equator, each line of latitude is identified by a number between 0 and 90 and by the word North or South (or the abbreviations N or S) to indicate its location north or south of the equator. Thus, the line that is 20 degrees north of the equator is referred to as Latitude 20° North (or 20° North Latitude). It would be misleading and incomplete to just call this line "Latitude 20" because another line of latitude south of the equator could also be called "Latitude 20."

- » Only one line of latitude — the equator — is a *great circle,* a line that divides Earth in half.

WHAT'S WRONG WITH THIS MAP?

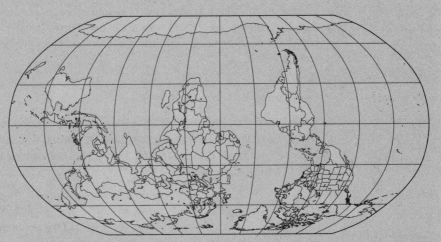

(© John Wiley & Sons Inc.)

The answer to the headline is this: Nothing much, really. You could say the map is upside down, and you would be right to a point. After all, nowadays maps commonly have north at the top. But considered as a planet in the multi-dimensional vastness of space, Earth has no "right side up." Thus, no compelling scientific reason exists as to why you can't make a map with south toward the top — other than that it would look strange and confusing to most people. Indeed, in olden times maps were oriented every which way. Chinese maps tended to have south at the top, and early Christian maps had east on top with Jerusalem at the center.

(continued)

(continued)

> Suffice to say, there are a number of competing ideas about how north got on top. The discovery of magnetic north? The preference of ninth-century Frankish emperor Charlemagne, who as powerful king would have held a bit of sway? The widespread use of the Mercator projection by navigators (see Chapter 4 to find out more about this)?
>
> Regardless of how it came about, north at the top tends to be the default, but there are exceptions. If you have a bit of free time, go online and search for Chile's Directorate General of Civil Aviation. Their logo includes a map that bucks convention.

Longitude

REMEMBER

The system of longitude lines has the following characteristics:

- » Lines of longitude run from the North Pole to the South Pole (top to bottom of the map) and are called *meridians.*

- » As opposed to latitude, no two lines of longitude are parallel to each other. Rather, successive lines of longitude are about 70 miles apart at the equator, but from there they slowly converge until they come together at the two poles (see Figure 3-2).

- » The prime meridian (Longitude 0°) divides the world into the *Eastern Hemisphere* and the *Western Hemisphere.*

- » Starting from the prime meridian, every line (degree) of longitude is numbered consecutively to the east and to the west half way around the world. Because Earth is 360 degrees around, 180 degrees of longitude lie east and west of the prime meridian.

- » Every line of longitude (except the prime meridian and the 180-degree line) is identified by a number from 1 to 179, and by the words East or West (or the letters E or W) to indicate its location east or west of the prime meridian. Thus, the line that is 20 degrees east of the prime meridian is referred to as Longitude 20° East. It would be misleading to call this line "Longitude 20" because some another line that is 20 degrees west of the prime meridian also could be called "Longitude 20."

- » Every line of longitude is a *great circle* — a line which, if continued around the world, would divide Earth equally in half.

Graticule

TECHNICAL STUFF

As far as geographers are concerned, latitude and longitude make for a very special grid that deserves a special name, *the graticule*, to distinguish it from every other kind of grid. Indeed, this name is so special that many dictionaries and computer spell-check programs do not recognize it. But geographers do, and they are extremely impressed if they hear it used by a layperson.

But more important than saying "graticule" is the ability to use it properly. That means, among other things, correctly identifying the grid coordinates (latitude and longitude) of locations indicated on a map. With that in mind, take a look at Figure 3-3, which represents a portion of the graticule. Note that lines of longitude are shown parallel (when in reality they converge toward the poles) and that only every tenth degree-line of latitude and longitude are indicated. World maps typically "skip" lines in a similar fashion, lest they become cluttered by the graticule. But what I really want you to focus on are the three dots lettered A, B, and C. See if you correctly can identify the coordinates of each dot, keeping in mind the following rules:

1. When reporting coordinate locations, always give the latitude first, and then give the longitude. (Why? Because latitude comes first alphabetically? Because latitude was accurately measured before longitude? It doesn't really matter. It's just the rule and the system only works if everyone uses it the same way.)

2. Correct reference to latitude must specify whether a location is north or south of the equator (Latitude 0°), assuming the location is not on the equator itself.

3. Correct reference to longitude must specify whether a location is east or west of the Prime Meridian (Longitude 0°).

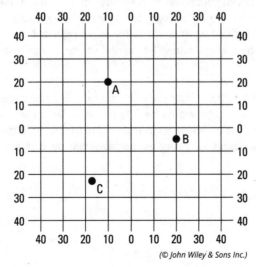

FIGURE 3-3: A representation of a portion of Earth's grid.

(© John Wiley & Sons Inc.)

The correct locations of the dots are as follows:

A	= Latitude 20° North, Longitude 10° West
B	= Latitude 5° South, Longitude 20° East
C	= Latitude 22° South, Longitude 17° West°

Minutes and seconds that don't tick away

On Earth's surface, adjacent lines of latitude and longitude may be several miles apart, and that creates a potential problem if you wish to state the absolute location of a spot that is "between the lines." For this reason, the graticule contains a couple of levels of refinement (see Figure 3-4).

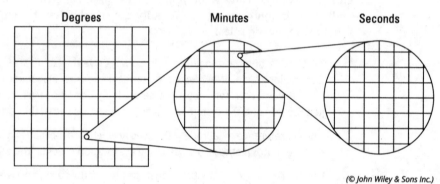

FIGURE 3-4: Degrees, minutes, and seconds.

(© John Wiley & Sons Inc.)

First, the space between successive degree lines may be subdivided into 60 equidistant units called *minutes* ('). Second, the space between successive minute lines may be subdivided into 60 equidistant units called *seconds* ("). And if more exactitude is needed, then seconds may be carried out to as many decimal points as may be necessary.

Doesn't this sound familiar? Sixty seconds in a minute? And for good reason. The system that you use to tell time goes back to the same Sumerian base-6 arithmetic that Hipparchus used to divide up a circle and also the world. Hmmm . . . there are 24 hours in a day. Think 24 being evenly divisible by 6 is just a coincidence? No way.

> **IN THIS CHAPTER**
>
> » Stretching the truth
>
> » Understanding how maps are dishonest
>
> » Weighing the pluses and minuses of globes and flat maps
>
> » Analyzing different maps
>
> » Looking out for *really* bad maps

Chapter **4**

Truthiness in Mapping

Imagine a million-dollar map contest. The only thing you have to do to win is to supply an exact map of the entire Earth that's flat. Here's how to enter!

1. Get your hands on a globe.
2. Peel off the surface layer in such a way that you end up with one big piece of map peel. (You may want to use somebody else's globe because this procedure results in the globe's complete ruin.)
3. Lay the map peel on a flat surface so that the two surfaces are completely in contact but without distorting the original map in any way. You can cut the map if you want, but pulling and stretching it is prohibited.

You are absolutely right if you think it's going to be tough to submit a winning entry. Actually, it's impossible. You can't take a sphere-like surface (see the sidebar "Earth's shape: Sphere-like, not spherical" for more on this), such as Earth, and lay it down flat without distorting the original image. This fact, however, hasn't deterred people from making flat maps of the world or parts thereof. And, to do that, the mapmaker has to figuratively *pull it here* and *stretch it there*. The result is a map that's full of distortion. Full of distortion? Well, simply put: Maps that lie flat lie!

> **EARTH'S SHAPE: SPHERE-LIKE, NOT SPHERICAL**
>
> People often say that Earth is a sphere. Not so. By definition, a *sphere* is a curved solid whose surface is always the same distance from its center, no matter at what point of the surface. Technically, Earth doesn't fit that definition. Instead, Earth is an *oblate spheroid,* meaning it is somewhat flattened at its poles, or, if you prefer, it bulges somewhat around the Equator. The average distance from Earth's center to the Equator is about 26 miles farther than the average distance from Earth's center to the poles. Compared to the size of Earth, 26 miles isn't a great distance, but it's enough to make Earth not a real sphere. It's better to say Earth is sphere-like, or an oblate spheroid.
>
> Earth's rotation causes its oblate-ness. The speed of Earth's rotation is much faster at the Equator than near the Poles. This difference in speed may not be obvious, so think of it this way. Earth's circumference measured along the Equator is about 25,000 miles. If you stand at a spot on the Equator for one day — for one full rotation — you'll travel 25,000 miles. In contrast, if you stand a foot or two from the North Pole for one rotation, you'll only travel a few yards. Obviously, somebody who travels 25,000 miles in one day is moving much faster than somebody who travels a few yards in the same time. So, the area near the Equator is spinning much faster than other parts of Earth. The outward, or *centrifugal,* force the high speed of rotation causes is so great that Earth bulges around the Equator as a result.

Maps of the world are among the most basic aids to geographic learning. Many people take it for granted that they are truthful. But in reality, all flat maps of the world lie — they simply cannot help it. If you're new to thinking geographically, it is important that you appreciate that simple fact and understand the ways in which maps distort their portraits of your Earthly home. This chapter shows just how flat maps lie.

Seeing the Light: Map Projections

Accordingly, this chapter is about mapmaking with emphasis on the distortions that are inherent in flat maps of the world. But first, some basic vocabulary is in order. A *map* is a representation of all or part of Earth's surface. *Cartography* is the field of mapmaking, and a *cartographer* is a person who makes maps. Way back when, cartography was pure freehand, and I do mean way back. The oldest known map is a 5,000-year-old clay tablet that shows physical features of Mesopotamia. Later, cartography became associated with instruments and techniques that most people think of as *drafting.* Nowadays, most cartography is done using a computer.

REMEMBER

Flat maps are called *projections* because, theoretically, making a map of the world or a large part of it involves projecting a globe onto a piece of paper or similar flat surface. Imagine, as shown in Figure 4-1, a clear plastic globe with a light source at its center. When the bulb is turned on, light passes through the glass sphere and *projects* the lines from the globe's surface onto a receiving flat surface. The result is a flat map of Earth — a projection.

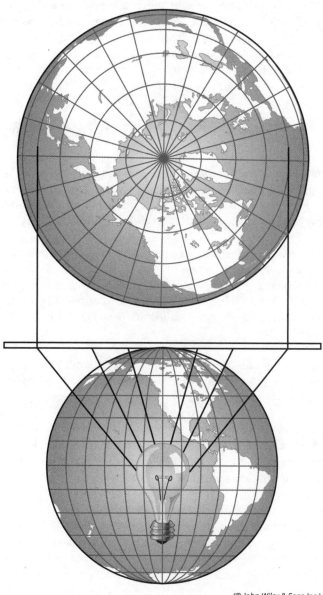

FIGURE 4-1: Map projection.

(© John Wiley & Sons Inc.)

REMEMBER

Projection has two meanings. On the one hand, it refers to the process of transferring a globe to a flat surface. On the other hand, projection refers to the map itself, the result of the transferal. One could say, then, that projection (transferal) results in a projection (flat map).

The diagram that shows the globe and light bulb is a simple model that most people find helpful in visualizing how projections are made. In reality, projections aren't made with a glowing light bulb in the center of a globe. Instead, projections are products of mathematical formulas, trigonometric tables, and things of that ilk. The specifics are pretty tedious; fortunately, trying to explain it all in language that even I can understand is beyond the scope of this book. It will be sufficient for you to appreciate that different projections exist, but none are totally truthful.

Realizing Exactly How Flat Maps Lie

TECHNICAL STUFF

The business of making map projections requires a somewhat deviant personality. Cartographers know that maps that lie flat lie. They know for certain before they begin a project that it's absolutely impossible to create a flat map that looks exactly like the world. Does that deter them? Nope. No way.

Cartographers have developed literally dozens of different kinds of map projections over the years. Each one contains some degree of misinformation. If you're like most people you've given little or no thought to map projections nor have you suffered from not doing so. Or have you? (For another perspective on why this matters, see the sidebar "Applied Geography: Putting your best projection forward.")

Understanding the facts about maps can't help but make you a better-informed person. Maps are a common means of communicating information. They pop up in internet articles, magazines, books, TV programs, and elsewhere. Because mainstream media is in the business of providing factual information, people may understandably assume that the maps they're looking at are accurate. But maps that lie flat lie, and there's nothing anybody can do about it — except maybe understand the nature of the distortions and appreciate that flat maps should be interpreted with a certain amount of caution.

Cartographers know projections lie, so their objective is to get as close to reality as possible. But enough of this blabber about maps that lie, it's time to consider a practical example that involves some honest-to-goodness maps. Or rather, some not-so-honest-to-goodness maps.

Singapore, please. And step on it!

Suppose you live in New York City and are preparing for a trip to Singapore, almost halfway around the world. In planning your trip, you decide to minimize your flying time and also to stop somewhere for a day or two, just to break up your travels. A friend suggests a stopover in Rome, Italy. But another friend tells you to layover in Helsinki, Finland. You have no idea which choice is best, so you decide to find out by plotting the two cities on a map (see Figure 4-2).

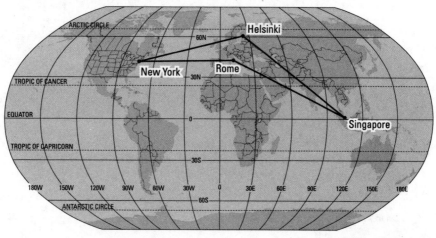

FIGURE 4-2: New York City to Singapore: Map # 1.

(© John Wiley & Sons Inc.)

Accepting the principle that a straight line is the shortest distance between two points, the map seems to make your choice pretty clear, doesn't it? The itinerary to Singapore via Rome is apparently much shorter than the route via Helsinki. As a result, you call your travel agent and make the appropriate bookings.

Upon hearing your travel plans, your second friend is shocked. "You're not going by way of Helsinki?" To show your friend the wisdom behind your choice, you take out your map and note the obvious: The linear distance from New York to Singapore is shorter via Rome. Whereupon your friend produces a map of her own (see Figure 4-3).

Looking at the map in Figure 4-3, three things are suddenly obvious.

- » First, the global view in this map is much different than in Figure 4-2.
- » Second, the results are different, too. In Figure 4-3, going to Singapore via Helsinki appears much shorter than the route via Rome.
- » Third, one of these maps is lying, but which one?

CHAPTER 4 **Truthiness in Mapping** 45

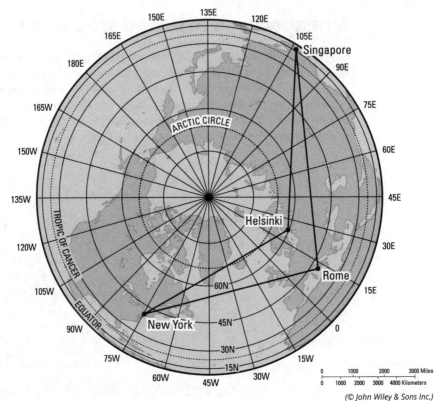

FIGURE 4-3: New York City to Singapore: Map #2.

(© John Wiley & Sons Inc.)

APPLIED GEOGRAPHY: PUTTING YOUR BEST PROJECTION FORWARD

Figures 4-2 and 4-3 provide different perspectives on air routes between New York City and Singapore. While this may seem a strictly academic exercise, airlines that compete on long-range international itineraries take the matter very seriously. There's an old saying: "Time is money." And for that reason, many business travelers (if they have a choice) prefer the shortest route to get them where they're going. Airline executives know this. Accordingly, marketing strategies sometimes involve making maps that present the airline's route system in the best light possible. And doing that, of course, involves choosing the best possible projection.

If you have a globe handy, you can determine the shorter of the two itineraries from New York City to Singapore. Get a string, pull it taut, and place it on the map so that the string connects New York City and Singapore. What you observe is that the string passes over the Arctic Circle and shows that a stopover in Helsinki is a minor detour, but a stopover in Rome is a major detour. If you don't have a globe, you can't do this demonstration, can you?

Wading through lies in search of the truth

The maps in both Figure 4-2 and 4-3 are lying. But the map in Figure 4-3 provides the most accurate — that is, most globe-like — perspective regarding the shortest route between New York and Singapore. I'd really love to be able to prove that to you right here on the page of this book, but therein lies the problem — literally. This page is flat. To find out which route is shortest, you need a map that really looks like the world itself. That is, you need a globe.

Because a globe doesn't come with this book, you have to come to grips with the four ways in which maps can lie: distance, direction, shape, and area.

TECHNICAL STUFF

Most flat maps lie with respect to at least two characteristics, and some lie in all four aspects. In modest detail, here is the lowdown on exactly how and why these fibs occur.

Distance

Theoretically, transferring a curved Earth to a flat map involves selectively stretching some parts of Earth's surface more than others. For example, imagine two cities are 1,000 miles apart and the land between them gets stretched a great deal during the map-making process. Now imagine that elsewhere on Earth, two other cities are also 1,000 miles apart, but the land between them gets stretched just a little to make the very same map. On the resulting maps, the distance of 1,000 miles isn't portrayed the same.

Direction

The situation with direction is pretty much the same as with distance. By stretching a globe to make a flat map, true directions become incorrect. If some parts of the globe are stretched more than others, then a north arrow placed on one part of the map may point in a different direction than a north arrow placed elsewhere.

Actually, it's possible to make a map that keeps true directions throughout its surface. The Mercator Projection, a rather famous map introduced later in the chapter, is an example. But maintaining true direction can only be achieved by distorting something else. As the Mercator Projection shows, that something else is distance and area.

Shape

Shape refers to the outline of objects on Earth's surface. In the process of projection, you can transfer a continent or island from a globe to a flat surface while keeping its shape pretty intact. Then again, you can make a complete mess of things because stretching here and pulling there is part and parcel to the projection process and may play havoc with shape.

For example, compare Greenland in Figures 4-2 and 4-3. Notice that the island appears very differently in the two maps. Greenland's shape is virtually correct in Figure 4-3 because the lines of longitude meet at the North Pole, just as in reality. In Figure 4-2, however, Greenland is seriously misshapen because the lines of longitude do not meet at the North Pole but are instead spread apart in the polar area. The result is a greatly distorted Greenland.

But before we sing the praises of Figure 4-3, compare the shape of Northern Africa on both maps. Africa appears much more accurately in Figure 4-2 because in that map, the spacing of North Africa's lines of latitude and longitude are pretty much true to life. In Figure 4-3, however, North Africa appears to have become an accordion. It has been stretched laterally out of proportion to its true shape. That happens because as the lines of longitude extend outward from the center point — the North Pole — the projection excessively stretches the distance between them. As a result, North Africa has a flattened appearance.

Area

Area refers to the size of objects on Earth's surface. As is the case with shape, you can transfer (project) some features from a globe onto a flat surface while keeping sizes accurate relative to other objects on Earth's surface. Then again, you can make a complete mess of things. As to the reasons why, well, I apologize that this is sounding like a stuck record, but the simple fact is that stretching here and pulling there to make a flat map screws up the relative sizes of continents, oceans, and everything else on Earth.

Isn't there a truthful map anywhere?

Many maps are honest. But before I point some of them out to you, let me re-emphasize that flat map untruthfulness is related to Earth's curvature. Obviously, big portions of Earth involve more curvature than small portions.

A flat map of the entire world is going to lie a lot because so much curvature is involved. In contrast, a flat map of the United States has the potential of being more truthful (strictly geographically speaking) because the area of the United

States has less curvature than the entire world. A flat map of the town or area in which you live — well, now we're talking little fibs as opposed to big lies because your local surroundings do not have *that* much of Earth's curved surface. And if we're talking about a map of your backyard that could be an absolutely honest map because Earth's curvature over such a small space is virtually nil.

So, yes there are honest maps, but only ones that involve relatively small portions of Earth's surface. Geography, however, involves study of the whole Earth or portions of it that typically are bigger than your backyard. That means curvature is involved and therefore the likelihood of dealing with dishonest maps.

The one and only honest map: The globe!

A *globe* is a spherical map of the world. I'm almost embarrassed to write that because everybody knows a globe when they see one. But over the years, I've been amazed at the number of people who tell me that a globe isn't a map because, according to them, maps are by definition flat. Not so. A globe is a representation of Earth; so, by definition, it most definitely is a map.

The globe is the one and only honest map of the world. Because the globe has the same shape as Earth, the appearance of Earth on a globe is free of distortion. Put differently, a globe doesn't lie flat so it doesn't lie at all (except for maybe the information displayed on or left off the globe). On a side note, globes are very attractive and fun to look at. Place one conspicuously in your home and guests are likely to think you have good taste and are very intellectual.

Honesty is the best policy, except . . .

Globes are truthful and the truth counts, but globes have four major disadvantages relative to flat maps.

Limited field of view

No matter how you look at a globe, you can never see the whole world at once (unless you're in a room full of mirrors, but forget that as a practical solution). Indeed, when you calculate the geometry, you cannot see even half of the world at once on a globe. However, it's often desirable to view Earth in its entirety or to visually compare far away parts of the world. These perspectives aren't possible on a globe but are possible on flat maps.

High cost

TIP

Globes are comparatively more expensive than maps. I checked the website of a well-known company that makes wall maps, atlases, and globes. The basic globe (12-inch diameter) sells for about four times the price of the basic world wall map and about twice as much as a really good world atlas. Want a map of the world without paying the world for it? Buy a flat map.

Before going any further, why is this book of maps called an atlas? See the sidebar "Why is an atlas called an atlas?" (later in this chapter) to find out more.

Lack of detail

Because globes entail the whole world, they tend to show less detail. Next time you're face-to-face with the typical desktop globe, look for the region in which you live. Unless you are a resident of a big city, there's a good chance the globe doesn't show your hometown. And suppose you wanted a detailed map of your home area. How big would a globe have to be to include that kind of information? Probably as big as the Empire State Building. Globes are good for giving you the big picture, but if you want to view an area in detail then you better get a flat map.

Inefficient data storage

Two paragraphs ago, I mentioned a globe with a 12-inch diameter. If you want to take it somewhere, you can't fold it up and put it in your pocket. It probably won't even fit in your backpack. In contrast, I have an atlas that is 12 inches long, 8 inches wide, 1.5 inches thick and contains more than 100 maps. Better still, you can find all kinds of different maps via the internet. By comparison, globes are very inefficient when it comes to data storage. (Besides, it's very difficult to walk around carrying a globe and look cool at the same time.)

How serious are these disadvantages? So serious that you'll need to amend a pearl of wisdom you learned as a kid. Honesty is the best policy except when it comes to globes. Globes are truthful, but the truth in this case comes at a very high and bulky price.

Telling the truth, but telling it skewed

It's certainly true that geography seeks to provide accurate information about Earth. It's also true that flat maps are inaccurate and therefore counterproductive to the pursuit of truth — at least in a limited sense. But the four disadvantages of globes are so serious that geographers prefer dispensing with honesty (globes) and using flat maps even though they lie. Indeed, those disadvantages of globes may be recast as advantages of flat maps:

- » **Unlimited field of view:** You can show as much or as little of Earth as you want on a flat map.
- » **Low cost:** Flat maps cost much less than globes. In fact, a good-sized atlas containing hundreds of maps may cost less than a single globe.
- » **Accommodates detail:** Want to show a small area in great detail? Not a problem on a flat map.
- » **Efficient data storage:** You can fold up a flat map and put it in your pocket. Or you can put the equivalent of a hundred globes in a single atlas and carry it in your hand or stick it in your backpack. Or you can search online via your smart phone for a map of about just any place you want. That's better than trying to carry 100 globes, right?

The bottom line is that it's okay if flat maps lie, as long as you know you are being lied to and understand the nature of the lie.

Different Strokes for Different Folks: A World of Projections

If you are a veteran map-gawker, you know that all world maps don't look the same. And if you're not, then look again at Figures 4-2 and 4-3. Figure 4-2 looks something like a rectangle, shows the entire Earth, and is centered on the intersection of the Equator and Prime Meridian. Figure 4-3 is a circle, shows only the Northern Hemisphere, and is centered on the North Pole. As mentioned earlier, the two maps offer contrasts with respect to the ways maps lie: distance, direction, shape, and size.

The appearances in the maps differ because of different kinds of projections. That is, the maps are products of different methods of transferring the curved globe to a flat surface. Over the years, cartographers have developed literally dozens of different projections. Most maps are accurate and/or visually pleasing in some respects, although inaccurate or visually displeasing in other respects.

At this point, you may feel like saying, "Look, Jerry, why don't you spare me the details? Just tell me which projection is the best one so we can move on to the next chapter." I wish it were that simple; I really do. But the simple fact is that a winning projection doesn't exist. Every projection has good points and bad points. The trick is to know the pluses and minuses of particular projections so that choosing the best map for specific purposes is easier. It really is a case of different strokes for different folks, or at least different projections for specific situations.

If you're starting to think that this is a somewhat arcane field of study, well, you're right. As a new geographer, you don't need to commit map projections to memory. (I know several professors of geography who don't go near this stuff!) What is important, however, is that you appreciate the variety and complexity of map projections and understand that even though all flat maps lie, some do a pretty good job of showing all or part of Earth.

All in the (map) family

Generally speaking, map projections belong to one of three families: azimuthal, cylindrical, and conic (see Figure 4-4).

>> **Azimuthal (or planar):** A flat piece of paper (or plane, hence *planar*) is placed against the globe. The globe is then projected onto the flat paper, rendering a flat map.

>> **Cylindrical:** A paper cylinder is placed over a globe. The globe is projected onto the paper. The cylinder is then cut vertically and unwrapped from the globe, yielding a flat map of the world.

>> **Conical:** A conical paper *hat* is placed on the globe. The portion of the globe under the hat is projected onto the paper. The paper is cut in a straight line from its edge to the tip of the cone. The cone is then opened up and put down flat.

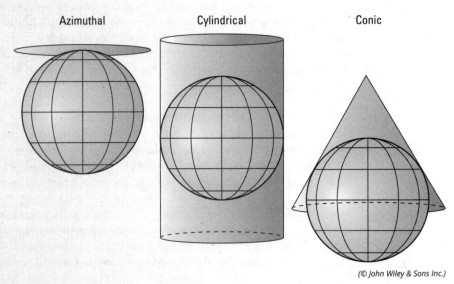

FIGURE 4-4: Families of map projections.

(© John Wiley & Sons Inc.)

REMEMBER

This reminds me to remind you that the process of projection does not literally involve projecting a globe onto a flat surface. Instead, mathematical formulas are used to plot the locations of lines (latitude, longitude, continental boundaries, and so on) on maps. Thanks to satellite imagery, you can now check the accuracy of your work in a way that was never possible before.

Five noteworthy liars

TIP

Here are five rather well-known projections that represent the range of formats shown in Figure 4-4. There will not be a test over this. I repeat, there will not be a test. So don't try to memorize this stuff, but instead, just sort of let the maps visually soak in to give you an appreciation of the variety of projections that are available.

The Mercator projection

Gerhard Kremer, who's much better known by his adopted Latin name, Gerardus Mercator, developed the Mercator projection in 1569. This cylindrical projection (see Figure 4-5) is easily the most famous world map of all time. Mercator crafted his projection to aid navigation, and in that regard, the map is a gem. Straight lines on this map correspond to true compass bearings so a navigator could use it to plot an accurate course. This achievement was a very big deal in the late 16th century, and by the middle of the 17th century, a majority of Western European navigators swore by this map.

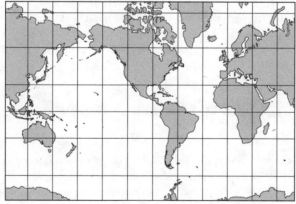

FIGURE 4-5: The Mercator projection.

(© John Wiley & Sons Inc.)

CHAPTER 4 **Truthiness in Mapping** 53

Because of its seafaring fame, the Mercator Projection later came into widespread use as a general-purpose map. That is, it found its way into classrooms as wall maps and into books and atlases. It became more or less *the* official world map, which is unfortunate because, although the shapes of landmasses are fairly accurate, the projection is extremely distorted with respect to size.

Notice that the lines of longitude on the Mercator projection don't meet at the Poles, as is the case in reality. Instead, the map shows the lines of longitude as parallel lines. This means that the North and South Polar regions have been stretched and become lines (the top and bottom borders of the map) that are as long as the Equator — 25,000 miles. One result is that land areas become disproportionately enlarged the closer they are to the areas of maximum distortion — the Poles. Alaska and Greenland are good examples. Alaska appears much larger than Mexico, while Greenland appears much larger than the Arabian Peninsula. In reality, Mexico is larger than Alaska, and the Arabian Peninsula is bigger than Greenland, but you'd never know by looking at the Mercator projection.

TIP

Keep in mind that there is nothing wrong with this projection. It is a representation of Earth, nothing more. Are there better representations for showing the size of Earth features? Sure. But this projection shows shapes quite well. Just as a hammer is great for striking a nail, but poor for drilling a hole, the same idea is true for projections. Some tools are better used for some purposes than others. It's up to the user to be wise about that choice.

Quite famously about two decades ago, a very reputable news magazine was not so wise. Hoping to portray how far North Korean missiles could travel, they drew a set of concentric circles atop a Mercator projection. This of course ignored the distortion toward the poles and made the missiles appear to have a much shorter range than reality. The implication? Hand a wrongly made map to a policy maker and you could have decision making that *does* amount to a whole world of trouble.

WHY IS AN ATLAS CALLED AN ATLAS?

An *atlas* is a book of maps. For the longest time, maps were published singly and tended to be stored as rolled-up scrolls standing in a corner. Gerardus Mercator was apparently the first person to compile a book of maps. His publisher decided to decorate the cover with a likeness of Atlas, the legendary Greek giant who supported Earth and the heavens on his shoulders. Other books of maps copied Mercator's idea and the image of Atlas on the cover or title page became standard — which is why such volumes are called atlases.

The Goode's Interrupted Homolosine projection

Noted American cartographer Dr. J. Paul Goode (1862–1932) developed this cylindrical projection (see Figure 4-6). It's an *equal area projection*, which means that the land areas are shown in their true sizes relative to each other. In that respect, Goode's projection is far superior to Mercator's. *Interrupted* refers to the map's outline. Earth is *cut into* once above the Equator and three times below it. Therefore, the Northern Hemisphere appears as two lobes and the Southern Hemisphere as four lobes.

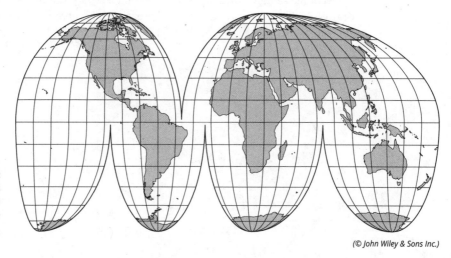

FIGURE 4-6: Goode's Interrupted Homolosine projection.

(© John Wiley & Sons Inc.)

As a result, the map's outline is not a rectangle or some other compact form, but instead is *interrupted*. The word *homolosine* reflects the fact that Goode's map is a combination of two other projections: the Mollweide homolographic and the Sinusoidal. (Whether or not you ever learn what that means, I will be happy to give you extra-credit for correct spellings.) Although Goode's projection appears in various atlases and despite its desirable equal-area attribute, many people are visually uncomfortable with its interrupted format.

The Robinson projection

Dr. Arthur H. Robinson, a noted American cartographer, introduced this cylindrical projection in 1963 (see Figure 4-2). If you lie really well, people may not notice. In fact, they may love you because of it. With all due respect and admiration to the good doctor, his map lies really well!

Although the projection contains distortion with respect to size and shape of land areas as well as to distance and direction, it has good overall balance with respect to these elements. In particular, the high latitude land areas are much less distorted than in the Mercator projection. Furthermore, Robinson's format does not have the interruptions of Goode's map. As a result of these pluses, the Robinson projection has become one of the popular choices among publishers of atlases and classroom wall maps.

The Lambert Conformal Conic projection

Johann Heinrich Lambert (1728–1777), a noted German physicist and mathematician, developed the Lambert Conformal Conic projection in 1772 (see Figure 4-7). Projections cannot correctly show the shapes of large areas, but they can be drafted such that the shapes of small areas closely *conform* to reality. That is what the Lambert Conformal Conic Projection achieves.

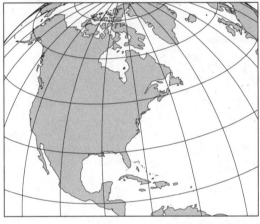

FIGURE 4-7: Lambert Conformal Conic projection.

(© John Wiley & Sons Inc.)

Accuracy of shape (*conformality*) is most closely achieved where the cone, which is intrinsic to a conic projection, touches the globe. If you refer back to Figure 4-4, you can see that the conic projection makes contact in the latitudinal vicinity of the United States. For Americans, therefore, this projection is noteworthy because it is commonly used to make maps of their country.

The Lambert Azimuthal Equal Area projection

The same Herr Lambert who developed the conformal conic projection (see the preceding section) presented the Lambert Azimuthal Equal Area projection in 1772 (see Figure 4-3). Because it's an azimuthal projection, as shown in Figure 4-4, it portrays only a hemisphere, as opposed to the entire world. On the other hand, it has two positive aspects: Areas are shown in true proportion to the same areas on Earth and, as revealed in my New York-to-Singapore exercise (see "Singapore, please. And step on it!" earlier in this chapter), long-range directions are depicted with a fair amount of accuracy.

Mapping a Cartographic Controversy!

If you're under the impression that the world of map-making is rather staid and geeky, you're right. In recent decades, however, a map known as the Peters projection has come along and stirred things up. An episode of the television series *The West Wing* showed just how geeky, wonky, and controversial map projections like Peters can be. Although this projection is controversial to some, it serves as an excellent example of why average citizens and novice geographers ought to know the facts about flat maps.

The Peters projection was introduced and subsequently promoted in 1972 by Arno Peters, a German historian (see Figure 4-8). It's also the subject of his book *The New Cartography (Die Neue Kartographie)*. As far as accurately showing the world is concerned, this map lies just like any other. The appearance of the continents has been likened to wet laundry hanging out to dry. The shape of land and water bodies is badly distorted, but size is maintained.

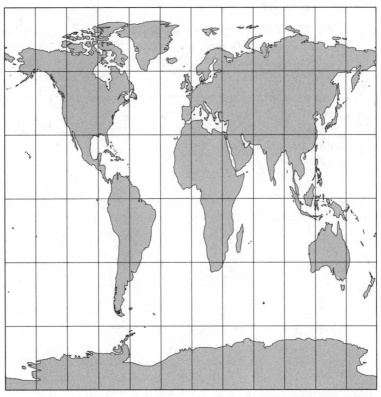

FIGURE 4-8: The Peters projection.

(© John Wiley & Sons Inc.)

This size issue is key for advocates of the Peters projection, saying that it renders an important measure of cartographic justice for tropical less-developed regions. They claim that by inflating the size of high-latitude regions relative to the tropics, the Mercator and some other projections present a Europe-centered view of the world that denigrates places in Asia, Africa, and South America. Proponents point out that the Peters projection is an *equal area projection* that shows tropical regions in their true size relative to, say, Europe and North America. As a result of such advocacy, several agencies with strong interests in these places have adopted the Peters projection as their official depiction of the world.

But is this projection really any better than the others? When you look at the facts of the matter, three things are obvious:

- First, the Peters projection terribly distorts shape (especially near the Poles and Equator) and distance.

- Second, there is a perfectly good alternative to the Peters that is an equal area map *and* depicts shape of tropical regions with considerable accuracy — the Goode's projection.

- Third, there is nothing new about *The New Cartography*. The Peters projection is a knock-off of a projection that was developed by James Gall in 1885. As result, it is usually called the Gall-Peters projection nowadays.

REMEMBER

So, is there one perfect world projection that lies flat without lying? No. Is there one perfect projection for all situations? Again, no. Representing Earth cartographically can be as crazy as all the physical and human features contained within it. Do your best!

IN THIS CHAPTER

» Knowing what a map is showing

» Measuring distance and size

» Taking a look at graphics

» Using symbols to depict reality

» Finding ways to gather information

» Harnessing new technologies

Chapter **5**

Telling a Spatial Story

I collect old maps, particularly ones of places I once lived like Maryland, Pennsylvania, California, and so on. So, I'm a geek, I guess. Before you think of yourself as being too cool, remember that no one made you pick up this book. So, I guess I'm not the only geek out there!

Of those maps, one is my most prized and until I found a copy in a dusty shop in Savannah, Georgia, I had not seen one like it. It's a map of major world rivers and mountains published in *Gray's Atlas* in 1874. But it isn't laid out like a traditional map. Each major river is positioned side by side from longest to shortest, and each major mountain is in a row by continent from shortest to tallest. I find the comparisons and different layout fascinating and spatially pleasing for some reason only geeks like you and me can understand. A neat added feature in one corner is a smudged fingerprint. I love the idea of someone like me — or you — exploring physical Earth space with this map nearly 150 years ago.

Over the years, and much to my relief, I have met numerous other geography geeks (geography teachers and students among them) who, however meekly, admitted to similar map affections. Indeed, such behavior turns out to be perfectly normal for people who, whether or not they know it, have a yearning for geography. No doubt, that is because the map is the most basic geographical tool.

Complementing the previous chapters on Earth's grid and the properties of projections, this chapter focuses on ways in which maps communicate information and how some of that information is obtained. Basically, this chapter is about map reading and map information for the budding geographer. Therefore, if you are, in fact, one of those people who can stare at maps for hours, then you can probably skip this chapter. But if maps confuse you or seem overwhelming, or if you have never been taught the fundamentals of maps and map reading, then this chapter is for you. While you probably won't master all there is to know, you can familiarize yourself with enough fundamentals so that you get the message of maps.

Why We Need Tal(l) Dogs

The basic function of maps is to show how particular phenomena are distributed over all or part of the world. *Cartographers* (mapmakers) communicate these and other kinds of information in part by incorporating into their maps a standard set of elements whose purpose is to help the map-reader get the message.

An easy way to remember these elements is with the acronym TALDOGS:

- **Title:** The title conveys the subject of the map and is the first thing a map-reader should look for. Ideally, its wording is simple and accurate. If the title confuses you, then that is probably more of a comment on the cartographer's communication skills than your intelligence quotient.

- **Author:** Knowing who made the map might say something about the map's credibility. Are you more likely to trust a map made by the National Geographic Society or one you found online created by joebubba12@email.com (that's right, we just made that email address up, so no writing in, Joe!)?

- **Legend:** Maps commonly convey information with the aid of symbols whose meanings may be uncertain. Thus, the cartographer always provides a legend (or *key*) that contains and defines the symbols found on the map.

- **Date:** Every map in print is out of date once printed — the world changes that fast. Newer digital mapping technologies are making this a problem of the past as we can update spatial data more quickly. However, knowing *when* data is collected can help us in making spatial and temporal comparisons.

- **Orientation:** *Orientation* is the alignment of the map with respect to *cardinal directions*: which way are north, south, east and west? The standard rule is that north is towards the top of the map, but not every layperson knows the rule and not every map follows it. Accordingly, many maps include a direction indicator, minimally a north-pointing arrow.

60   PART 1 **Getting Grounded: The Geographic Basics**

- » **Grid:** Many maps contain a couple of labeled grid lines of latitude and longitude (see Chapter 3) in order to convey the global context of the mapped area. If the cartographer has reason to believe that the map reader is intimately familiar with the mapped area, or if previous maps have indicated the global context of the mapped area, then grid lines may be omitted.

- » **Scale:** Scale (described more fully in the following section) provides information about the actual size of the area shown on the map. Typically, this is achieved with a small ruler-like entry on a map that equates distance in miles and/or kilometers with measurement in inches and/or centimeters.

Taking It to Scale

REMEMBER

Scale is the relationship between a distance as measured on a map and the corresponding actual distance on Earth's surface. Calculating distance between locations and comparing the size of areas are two of the more important functions of maps.

Going the distance

The scale of a map may be stated in three rather different ways, described in the following sections. Figure 5-1 shows you what the three ways look like. Some maps include just one of them. Others include two, and still others all three. Perhaps the most important thing to remember is that every map has a single scale, but a cartographer has three ways to tell you what it is. If, therefore, a single map contains two or three of the scale-types, then each is saying the same thing, albeit in a different way.

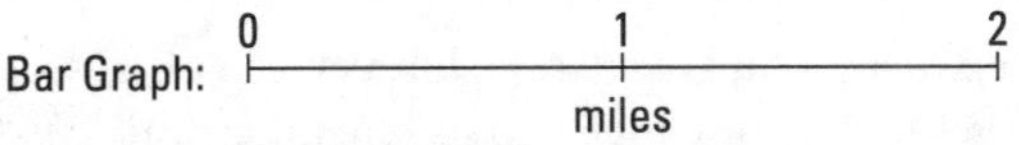

FIGURE 5-1:
Three ways to indicate the scale of a map.

Bar Graph: |—0—————1—————2—|
                    miles

Statement: One inch represents one mile

Representative Fraction: 1/63,360 or 1:63,360

(© John Wiley & Sons Inc.)

Scale bar

A *scale bar* looks like a miniature ruler. But whereas the ruler you use may show inches and millimeters, the one on the map shows miles or kilometers, (as shown in Figure 5-1). The principal virtue of the bar graph is that it provides a clear visual reference to the size of the area portrayed on the map. For actual measurement, however, it may be a bit unwieldy because you can't pick it up like you can a real ruler.

Verbal scale

A *verbal scale* (also called *statement of scale*) communicates the relationship between map distance and real-world distance in a sentence or sentence-like format. In Figure 5-1, "One inch equals one mile" is the example. (By the way, if you want to know why a mile is a mile, see the nearby sidebar "Whence comes the mile?")

As far as most people are concerned, the verbal scale is particularly convenient for measuring distances on a map, provided a ruler is available. In the case of "one inch equals one mile," one need only measure the number of inches between two points to arrive at the number of miles that separate them on Earth. If, on the other hand, the verbal scale on another map reads "one inch equals 20 miles," then the number of inches between the two points on the map needs to be multiplied by 20 to render the actual distance.

REMEMBER

Maps come in different scales. Thus, the scale you use to calculate distance on one map may not be the same for the next map. Always check the scale before you calculate distance.

Representative fraction (RF)

TECHNICAL STUFF

The area shown on a map is a fraction of its actual size. Appropriately, therefore, scale may be indicated as a *representative fraction (RF)*, which states the ratio between a unit of distance on the map and the same distance measured in the same units on the ground. As far as most people are concerned, this is the most confusing scale-type and the most difficult to explain. OK, here goes.

Check out Figure 5-1 again. The RF shown is 1:63,360. That means the map is 1/63,360th the size of the area it shows. Stated differently, a distance of one inch on the map equals 63,360 inches on the Earth's surface.

Once more, a given map has a given scale, but you can express it in different ways. In the example, therefore, "One inch equals one mile" and "1:63,360" must mean the same thing. And, indeed, they do. Proof is obtained by calculating the number of inches in a mile. To do that, multiply the number of inches per foot times the number of feet per mile (12 × 5,280). The answer is 63,360, so the statement of scale and the RF are, in fact, the same.

Comparing Earth at different scales

REMEMBER

Maps come in different scales; and because they do, the amount of area and degree of detail shown on one map may be very different from another. This is demonstrated in Figure 5-2, which shows three maps that have identical dimensions and progressively "zoom in" on Chicago. Specifically:

TECHNICAL STUFF

WHENCE COMES THE MILE?

A mile is a unit of linear measurement that equals 5,280 feet. While most of the world has adopted metric units (kilometers), Americans continue to express distance in miles, which, therefore, commonly appear as units of measurement on maps made in the U.S. But exactly what is a mile? And why does it consist of 5,280 feet instead of a more convenient figure, like 5,000?

"Mile" comes from the Latin *milia,* meaning thousand. In Roman times, a unit of linear measure called the *milia passum,* or thousand paces, was common. Somehow, somebody's thousand strides became a standard Roman mile, equal to about 1,650 yards. This measurement became widely used in Britain following the Roman's invasion. After the Empire's demise, however, the *milia passum* fell into disuse, although "mile" endured in the British vocabulary as a word applicable to a substantial distance.

The mile's present length has its origins in medieval English agriculture. Back then, a team of oxen was used to pull a heavy wooden plow. The farmer walked behind, making liberal use of an ox goad — a big stick — to influence the animals' behavior. The stick was known as a *rod,* and at some point its length was standardized to 16.5 feet. The length of a parcel of farmland was "a furrow long," or *furlong.* That was the distance the oxen could pull the plow before the farmer had to stop and rest them. Naturally, that length varied. In time, however, the furlong was standardized to a distance of 40 rods (660 feet or 220 yards). Sometime later, a distance of 8 furlongs (5,280 feet or 1,760 yards) became the standard mile, and remains so to this day.

» In Figure 5-2a, 1 inch represents 630 miles. As a result, this map shows a comparatively large area that includes most of the Great Lakes, Upper Midwest, a handful of major cities, and a portion of Canada.

» In Figure 5-2b, 1 inch represents 190 miles. What is shown now is a much smaller area that includes parts of Lake Michigan and Midwest states, a few medium-size towns, and a few major regional highways.

» In Figure 5-2c, 1 inch represents 64 miles. Now we have "zoomed in" to the extent that the map shows Greater Chicago, southern-most Lake Michigan, more municipalities, local highways, and several streets.

Notice that as the area shown on these maps decreases, the amount of detail increases. And if you think about it, that makes a great deal of sense. When 1 inch represents 630 miles — a large area — only very large surface features (such as the Great Lakes) can be shown. But when 1 inch represents 64 miles — a much smaller area — then comparatively small surface features (such as roads) can be effectively shown.

FIGURE 5-2: These three maps have different scales, and therefore differ in area and detail.

TECHNICAL STUFF

In the lingo of cartography, *small-scale* maps show large areas in little detail, while *large-scale* maps show small areas in big detail. Figure 5-2a has a comparatively small scale. In contrast, Figure 5-2b has a somewhat larger scale, while Figure 5-2c has the largest scale among the three maps. And indeed, as the scales of these maps get larger, the degree of detail increases. Calling the first map a small-scale map makes sense as the representative fraction, if actually divided, would result in a much smaller number than the other two maps.

Showing the Ups and Downs: Topography

All points on Earth have an elevation with respect to sea level. Altogether, they constitute "the lay of the land." (Keep in mind that elevation also pertains to points on the ocean's bottom.) *Topography* is the art and science of depicting heights and depths on a map. Like scale, topographic information is a basic feature of many maps and is commonly represented in three ways as indicated in Figure 5-3. The following sections discuss the three ways of showing topographic information.

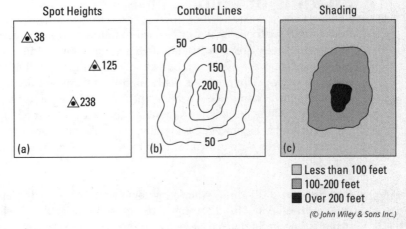

FIGURE 5-3: The terrain of a make-believe place, as depicted three ways.

(© John Wiley & Sons Inc.)

Spot heights

A *spot height* is a symbol (typically a tiny dot, plus sign, or triangle) accompanied by a number that indicates the elevation of a given point in feet or meters (see Figure 5-3a). Sometimes a cartographer wishes to emphasize something other than topography on a map, yet provide elevation information for a few selected points in order to convey the lay of the land. Spot heights serve this purpose.

Contour lines

Contour lines connect points of equal elevation. In so doing, they convey the shape (hence, "contour") of the land they depict. Near the top of Figure 5-3b is a thin line labeled "50," which connects points that are 50 feet above sea level. Farther inland is a line labeled 100, which connects points that are 100 feet above sea level. Thus, a walk from the water's edge to a point on the second line involves a 100-foot gain in elevation. These elevation line features are a major component of topographic maps published by the United States Geologic Survey.

One important feature of contour lines is also their ability to tell us about slope steepness. Contour lines very close together means a greater elevation change over less distance. In other words, in that place the slope would be steep.

Other maps have similar lines that connect equal data points using an isoline (*iso* means equal). Want to connect areas of equal temperature on a map instead of elevation? Using the same process, now you have an *isotherm* instead of a contour line. All maps that use isolines to connect features of the same value are called isopleth maps.

Shading or Color

Colors and gray tones may also be used to indicate elevation above sea level. On color maps, deep green is usually used to depict low-lying coastal land. Light green and yellow are used for progressively higher lands, followed by light brown and dark brown. The peaks of really high mountains are often shown in white (just like snow). Map readers need to understand these color gradations and what they signify; none of us is born knowing that brown means "higher elevation." Our brain does help, however. Generally, we think of darker colors as representing greater values, so dark brown to show higher elevations makes sense in many of our heads.

Like just stated for color, when gray tones are used in cartography, the general rule is "the darker the gray tone, the greater the value of whatever is being mapped." Accordingly, and as seen in Figure 5-3c, the lightest shade indicates the lowest-lying land, while deeper shades signify progressively higher elevations.

While spot heights and contour lines identify the precise elevations of precise locations, shadings refer to a range of elevations over an area. Thus, the lightest gray tone on Figure 5-3c signifies land that is anywhere between sea level and 100 feet above sea level.

DISTORTION FOR A PURPOSE

A *cartogram* is a map in which different areas are distorted in proportion to numerical data. Following are two maps of Australia. The one on the left shows the true shape of the continent. The one on the right is a *cartogram* in which the sizes of Australia's states and territories are distorted in proportion to their populations. As a result, the cartogram looks much different than the "real thing." New South Wales, which is home to Sydney (the nation's largest city), contains about 10 percent of the country's territory, but about 32 percent of its population. On the cartogram, therefore, New South Wales appears bloated. In contrast, Northern Territories accounts for 17 percent of the country's territory but only 1 percent of its population. On the cartogram, therefore, Northern Territories is quite small. These extremes visually highlight Australia's uneven population geography. Usually, cartographers seek to minimize map distortion. In the case of cartograms, however, distortion is the purpose of the exercise.

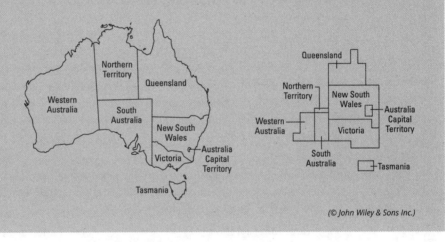

(© John Wiley & Sons Inc.)

TIP

But that's just this one map. On a different map, the same gray tones may mean something very different. Similarly, a light brown color may signify a particular elevation on one map, but a very different elevation on a different map. Always check the legend to make sure of the meaning of particular shades.

Using Symbols to Tell the Story

As highlighted by the discussion of topography, maps commonly show things by means of point, line, and/or area symbols. Each category, in turn, may display either qualitative or quantitative information. That is, each can simply show where something is located, or how much of something exists at a particular location or over a particular area.

Point symbols

Point symbols are used to locate discreet phenomena on Earth's surface. Most fall within one or more of the following categories:

» **Nominal icons** are tiny likenesses or symbols they name (hence, nominal) and indicate the locations of particular landscape features. Thus a tiny black dot (•) may be used to symbolize a residence while a cross (†) may be used to locate a cemetery or a church. Whatever the symbol, the cartographer must explain its meaning in the map's legend.

» **Ordinal icons** are very much like nominal icons except that they come in different sizes that suggest comparable size or order (hence, *ordinal*). On some maps, for example, a tiny airplane might be used to symbolize a small airport, while a larger airplane is used to indicate a major airport. Similarly, a lower case *u* might be used to pinpoint a minor uranium deposit while a capital *U* locates a major one.

» **Dots** are often used to show how the distribution of something varies numerically from place to place. Thus, for example, a map showing the geography of dairy cattle might use a series of dots, each one representing, say, 100 head of cattle. Similarly, a map of tobacco farming might use a series of dots, each representing, say, 100 acres of land in cultivation. As each dot represents some measurable quantity in a place, we call these dot density maps.

TIP

Be careful here. Using dots could be a bad representational choice. If we used one dot for each person in the United States on a map, you'd have trouble seeing the base map and each dot would overlay another. So dots are not always a good way to convey information.

» **Proportional symbols** vary in size in direct relation to numerical values. Thus, circles whose areas are proportional to population may indicate the locations and sizes of cities (Figure 5-4).

Line symbols

A number of important features on Earth's surface are linear in nature, meaning they look like lines, such as roads or railways. Likewise, migration, travel, trade, and other movements of interest to geography are basically linear phenomena that connect points. Accordingly, line symbols are common features on maps and take one of the following forms:

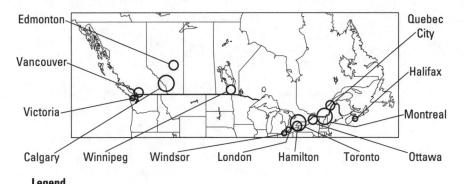

FIGURE 5-4:
This map uses proportional circles to indicate the size of cities.

(© John Wiley & Sons Inc.)

>> **Nominal lines** note the locations of particular linear features, such as roads, railways, rivers, and borders. They may appear as solid, dashed, or embellished lines, the standard symbol for railroads being an example of the latter. Colors may also be employed. Blue lines, for example, are commonly used to indicate rivers.

>> **Ordinal lines** vary in thickness or color to indicate relative importance. On many maps, for example, city, state, and country boundaries are progressively thicker so as to indicate the relative importance of the political units they mark. In Figure 5-4, the line that separates the United States and Canada is thicker than the lines that separate the states and provinces. Similarly, lines that symbolize roads often vary in thickness in proportion to the width of the highway or number of lanes.

>> **Flow lines** indicate movement, travel or trade along a given route or between two points. On some maps, the thickness of the lines varies in direct proportion to the quantity or volume of the flow. Thus, on a map of immigration, arrows of varying widths may be used to indicate the volume of movement between sender and receiver regions (as shown in Figure 5-5).

>> **Isolines** connect points of equal value with respect to a certain phenomenon. Didn't you just hear this recently? Oh yes, the contour lines shown in Figure 5-3b are an example (flick back if you missed that part of this chapter). Similarly, daily weather maps often contain isolines that connect points with identical atmospheric pressure or the day's projected high temperature or precipitation.

CHAPTER 5 **Telling a Spatial Story**

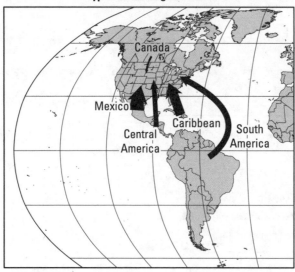

FIGURE 5-5: This map uses flow lines of different widths to indicate hypothetical migrant flow.

(© John Wiley & Sons Inc.)

Area symbols

Area symbols use gray tones or colors to depict phenomena that characterize areas as opposed to points or lines and are separated into two basic varieties:

>> **Nominal symbols** identify qualitative characteristics or phenomena that pertain to areas or regions. Figure 5-6, for example, uses nominal symbols to identify official languages of South American countries.

>> **Choropleth maps** (from the Greek *choros* and *pleth,* meaning place and value respectively) use colors or gray tones to show how the quantity or numerical value of something varies from one area to the next. Figure 5-7 uses gray tones to depict population density in South American countries. Choropleth maps always process data, and do not map raw numbers. These maps always show data *per* something else. Cows *per* square mile. Income *per* capita. Doing so removes the influence that the area size has on the data representation. For example, wouldn't you expect Texas to have a lot of tornadoes simply because Texas has so much land area? More informative is how many tornadoes that state has per square mile, say to Florida, for direct comparison purposes.

FIGURE 5-6: This map uses nominal area symbols to identify the distribution of Primary European Languages of South America.

(© John Wiley & Sons Inc.)

FIGURE 5-7: This choropleth map uses shades of gray to show population density in South America.

(© John Wiley & Sons Inc.)

CHAPTER 5 **Telling a Spatial Story**

New Ways of Seeing: How Technology has Changed How we Make and Use Maps

Map-making technology has come a long way from chiseling on stone, through ink on papyrus, through pen on paper splayed across a drafting table, and into the modern world. In the past three decades we have seen a revolution, a geospatial revolution, in how Earth data is collected, measured, and analyzed. That last part — analysis — is crucial. Data analyzed becomes information. Information can be acted on. This action takes the form of problem solving. And with this geospatial revolution, we are acting faster, more accurately, and hopefully for the betterment of all people than ever before. Three of these technologies — separate but integrated — are geographic information systems (GIS), global positioning systems (GPS), and remote sensing.

Geographic Information Systems

Today nearly all cartography at the professional level is done on a computer. The maps in this book are an example of this. Special kinds of software are available that allow cartographers to make maps with a degree of speed, accuracy, and data management that were unimaginable three decades ago. These qualities have also served to make mapmaking a powerful tool for a variety of businesses and planners. And in that regard, the most significant, cutting-edge field in contemporary cartography is the *geographic information system (GIS)*.

Giving you the complete lowdown on GIS would involve a lot of techo-babble that you don't want to read and I don't want to write. So perhaps the best way for me to describe GIS begins with a description of what it has replaced.

If you had poked around a city or regional planning office years ago, you'd be sure to find a huge table someplace with a huge base map that showed the streets and roads of the city or region in question. There would also be numerous overlays of different phenomena drafted on individual pieces of transparent film. For example, one transparent overlay might show the location of property boundaries. Others might show land use, sewage pipes, water mains, building characteristics, telephone lines, school districts, voting precincts, contour lines, wooded areas, and anything else that may be deemed useful for planning purposes.

Again, each characteristic would be on its own piece of transparent film — that is, its own map. So, if a planner wanted to see how two phenomena coincided geographically, the respective transparent films would be manually overlain on the base map and comparisons manually noted. Of course, the landscape changes. Thus, every so often a particular overlay would have to be manually updated or manually redrafted from scratch.

If all of this sounds a bit tedious, then you get the point.

With the advent of GIS, all of those physical base maps and pieces of transparent film have been replaced by layers of information that exist in computer memory (see Figure 5-8). This permits multiple layers, or even parts of layers, to be compared electronically, which is to say instantaneously. But the bottom line is that GIS has given geographers and planners the power to map and compare phenomena with great speed and accuracy. Indeed, remotely sensed images (I tackle that in the next couple of pages!) can be directly "fed into" a GIS, reducing to minutes and seconds a process of field observation and mapping that used to take weeks and months.

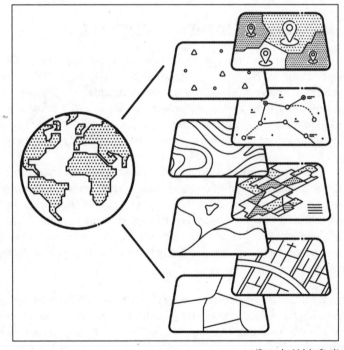

FIGURE 5-8: Geographic Information Systems (GIS) use layers of locational and attribute data about places for analysis and decision-making.

(© naschy / Adobe Stock)

REMEMBER

The beauty of GIS is the ability to ask questions, or query, the various layers. Let's say you want to locate a new grocery store. "Where" is a pretty important question. You might pull up a layer of population density, income levels, land for sale, established infrastructure such as sewers, competing grocery stores, and so on.

CHAPTER 5 **Telling a Spatial Story** 73

Collectively overlain, each of these layers helps you to eliminate areas that don't quite meet your criteria. The business applications are near endless, but think about what else we can do: Track disease! Establish school bus routes! Identify hurricane storm surge zones! It's endless!

Most of you have probably not used a GIS, but you've come close. Whenever you use an online mapping service to find your way between two cities, you've had a simple GIS-like product at your fingertips. There's a base map, there's data — say restaurants or hotels — and you have the ability to "query." Encoded into that map is not just locations but whether a street is one-way or not, and so on. As a result, your query about directions results in a selection of best paths for you to take, usually based on distance or time.

In the 1967 movie, *The Graduate*, moviegoers were told that the future is in plastics. Book mark this page now. The real future is in geospatial technology with GIS at the lead!

Global Positioning Systems

Few things are more important in cartography than the positional accuracy of mapped objects. Historically, this was accomplished by field observation. That is, explorers or surveyors would travel to a particular area, observe locally important features, and map their locations. Nowadays, GPS (global positioning system) technology has greatly contributed to positional accuracy. Think about exactly how accurate our spatial data needs to be for self-driving cars to work!

While you may think of GPS as the nice voice that gives you directions in your car, have you ever thought about the system that makes it all happen? The United States has launched a series of satellites (31 operational at the time of writing) that talk to a GPS receiver, often now in your smart phone. The United States is not alone here. Russia, China, India, Japan, and the European Union all have systems, so geographers are now likely to refer to a Global Navigation Satellite System to encompass it all.

Using *trilateration* — or measuring distances — the receiver and satellite bounce signals between each other and record the time the signal takes to be read. With three satellites doing the same thing, we can accurately locate you on Earth by putting you in the middle of a satellite Venn diagram. Add a fourth satellite and we can determine your elevation.

There are a number of really neat smart phone apps that allow you to collect data with accurate locations and other ones that allow you to get outside and play (and I mean, play!). Try *geocaching* some time, where you use GPS to undertake little treasure hunts. You'll learn something about your local environment, engage with cool technology, and probably get some exercise at the same time. What could be better?

Remote sensing

Getting information *remotely* is the game, here. In other words, someone — or rather, something — is doing the work for you. In remote sensing we are talking about gathering information at a distance, and mainly about Earth from above. This might be aerial photography from a balloon all the way to infrared imagery from a satellite hundreds of miles above you, and much of this information now makes up the base maps used in a GIS.

Numerous satellites monitor and provide map-ready information about Earth's surface and atmosphere. Virtually all of them utilize non-photographic scanners that produce thermal, infrared, or radar imagery. This information is processed and assembled into photo-like images. But we also use digital photography taken by aircraft or drone, and each type of sensor has special advantages.

Aerial photography

Aerial photography refers to photos of Earth's surface taken from aircraft. Today many maps produced under government approval at all levels, municipal through federal, are directly derived from aerial photography. Black and white film was a widely used medium historically, but today almost all of this is done with digital cameras.

Infrared photography is also very popular. Infrared energy is contained in the sunlight that strikes Earth and reflects off its surface. You and I can't see it, but special types of sensors can. Infrared energy readily passes through haze and air pollution, resulting in crisp images even on days when the atmosphere is far from clean.

Because of that very desirable characteristic, infrared photography is widely used in aerial surveys.

TECHNICAL STUFF

The gray tones and colors on an infrared photograph may be very different than those observed on regular black and white or color film. Because of that, the term *false color* is widely applied to infrared film and photographs. Most bizarrely, vegetation appears red. Indeed, the more lush or healthy the vegetation, which appears downright green to you and me, the redder it appears on an infrared photo. Differences in redness may indicate different kinds of crops or forests, or indicate plant life that is stressed because of disease or drought.

CHAPTER 5 **Telling a Spatial Story** 75

Non-photographic imagery

TECHNICAL STUFF

Like infrared photography, other remote sensing technologies record surface features in ways that are beyond the capabilities of human eyesight and normal cameras and film. Virtually all of them make use of sensors that scan Earth and record surface information electronically. In the lingo of remote sensing, you have aerial *photographs* (terminology left over from using film) and non-photographic *images*. Several image-types are widely used:

» **Radar imaging:** In radar imaging, a sensor emits continuous beams of energy that bounce off Earth and return to the sensor, which records them. Because the emitted beams travel at a known and uniform speed, the time that it takes them to make the round trip is a function of the elevations of the locations where the beams reflect. For example, a beam that bounces off a mountaintop takes less time to return to the sensor than one that reflects off a valley bottom. This information can be used to produce detailed images of terrain and very exact topographic maps.

Radar beams can penetrate clouds and fog with no loss of strength. Thus, radar imaging is extremely useful for monitoring and mapping Earth's surface in regions where atmospheric characteristics inhibit aerial photography (such as characteristically cloudy equatorial areas). It may also be used at night to the same effect as day. The same, of course, cannot be said of regular film.

» **Thermal imaging:** Thermal scanners (a form of infrared imaging) record heat differences on Earth's surface. This is particularly useful for mapping ocean surface currents (whose temperature variations have a major effect on weather and climate) as well as for identifying and mapping different kinds of pollution. It has also proved very useful in mapping and monitoring forest fires and other fire-related phenomena, especially in situations in which smoke prohibits analysis by means of standard photography.

» **LiDAR (Light Detection And Ranging):** This has really come into its own over the past two decades, primarily for its ability to make very high-resolution maps. Using laser pulses, the system has helped us better map ocean bottoms (*bathymetry*) and helped archaeologists find sites under vegetative cover. LiDAR highlights that mapping is never really done. Sure, there are always new things to map but as LiDAR shows us there are always new, better, and increasingly more accurate ways to do so.

Whether we are gathering photographic or imaged information, there's one other game changer in this whole process: the UAV. The what? An unmanned aerial vehicle, that's what. You probably call it a drone. While many people are playing with these and snapping pictures over their backyards, we know better. They are really a new army of remote sensors. Put one of those UAVs into the hands of a credentialed pilot with a powerful LiDAR sensor on board and look out. We can gather Earth data in really inaccessible places and make a lot of change for the better as was done in my state, South Carolina, by monitoring old agricultural dams that can cause big problems downstream if they fail.

So maybe the future is plastics. And metal. And glass. And circuitry. All the things that collectively make up the computing, sensing, and locational hardware behind GIS, GPS, and remote sensing.

Making Maps Yourself!

Just about anyone can make a map these days. There are a number of online software options that allow you to drop points on a map or make full-blown presentations of stories atop a base map of remotely sensed imagery.

But there's still one big problem: These new mapmakers have not read this book. As a result, they use the wrong symbols, don't process their data for choropleth maps, and use crazy color schemes that do everything but illuminate. Perhaps worse than having no map at all is one that is flat out misleading. That is, of course, only a problem if you want your maps to be truthful. Not everyone does. History is replete with examples: from World War Two Nazi maps (propaganda to show how much *lebensraum* (living space) Germany needed, thus a reason to conquer Europe) to Cold War Russians "hiding" cities by moving them from one map to another over time (why? To avoid American missiles.).

Be excited about your ability to map, but take care, too.

2 Let's Get Physical: Land, Water, and Air

IN THIS PART...

The natural world is big and complex. People want to make sense of it. That, as it turns out, is one of the principal tasks of geography.

Earth's surface is a mosaic of landforms covered by a rich variety of natural vegetation that is produced by diverse climate-types. Complementing this is a world of water, most of which is out of sight (and usually out of mind as well), on which life as we know it depends. None of these phenomena "just happen." Instead, mountains, plains, forests, climates, precipitation, oceans . . . every aspect of the physical world is the result of one or more natural processes that help explain the world we see and live in.

In this part, you will discover the key concepts and concerns of physical geography, which describes and analyzes the distribution of natural phenomena over Earth's surface. Yes, Earth is a big complex world. And as you will see, physical geography makes sense of it.

IN THIS CHAPTER

» Moving continents

» Giving rise to mountains

» Shaking and baking with earthquakes and volcanoes

» Considering terms: natural hazards or natural processes?

Chapter 6
Shape-shifting Earth

Earth originated about 4.7 billion years ago as a molten fireball and has been slowly cooling ever since. As a result, and after so many years, the outermost portion has hardened into a layer of rock called the *lithosphere* (from the Greek *lithos,* meaning stone). Most of this layer is so hot that it would literally fry your feet, along with the rest of you.

Fortunately, however, the outermost portion of the lithosphere is relatively cool. This sub-layer, called the *crust,* is no more than between 5 and 40 miles deep, so it accounts for a very small portion of planet Earth. But the crust has a degree of importance that is out of proportion to its volume because you live on it. The crust is your home.

The crust is also home to every kind of landform you have ever seen or will see — mountains, valleys, plateaus, plains, and so on. These features give character to different parts of Earth and are among the first things that come to many peoples' minds when they think about geography. And indeed *geomorphology*, the study of the nature and origins of landforms, is an important sub-field of geography. I'm just throwing in a new -ology for you there, free of charge.

A world war of sorts is engulfing your "crusty home." The combatants are two powerful opposing sets of forces that shape and reshape Earth's surface. On the one hand, and the subjects of this chapter, *tectonic* forces (from the Greek *tekton*, meaning "builder") build up Earth's crust. The pressures involved here are mighty enough to literally make mountains out of molehills, and also cause earthquakes and volcanoes to occur. Tectonic force has been modifying the crust for as long as crust has existed, and will continue to do so for billions of years to come. Thus, I can say with complete confidence that the force will be with you always.

On the other hand, *gradational* forces wear down the crust. Given enough time, they can transform today's mountains into tomorrow's molehills. Gradational forces are the subjects of Chapter 7.

Starting at the Bottom: Inside Earth

It would be great if you could go deep into Earth and see what's going on, but that's impossible. The average distance from Earth's surface to the center is 3,960 miles, and no human has ever come close. Several books and movies have portrayed such fanciful feats, but the truth is that people have barely penetrated the crust. The record is held by Russian researchers who managed to drill almost 8 miles deep near Murmansk from 1970 until 1989. They had to stop when Earth's heat began warping drill bits and rock behaved more like plastic. But don't worry about falling in yourself! The hole is only about 9 inches wide. So instead of going on a fantastic journey to see Earth's innards, you must settle for a diagram (as shown in Figure 6-1) based on informed speculation. Looking at it may cause you to wonder, "Well, if nobody's ever been down there, then how do you know what it looks like?" Great question! Check out the "How do we know what's down there?" sidebar for the answer.

The composition and temperature of Earth's interior are the reasons nobody has ever gone there and probably never will. Most of that realm is molten or almost molten. Thankfully, not only is it out of sight and out of mind, but also out of touch. Were it not for the insulating crust, life as we know it simply would not exist.

TECHNICAL STUFF

Directly beneath the lithosphere lies the *asthenosphere*. Measured in the thousands of degrees Fahrenheit, its rock assumes a plastic, almost molten quality. Directly beneath the asthenosphere is a vast volume of somewhat stronger rock and below that liquid iron of the outer core and solid iron of the inner core that is hotter still (as shown in Figure 6-1).

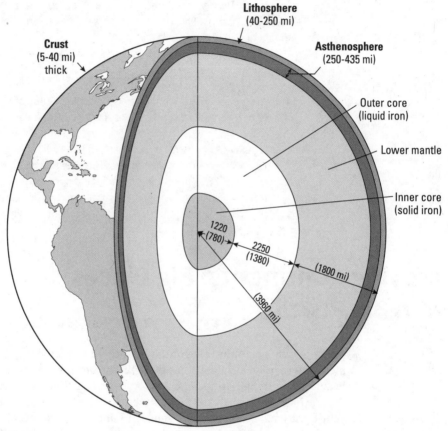

FIGURE 6-1: A cut-away view of Earth.

(© John Wiley & Sons Inc.)

Altogether, that vast volume of incredibly hot stuff is a powerful source of pressure — tectonic force. Indeed, it is mighty enough to create and rearrange continents, and in the process build mountains and cause earthquakes and volcanoes to occur. This knowledge has been available for only a couple of decades. But the idea of a force powerful enough to move continents has been around for centuries.

TECHNICAL STUFF

HOW DO WE KNOW WHAT'S DOWN THERE?

Our understanding of Earth's interior rests on a combination of inference, analysis of alien objects, sound waves, and rocks and minerals. By alien objects, I do not mean UFOs, but instead meteorites and such that have fallen to Earth. These uniformly reveal a high percentage of iron. Because these alien objects are the result of the same process of planetary formation that produced Earth, the assumption is that the proportion of iron in these objects is probably about the same for planet Earth. That suggests an incredible amount of iron beneath your feet.

Earthquakes produce sound waves. Over the past several decades, *seismologists* (people who study earthquakes) have placed within the crust hundreds of "listening devices" that record and analyze sound waves made by earthquakes. Some of these waves, it turns out, have peculiar characteristics: They cannot penetrate liquids, or liquids deflect them, or they travel at different speeds through liquids and through solids with different characteristics. Analysis of the tracks and characteristics of literally hundreds of such waves, plus the previous inference concerning iron, provide much of the input for Figure 6-1. Also, geologists have studied lots of rocks and minerals that have been thrust up through Earth's crust. Analysis of these materials reveals a relative scarcity of iron, which suggests this substance must be concentrated deep within Earth.

Moving Continents: Big Pieces of a Big Puzzle

The continents on either side of the Atlantic Ocean can easily be viewed as giant pieces of a jigsaw puzzle. If that thought has never occurred to you, then take a look at a world map, focusing on the Atlantic coastlines of Africa and Europe on the one hand, and North and South America on the other. Imagine you could place your right hand on the former, your left hand on the latter, and push the pieces together. It really does look as if they would fit together, doesn't it?

TIP

Even little pieces seem like they can be merged with larger ones! Go on, take a look at Madagascar and tell me its cousin isn't right next door in southeastern Africa.

People began puzzling over this almost as soon as those shorelines were accurately mapped. Many, such as Abraham Ortelius (way back in the 16th century!), were convinced the "fit" was too good to be mere coincidence. The implication, of course, was that there was once a super-continent that broke into big pieces that subsequently moved apart. But how could this happen? What gigantic force was responsible? And where did it come from?

Where have you gone, Gondwanaland?

Various explanations were proposed. In the 1850s, Antonio Snider-Pellegrini suggested that during Noah's time, Earth had several deep volcano-related cracks. Water pressure during The Flood exacerbated these cracks, created the continents, and moved them apart. A few years later, Eduard Suess proposed that a supercontinent he dubbed *Gondwanaland* (after a geological area in India) had fragmented and broken apart for reasons that he did not identify or endorse.

Alfred Wegener, mover and shaker

REMEMBER

The greatest theorizer of all was Alfred Wegener, a German geographer and meteorologist. In 1915, he proposed a theory that explained not only the shapes and locations of continents, but also the geography of mountains. According to Wegener, Earth's surface once consisted of a single supercontinent called *Panagea* ("all the Earth") and a single world ocean called *Panthalassa* ("all the seas"). Pangea subsequently broke into two pieces of roughly equal size: a northern component called *Laurasia*, and a southern component called *Gondwanaland* (borrowing from Suess). Both of these, in turn, later broke up. Pieces of Laurasia became North America, Central America, Greenland, Europe, and Asia. Pieces of Gondwanaland became South America, Africa, Australia, and Antarctica

These pieces, Wegener suggested, subsequently "drifted" apart, hence the popular name for his theory, *continental drift*. These continental "rafts" did not float on water, but instead bulldozed their way over other firmament (ocean bottom, typically). Now, at some time in your life you have probably watched a bulldozer do its thing. It scrapes the surface and produces a pile of debris as it moves forward. As Wegener saw it, that is exactly what the continental rafts were doing: scraping Earth and producing mountains and mountain ranges along their leading edges. Thus, the "rafts" containing North and South America were drifting westerly, bulldozing as they went, the result being the chain of mountains that today extend along the west coast of the Americas from Alaska to the southern tip of South America.

Wegener's theory had a number of partially correct elements, but it remained unproven until after his death because the matter of how Pangaea was broken up was unresolved. Obviously, a mighty force was required to break up Pangaea and cause its pieces to move. But what could that force be, and where did it come from? Could it be wind driven? Tidal? Related to Earth's rotation? Again, that was the puzzle within the puzzle.

Puzzle solved!

Imagine a mountain range about 10,000 miles long and nobody knew it existed until the latter half of the twentieth century. Sound bizarre? Well, it really happened, and of course, there's a catch of sorts. The range in question is the Mid-Atlantic Ridge. As the name implies, it pretty much runs down the middle of the Atlantic Ocean. It was virtually unknown until ocean-floor mapping (a very deep subject!) revealed its presence.

Through use of remotely-controlled submersibles that carried cameras and other instruments, it was learned that this mountain range is basically a 10,000-mile long active volcano. The Mid-Atlantic Ridge is the product of *magma*, molten material from beneath the asthenosphere. A series of deep cracks, or *fissures*, in the lithosphere run the full length of the Ridge. Over the years, magma has

- oozed upwards through the fissures in response to tectonic force
- piled up on the ocean floor
- hardened to form the Mid-Atlantic Ridge

But this oozing is ongoing. Slowly, but inevitably, other magma is rising up through the fissures, to cool and harden, and in doing so, elbowing to either side of the fissure the previously hardened magma. The result is a *spreading sea floor*. The Atlantic Ocean is getting wider. The New World and the Old World are moving apart.

Subsequent deep-ocean mapping revealed other spreading sea floors in other oceans. Here was the explanation of how continents had split apart and moved! Here was the solution to the great puzzle that had been driving people nuts for years!

But if Earth is growing along these ridges, why isn't the planet increasing in size? More on that shortly (see the section, "Subducting Plates: Volcano Makers").

Getting Down to Theory

In a manner of speaking, Earth's lithosphere is like an eggshell — a thin, hard, brittle outer covering that encases a big mass of goo. Due to tectonic force, that earthly eggshell is easily cracked.

REMEMBER

The Theory of Plate Tectonics explains what has been happening over the years. Here are a few points from that theory:

>> Tectonic force has broken up the lithosphere into seven major and eight minor pieces, or plates.

>> Tectonic force causes the plates either to move apart, collide, or slide by one another.

>> Mountains, volcanoes, and earthquakes result when plates collide.

>> Earthquakes may also result when plates slide by each other.

Taken together, these statements constitute much of the Theory of Plate Tectonics.

Tectonic force has broken the lithosphere into 15 plates that vary in size. The Pacific Plate, for example, covers millions of square miles. In contrast, the Juan de Fuca Plate, which borders the Pacific Northwestern States of the U.S., is barely visible on the map (as seen in Figure 6-2). Every plate is either on the move or is being affected by the movement of a neighboring plate. Arrows on the map indicate the direction in which different plates are moving. As geographers have seen, the Mid-Atlantic Ridge marks a boundary at which neighboring plates are moving apart. Other spreading sea floors also are evident on the map. But, if plates are moving apart along some of their boundaries, that means there must be other locales where they are meeting head on, sliding by each other, or sliding under each other (a process called *subduction*).

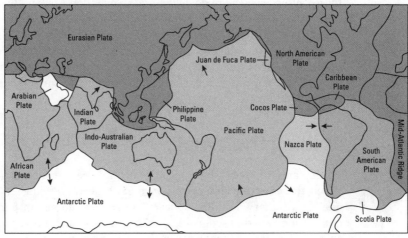

FIGURE 6-2: Plates and plate boundaries.

(© John Wiley & Sons Inc.)

CHAPTER 6 **Shape-shifting Earth**

Making Mountains Out of Molehills

Mountains and mountainous terrain are pretty hard to miss, and everybody knows them when they see them. Likewise, most people recognize the aesthetic appeal of mountains, their value as recreational resources, and the problems they pose for surface transportation systems, land settlement, and agriculture. In addition, however, mountains are of particular interest to geography for three reasons:

» **They are climate makers.** Mountains may cause some areas to have abundant moisture and others to be bone-dry. Thus, they are major factors in the geography of climate, as you will see in Chapter 9.

» **They are culture makers.** Mountainous terrain has historically tended to isolate people and impede their ability to share ideas and material things. Thus, as you will see in Chapter 13, they have tended to encourage development of separate cultures and act as barriers between cultures.

» **They are country makers.** Because they are such visible landscape features, mountains and mountainous features — such as ridges — have often been used to designate frontiers between countries and states. Thus, as you will see in Chapter 14, they are major factors in political geography.

For now, however, the focus is on the causes and consequences of mountain building *per se* rather than the climatic, cultural, or political effects. To get to those causes, we need to go down under. And I don't mean Australia.

Mountains that soar up into the sky have their origins way down below, well beneath Earth's crust. For that reason, tectonic forces are sometimes called *endogenous* forces. This comes from the Greek *endon*, meaning within, and another old word that is the source of "genesis." So endogenous forces have their genesis, or origin, within Earth.

Folding the crust

When plates collide head-on, a couple of outcomes are possible. One is folding, in which the crust buckles in response to the compression, and may eventually assume a rather wave-like appearance, as you can see in Figure 6-3. You can crudely simulate that sequence by doing the following: Put a piece of paper on the surface of a desk or table and place your right and left fingertips on opposite edges. Then very slowly move your hands together. Hopefully, the paper will assume a wave-like form as your fingers approach each other.

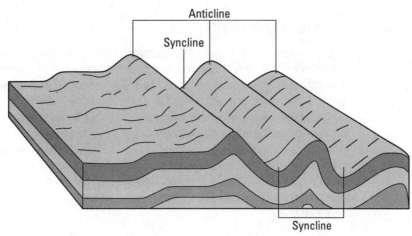

FIGURE 6-3: A folded landscape.

(© John Wiley & Sons Inc.)

Remember, however, that the Earth area represented by that paper contains hundreds of square miles, and that the convergence of your fingertips mimic plate movements that span millions of years. These folds are called *anticlines* (pointing up) and *synclines* (pointing down). It easy to remember which is which if you think about misbehaving people who "syn." They end up going down and spending eternity in a very hot place. The image in Figure 6-4 shows this work over millennia in Crete.

FIGURE 6-4: A photograph of exposed anticlines and synclines in Crete.

(© Nigel / Adobe Stock)

CHAPTER 6 **Shape-shifting Earth** 89

Former grandeur

The Appalachian Mountains of the eastern United States are a prime example of a folded mountain range. They do not, however, coincide today with a plate boundary. This tells geographers that plates and plate boundaries come and go over the broad expanse of geologic time. In some cases, therefore, mountains mark an active plate boundary where mountain making is in progress. In other cases, mountain ranges mark ancient (extinct) plate boundaries, and are themselves mere eroded remnants of what used to be. Today, the highest peaks in the Appalachians are in the 6,000 feet range. But orientation of various rock layers suggests to geologists that in their ancient heyday, the Appalachians towered 30,000 feet and more above sea level. That's higher that Mt. Everest, the tallest mountain on Earth today.

Making resources accessible

The geography of mining often coincides with the geography of mountains. The tremendous heat and pressure of mountain-building creates and reveals the presence of valuable ores and minerals, and facilitate their accessibility to humans. The coal reserves of the Appalachians are a case in point.

In the diagram of folded mountains (see Figure 6-3), assume that coal occupies the second layer, or *strata*, of rock from the surface. In the flat terrain on the left of the diagram, one can see no indication of a valuable resource underfoot. On the right, however, folding may reveal the coal. If not, then subsequent erosion — say, a river that "cuts through" the landscape — may reveal the presence of valuable strata. This, in turn, may give rise to economic activity that — such as in the case of coal mining in the Appalachians — is virtually synonymous with the region.

Resources created in the mountain-building process include marble, jade, and rubies. These are *metamorphic*, meaning that existing rock was changed into something new and in many cases much more valuable from a human point of view.

Whose "fault" is it?

In addition to folding, head-on collisions of plates may also produce mountains by *faulting*. In this case, a series of deep fractures develop through the crust. Over time, the immense pressure attendant to the slow-motion collision causes large-scale rock units to be raised (a *horst*) and sometimes lowered (a *graben*) along normal *fault lines* that mark the intersections of the fractures and Earth's surface (as you can see in Figure 6-5). Small landform features all the way up to mountains of significant size may result. And, actually, they have — the highest mountains on Earth, the Himalayan Range, are products of uplift along fault lines produced by collisions between the Indian Plate and the Eurasian Plate. In this range are found Mt. Everest, the highest mountain on Earth (29,028 feet), plus K-2, Kanchenjunga, Makalu, and every one of the twenty tallest summits on Earth.

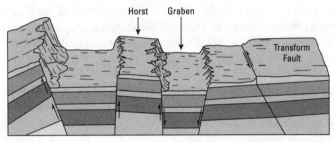

FIGURE 6-5: Types of faults.

(© John Wiley & Sons Inc.)

This impressive collection of peaks is only getting higher. That is the fault, quite literally, of the ongoing collision of the Indian and Eurasian Plates. As a result, Mt. Everest is growing by about one-half inch per year.

Another fault type is called *transform* (see Figure 6-5) where the motion is primarily horizontal compared to vertical as in a normal fault. One fairly large and long transform fault is already probably well-known to you. It goes by the name of San Andreas in California.

Plate tectonics: A four-letter word!

REMEMBER

That word is "slow." Mt. Everest is growing by about a half inch per year. And since the break-up of Pangea, it has taken the continents hundreds of millions of years to travel a few thousand miles.

Tectonic force has massive power, but the continents and crust have massive weight. That means massive *friction of resistance* needs to be overcome if anything is going to be moved. Looked at over long periods of time, therefore, alterations to the crust due to tectonic forces tend to happen real *slowly*. Or so things normally seem. As befits a four-letter word, things can get nasty. A dark side to the force is out there.

Experiencing Earthquakes: Shake, Rattle and Roll!

The study of plate movement rather naturally leads one into the field of *seismology*, the study of earthquakes. These are inevitable and dangerous consequences of plate tectonics. Why "inevitable?" Well, an *earthquake* is a sudden movement of Earth's crust. Given plate tectonics, earthquakes are inevitable. That is, sooner or later they're bound to happen.

During an earthquake, ton after ton of crust is moved. The amount of pressure required to do that is incredible; and it does not accumulate in a day. Therefore, at a given location, on an active plate boundary, earthquakes are not everyday ongoing events. Instead, it takes long periods of time — years and decades — for tectonic force to build up enough pressure to move a mass of crust.

The mechanics are crudely similar to inflating a balloon until it bursts. As the volume of air increases within, so does the pressure and tension along the surface of the balloon. Eventually, the pressure exceeds the balloon's capacity to contain it, and the balloon gives way . . . pop! Obviously, crust does not inflate and pop, but pressure does build-up slowly, especially along plate boundaries. Tension increases and keeps increasing over the years. And finally, perhaps after decades of pressure building, the crust just can't take it any longer. And so it suddenly gives. That is, it suddenly moves, releasing the built-up pressure, only to warn that a new cycle of tension build-up and release is underway.

Because tectonic force exists everywhere, an earthquake can happen anywhere. But given what you have read about plate tectonics, it should come as no surprise that the geography of earthquakes largely coincides with the geography of plate boundaries. Zones of spreading sea floors are prime candidate locales. So, too, are areas where plates collide. And so, too, is another possibility that has yet to be mentioned.

Splitsville in California

Sometimes neighboring plates do not diverge or collide, but rather slide by each other. The linear break in the rocks that marks the occurrence of this kind of movement is called a *transform* fault. California's San Andreas Fault, no doubt the most famous fault line in the United States, is an example of a transform fault. The land on the western side of the fault is part of the Pacific Plate and is slowly moving to the northwest. Meanwhile, the land on the eastern side, which is part of the North American Plate, is slowly moving towards the southeast. Pinnacles National Park in central California aptly demonstrates what has been going on for millennia; it's rocky origin exists in southern California nearly 200 miles to the south.

> ## THE NEW MADRID EARTHQUAKE(S)
>
> On December 16, 1811, perhaps the most powerful earthquake in the recorded history of the United States occurred near New Madrid, Missouri, located on the Mississippi River in the extreme southeastern part of that state. I say "perhaps" for two reasons. Scientific instruments that accurately measure earthquakes were not then available. And in the days and weeks that followed, two other major quakes (and literally thousands of much lesser ones) rocked the region. Either or both of those may have been stronger than the first.
>
> How powerful were they? Apparently, each may have registered an 8.0 or higher on our "old" Richter Scale. The Mississippi River changed course. An island in the river disappeared. New land rose. Forests were knocked down for miles around. The ground rolled in visible waves, wiping away houses, gardens, and fields. Fortunately, remarkably few people perished, mainly because the region was then only lightly populated.
>
> The quakes were not flukes, but instead the product of a fault zone that is minor in terms of its length, but major in respect to seismic potential. Geographically, what is most interesting is that New Madrid is more than a thousand miles from the nearest plate boundary. Thus, while the vast majority of earthquakes occur on the fringe of tectonic plates, the New Madrid episodes demonstrate that earthquakes (even very serious ones) can potentially happen anywhere.

Actually, and as has been seen, "sudden fits and starts" is more accurate than "slowly moving." Pressure slowly builds on both sides of the fault line. Every so many years, enough pressure accumulates to overcome the friction of resistance offered by tons of crust. At that point, parts of California slide by each other not as continuous slow movement, but rather in short and sudden spurts as earthquakes occur. As a result, California is slowly being torn apart.

People at risk

REMEMBER

Ultimately, geography investigates natural phenomena to gain information that is relevant to humans. Earthquakes are particularly significant because of their destructive potential and because so many earthquake-prone areas are densely populated. More important, therefore, than long-term scenarios such as the splitting of California are short-term consequences for Los Angeles, San Francisco, and several other major cities that are on or near a fault line. And California isn't the only place affected. Several other plate boundaries coincide with major metropolitan areas. That includes about a dozen or so major cities along the West Coast of the Americas from Mexico City to Santiago, Chile. The same applies to virtually all of Japan and many parts of Southeast Asia. Then consider the long interface between the Eurasian Plate and its southern "neighbors" that extends from the Himalayas westward through Italy. All told, several billion people live close enough to an active fault zone to be in harm's way. See "The New Madrid Earthquake(s)" sidebar for more information on other vulnerable places.

How earthquakes kill and maim

Collapsing buildings are the cause of most earthquake-related casualties. That is, most people who end up as statistics are either inside or next to a building that experiences structural failure. The walls give out. The roof caves in. The floors of a multistory building become something akin to a pile of pancakes. And people get crushed.

As a result, an adage of sorts goes, "Earthquakes don't kill people: *Buildings* kill people." Of course, those buildings didn't just up and collapse. It was the earthquake that caused the collapse, and the greater the magnitude of the quake, the greater the likelihood of casualties resulting from collapsing buildings. But rarely does a quake *per se* prove fatal. Certainly, the attendant terror has been known to induce heart attacks. But shaking earth *per se* is not a major killer. Usually a side effect does people in, the principal one being structural failure.

A matter of wealth and culture

Wealth and culture also play major roles in determining earthquake damage and casualties. Countries characterized by low average income and a traditional cultural environment tend to fare far worse than their wealthy, modern counterparts. Say, for example, that a strong earthquake measuring 7.3 on the Moment Magnitude Scale strikes a major American West Coast city and a city in a developing country. Chances are the toll in human lives and injuries would be far worse in the developing country.

The reason relates to differences in building construction and building codes. Buildings with walls of concrete, cinder block, brick, or adobe-like materials have a rather brittle quality, so they tend to "snap" and give way rather readily. It helps if they have a supporting skeleton of steel rods or wooden poles, but even these may prove grossly insufficient in the event of a really strong quake. Unfortunately, literally millions of buildings such as these are located in seismically suspect regions in developing countries. Each is a potential tragedy waiting to happen.

In contrast, superior construction is much more predominant in wealthier settings. Skeletal steel is much more common, as is implementation of the latest thinking regarding earthquake-resistant buildings (Figure 6-6). In that regard, an ideal model is the way a tree bends in a strong wind, absorbing the punch. The implication is to build structures that bend with the punch — or rather, sway with the earthquake. This is done by attaching girders in a way that produces a skeleton that is rather like your own — it bends.

FIGURE 6-6: A church in Chillán, Chile designed as one large arch – structurally strong today and a memorial to a 1939 earthquake.

(© Jerry T. Mitchell)

Building codes in earthquake-prone areas of the United States, and in other parts of the world, now mandate this kind of construction. And these laws clearly are having their desired effect. But modern construction is costly — much more so than the traditional masonry that continues to dominate much of the earthquake-prone world as a matter of tradition and inability to afford the state-of-the-art alternative.

Tsunamis

REMEMBER

When earthquakes occur under the ocean, the sea floor may rise or fall by a few feet over an area hundreds of miles on a side. This happens in a matter of seconds. With the sea surface suddenly a few feet too high or too low over a huge area, enormous volumes of water are set in motion to bring the sea surface back to level. This produces a long, low wave that moves out across the ocean in every direction at speeds up to 400 miles per hour. In the open ocean the wave is less than 3 feet high, but may be 300 miles across. Although the wave is moving quickly, because the wave is so broad, the 3-foot rise and fall takes 10 or 20 minutes and if it passed

CHAPTER 6 **Shape-shifting Earth** 95

under your ship at sea it would not even ripple the surface of your martini. You wouldn't even notice it. But as the wave encounters shallow coastal waters, it slows down and grows enormously in height, manifesting itself as a *tsunami*, or a large (misnamed) "tidal wave." The size of the wave is directly related to the magnitude of the earthquake. Strong quakes may produce waves 30 to 50 feet in height, and 100-feet monster waves are not unknown.

Naturally, the destructive potential of these waves in regard to coastal settlements is substantial. Given the seismically and volcanically active plate boundaries that border the Pacific Ocean, more tsunamis affect that ocean's shores than any other. Japan has been one of the most affected regions, which explains why the word "tsunami" is of Japanese origin.

When a major earthquake occurs on land, devastation is limited to a few tens or perhaps a hundred miles of the point where the quake occurred. The tsunami, however, may travel for hundreds or even thousands of miles with little loss of energy. By means of tsunamis, therefore, earthquakes can wreak havoc far away in places where the earthquake itself is never felt. Consider that in 2004 thousands of people perished from an Indonesian earthquake's tsunami as far away as India (which is 1,500 miles distant).

Fortunately, global earthquake monitoring now makes it possible to warn tsunami-prone cities, hopefully in time to evacuate people from low-lying areas. This know-how has been unevenly applied, however. Generally, affluent countries have been able to put in place good civil defense systems while poor countries have not. As was the case with earthquakes, therefore, the geography of wealth and poverty has much to do with the resulting human toll. Another way to reduce losses is through land-use planning (see sidebar "Applied Geography: Coping with tsunamis" later in this chapter).

A matter of magnitude

The amount of destruction that results from an earthquake depends on a couple of factors, one of which is its strength. Earthquakes vary remarkably in their power. Some can barely be felt. Others can knock you off your feet or buildings off their foundations. Two methods are used to measure the power of earthquakes. Because both make use of a series of numbers, they are referred to as *scales*.

The Richter Scale (maybe?)

TECHNICAL STUFF

Probably the more famous of the two scales is the *Richter Scale*, which was formulated in 1935 by a *seismologist* named Charles F. Richter. This scale indicates ground motion in an earthquake. It ranges from 0 to 9, but theoretically can go higher. The numbers are *logarithmic*. That means each whole number is 10 times greater than the preceding whole number. Thus, an earthquake that measures 7.0 on the

Richter Scale releases 10 times more energy than one that measures 6.0, and 100 times as much as one that measures 5.0. Of course, the ground motion of earthquakes is not limited to multiples of the number 10. Thus when earthquakes are reported in Richter terms, you may see numbers such as 6.3 or 4.8.

I said "maybe" in this section title as this scale isn't really used any longer. The Moment Magnitude Scale is used instead, though its number scheme was kept similar to Richter for familiarity purposes. *The Moment Magnitude Scale* measures energy release and is better for estimating the size of large earthquakes.

The Mercalli Scale

Less famous is the *Mercalli Scale.* This was devised in 1902 by a gentleman whose first name was Giuseppe and whose last name you can guess. This measures the intensity or violence of an earthquake, particularly in terms of damage caused to human-built structures. It is a subjective scale and more a measure of earthquake intensity than magnitude. Expressed by a series of Roman numerals, I to XII, the higher numbers reflect increasing damage.

APPLIED GEOGRAPHY: COPING WITH TSUNAMIS

Urban planners are using geographical knowledge and ideas to help coastal towns in tsunami-prone areas mitigate the impact of future ones. Cities and towns consist of different kinds of land use — parks, apartment buildings, schools, office buildings, and so forth. A key goal of urban planning is to allocate different kinds of land use to different parts of town in ways that benefit local residents. With respect to tsunamis, land adjacent to the coastline is dangerous, but the threat lessens as one goes progressively farther inland. The typical planning response is to allocate land use such that schools and hospitals are away from the danger zone and placing residences, apartment complexes, and office buildings away from the shore, while allocating warehouses, open space, and other low-population density land use to the immediate coastal setting. What may sound like plain common sense is, in fact, a geography lesson that several coastal towns have learned by fatal trial and error. But there is, of course, a constant tension. Ocean views and access to the beach are what draw people close. It is, therefore, a risk versus opportunity decision: What is the likelihood of a tsunami compared to the enjoyment of that beautiful location?

The lowest number, I, means "not felt" while "XII" is total destruction. Though not well-known by most people for its earthquakes, Charleston, South Carolina was decimated by one in 1886. It's Mercalli description was "X" or "most structures destroyed." With up to 90 percent of the city's chimneys toppled, it is amazing that we do not speak today of the Great Charleston Fire.

Subducting Plates: Volcano Makers

Sometimes when two plates meet head-on, one overrides the other in a process called *subduction* (see Figure 6-7). You may have read earlier in this chapter that Earth wasn't growing despite new material being deposited along the Mid-Atlantic Ridge? This is why. Earth is going back into itself. The plate that gets overridden is said to be *subducting*. Only the oceanic lithosphere created at mid-ocean ridges is dense enough to *subduct* (sink) very far into the mantle below. When it does it heats up, its upper surface or the asthenosphere overlying it partially melts. The result is a local surplus of sorts of molten material that seeks to rise through the lithosphere, and will do so if a convenient fissure or area of weakness (see "The Hawaiian 'hot spot'" sidebar) provides a path to the surface. When that is accomplished, the result is a volcanic eruption. Thus, the geography of subduction largely determines the geography of volcanoes.

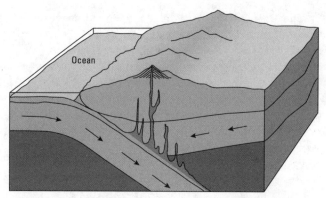

FIGURE 6-7: The process of subduction.

(© John Wiley & Sons Inc.)

"The Ring of Fire"

Zones of subduction occur in many parts of the world, but are especially prevalent around the shores of the Pacific Ocean. For that reason, the Pacific Rim has an extraordinarily high concentration of active volcanoes and is known as "The Ring of Fire." Locations include the western coast of South America, the western fringe of Central America, the U.S. Pacific Northwest, much of coastal Alaska from the Anchorage area westward through the Aleutian Islands, Russia's Kamchatka Peninsula, Japan, The Philippines, and Indonesia. As is the case with earthquakes, billions of people worldwide are directly or indirectly at risk, especially so in "the Ring of Fire."

Subduction: Another four-letter word?

REMEMBER

Subduction is another example of that four-letter word: slow. As a result, the build-up of pressure and the basic "stuff" of an eruption are slow. Therefore, volcanic eruptions happen infrequently, as do earthquakes. And in a sense, that's the really nasty thing about volcanoes: They erupt so infrequently.

Now you may say that is a really stupid statement. Why would anybody want eruptions to happen more often? My point is this: If eruptions happened with greater frequency, then people would get the message. They would more fully realize the dangers attendant to volcanoes and be more likely to avoid the danger zones.

But that's not how it works. Eruptions happen infrequently. Scientists who study hazards have a fancy term for this — *temporal spacing*. Subduction is slow, and there's nothing anyone can do to speed things up. So, the people who occupy volcanic environs either are unaware of the dangers because they happen so infrequently that they have no memory or personal experience. Or perhaps they just like their chances. Maybe they even employ that popular, cuddly term "sleeping volcano," and pray it doesn't wake up in their lifetimes. And indeed, most days it doesn't. But then one day it does.

The big blast

A volcanic eruption is perhaps nature's most spectacular show. In some cases, it consists of lava "fountains" and flows. More commonly, however, the event is a big blast accompanied by massive emissions of steam and hot rock particles of all sizes, rather than rivers of lava.

The power of eruptions is sometimes equated to so many atomic bombs. Obviously, nobody wants to be around a volcano when it goes BOOM! But few people tend to live in immediate blast areas, so BOOM! *per se* is not the big thing you may think, at least in terms of immediate human casualties. In that regard, two other side effects are of greater importance: ash and lahar.

THE HAWAIIAN "HOT SPOT"

Hawaii is clearly a "hot spot" as regards to tourism, but that moniker applies in another way, too. Hawaii is the most volcanically active place on Earth. In most cases, subduction is responsible for volcanoes. But in a relative handful of locales, including Hawaii, the cause is a "hot spot." In these instances, hot mantle rock is rising to the base of the lithosphere and then rolling back down, like water boiling in a pot. Some of the rock melts as it rises and the magma rises close to or onto the surface. The latter is the case with the Hawaiian Islands, and is the very source of their existence.

For the last several millions of years, the Pacific Plate has been moving to the northwest, passing over a hot spot. Over the eons, magma has come up the vent, issued onto the ocean floor and hardened, becoming a *seamount*. As magma continues to seep through the vent, the seamount grows, breaks the ocean surface, and becomes an island, which continues to grow as long as the connection with the magma-giving vent remains. Eventually, however, the moving plate severs that connection and the island stops growing. A new seamount — the forerunner of a future island — then begins to grow on the ocean floor. This sequence explains the southeast-to-northwest orientation of the Hawaiian Islands, as well as why all of the islands except for the Island of Hawaii, which is now over the hot spot, are volcanically extinct. It also explains why to the southeast of the Island of Hawaii a seamount exists, which, several thousands (if not millions) of years from now, will become the next island in the Hawaiian chain.

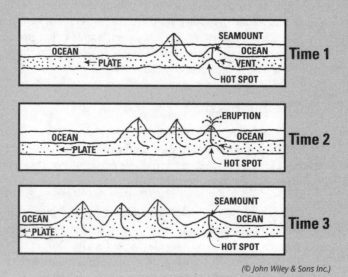

(© John Wiley & Sons Inc.)

Making an ash of itself

When explosive volcanic eruptions occur, tons of ash (tiny rock particles) are thrown into the air as a humongous, dense, and potentially suffocating cloud. Having weight, these particles eventually fall to earth over a wide area, coating crops, covering roads and houses, and potentially causing severe (and sometimes fatal) breathing problems for people and animals. The eruption of Mt. St. Helens provides an excellent case study of the possibilities (see the "A mountain blows its top" sidebar later in the chapter).

Lahars

REMEMBER

A *lahar*, a word of Indonesian origin, is a dangerous, fast-moving mudflow. During an eruption, vast quantities of emitted steam may cool rapidly, fall as rain, mix with ash, and form a mudflow. In the cases of very high volcanoes, such flows may be complemented by large amounts of water from rapidly melted snow and glaciers. The resulting lahar may race down flanking valleys for miles and miles from the volcano, burying everything in its path.

> ## A MOUNTAIN BLOWS ITS TOP
>
> Depending on the atlas you look at, you may get a different elevation for Mt. St. Helens in Washington state. One atlas may say that the summit is 9,677 feet above sea level. Another may give an elevation of 8,363 feet above sea level. That's a difference of 1,314 feet, which is close to the height of the Empire State Building. The reason for the disagreement is that one atlas would have been published before the volcano erupted (May 18, 1980) and the other published afterwards.
>
> But "erupted" is a bit of a misnomer. The mountain literally blew its top. The U.S. Geological Survey estimates 3.7 billion cubic yards of mountain got blown away. Another 1.4 billion cubic yards of ash got ejected, much of it in a cloud that reached 80,000 feet within 15 minutes. Accumulations of the inevitable ash fall averaged 10 inches 10 miles downwind, 1 inch 60 miles downwind, and 1/2 inch 300 miles downwind. Fifty-seven people perished (some from the blast, others from the suffocating ashfall), along with an estimated 7,000 big game animals and 12-million Chinook and Coho salmon fingerlings. As you can see, all kinds of numerical facts have been calculated and committed to print. What's really amazing to me is that the volume of forest that got blown down (4 billion board feet of timber) was enough to build 300,000 2-bedroom homes.
>
> Quirkier still? As an eleven-year-old growing up in San Francisco, I was treated to a souvenir container of ash as a promotional ploy. The purveyor? Kentucky Fried Chicken.

In November 1985, the destructive potential played out to its fullest following the eruption of the Nevada del Ruiz volcano in Colombia. Lahars as deep as 150 feet raced down the side valleys. Within four hours, locations as far as 65 miles away — seemingly well beyond the volcano's reach — were under mud. Hardest hit was the town of Armero, where some 23,000 people were killed and another 5,000 injured.

In a way, a lahar is to a volcano as a tsunami is to an earthquake — a mechanism by which the power of a major tectonic event may be fatally felt far away from the actual event itself. But being far away, the possibility exists for early warning systems that can significantly lessen the number of people who end up as statistics.

Categorizing Tectonic Processes

Ironically, beautiful mountains are products of powerful processes that can kill or maim. Due to the latter, earthquakes and volcanoes are rather commonly referred to as *natural hazards* — environmental events that are potentially harmful to humans and their handiwork, just as tornadoes, hurricanes, landslides, and floods can be. While some students of tectonics can't question the destructive potential of these various forces, they are rather put off by the "natural hazard" name, which in their view unfairly demonizes nature and conveniently absolves humans of any responsibility for fatal effects. Better, they say, to think of tectonic events as "natural processes" that only become "natural hazards" when people get in the way, as by building houses and cities in areas at risk. Some go further still and rename earthquakes as "class quakes" to further illustrate that not all people are impacted equally.

You may agree with that, or disagree with that, or have no opinion. But the statement underscores the interest of geography in this subject matter. Certainly, tectonic movements are natural processes. But when you consider the geography of these events in relation to the geography of humans and the things that they value, then tectonic forces may assume a degree of importance that goes well beyond their own immense power.

IN THIS CHAPTER

» Sculpting the planet

» Making soil for plants and food

» Depositing debris

» Living with flood plains and sea shores

Chapter 7
A Nip and a Tuck: Giving Earth a Facelift

Gradational force, which wears down Earth's crust and is the subject of this chapter, is the opposite of *tectonic force*, which builds up Earth's crust, and was discussed in Chapter 6. Indeed, you could say a competition of sorts is going on between those two powerful and opposing sets of forces which respectively wear down and build up Earth's crust, and thus create and alter the natural landforms that give character to Earth's surface. Gradational forces may not have the cataclysmic explosiveness of earthquakes and volcanoes, but their results, as shown in Figure 7-1, may be truly grand.

Listening to the reactions of people who are seeing the Grand Canyon in person for the first time is always extremely interesting. Some gasp at its enormity and color. Others say, "So we drove how far to see this hole?" Okay, so not everybody is mightily impressed.

But they ought to be! The beauty is spectacular, and the scale is grand. In Grand Canyon National Park, the featured attraction is one mile deep and 10 to 18 miles across, rim to rim, depending on from where you measure. By way of explanation, *carved* is a verb you see a lot — as in "The Colorado River *carved* the Grand Canyon." Not so! The river *carved* nothing. Instead, it *carried away* every last ounce of rock and soil that once occupied the space that is now the Canyon.

FIGURE 7-1: The Grand Canyon is a product of gradational force.

(© Amineah / Adobe Stock)

Now take a look at the Appalachian Mountains, even though it may seem like a complete change of subject. The Canyon "goes down" while the Appalachians "go up," but not very grandly. Indeed, while they sport a decent peak or two, nobody is going to call mountains that top out at 6,000 feet "The Grand Mountains." But they used to be. As noted in Chapter 6, geologists estimate some of the Appalachians were once 30,000 feet high. That's higher than Mount Everest, the highest mountain on Earth. Clearly a whole lot of earth was carried away here, too! This gradational force, mainly run-off from precipitation, made molehills out of some once grand mountains.

And so the Grand Canyon and the Appalachians — two very different landforms — turn out to have something fundamentally in common. A process of removal has shaped both of them. That is, both have been shaped by gradational force.

Getting Carried Away

Just how does gradational force work, as for example when it turns a 30,000-foot high mountain of yesteryear into a 6,684 foot high mountain of today? (That's North Carolina's Mount Mitchell, an excellent name for a mountain.) Basically, it's a two-part process. Gradational force is part of nature, which has at its disposal mechanisms (described in the next section) that can break great big rocks into tiny bits of rocks that are easily transportable. After that, another set of mechanisms picks up the tiny pieces and carries them away. These two sets of mechanisms are known respectively as *weathering* and *mass wasting*. It may take them hundreds or thousands or even millions of years to break everything down, but nature is in no rush. Indeed, it has all the time in the world.

Weathering Earth

Weathering refers to the natural processes that break rock into smaller and smaller pieces. Weathering can be broken down into two types, *mechanical* and *chemical*.

Mechanical weathering

TECHNICAL STUFF

Mechanical weathering is the disintegration of rock and other solid Earth material by physical means. Here are the major mechanisms used to accomplish this:

» **Frost action.** Some rocks allow water to occupy space between particles. If the water freezes, the resulting ice crystals expand and exert outward pressure that cracks the rock.

» **Hot-cold fluctuation.** In some locations, rock may be exposed to extreme temperature fluctuations. Over time, alternating heat and cold causes expansion and contraction, respectively, and may result in some parts of the rock breaking off.

» **Root action.** Fine, delicate roots may find their way into cracks or spaces within rock. As the roots grow, they may exert pressure sufficient to break the rock. Similarly, homeowners with basements know the power of roots to crack walls. So, too, have many plumbers been called as roots push through pipe cracks in search of moisture in water lines.

» **Abrasion.** Chips may break off when a hard object scrapes or rubs against rocks. During the Ice Ages, gigantic glaciers pulverized and ground up untold tons of bedrock, converting it to pebbles, gravel, and soil particles that are still with us. This action continues with today's glaciers, but on a much-reduced scale. Similarly, scraping or rubbing occurs when one piece of rock strikes another. This may happen as a rock rolls downhill, hitting other rocks as it goes; or when high winds send sand particles smashing into other rock particles; or when a swiftly flowing river causes one rock to hit another.

Chemical weathering

TECHNICAL STUFF

Chemical weathering is the disintegration of earth material by chemical means. Here are the major mechanisms used to accomplish this:

» **Rusting away** *(oxidation).* Lots of rock contain bits of iron. When oxygen combines with the iron, rust develops and destabilizes the rock, possibly leading to its break-up.

» **Dissolution.** Direct and prolonged contact between water and certain rock minerals may cause the latter to decompose and contribute to the ultimate break-up of rock.

» **Carbonation.** Atmospheric carbon dioxide may dissolve in rainwater and create a weak carbonic acid. Long-term, this may also contribute to dissolution of certain elements in rock and contribute to its ultimate break-up. Ever been to or seen pictures of Kentucky's Mammoth Cave (Figure 7-2)? Yep, that's carbonation at work. Limestone rock just can't stand up to the power of even that very weak acid.

By itself, weathering does not create a Grand Canyon or turn a mountain into a molehill. Instead, by converting large immobile pieces of rock into small transportable ones, it makes possible the movement of surface materials in ways that create or alter landforms.

FIGURE 7-2: Mammoth Cave is an example of chemical weathering. Note the size of the visitors for scale.

(© K.A / Adobe Stock)

GETTING DOWN AND DIRTY: SOIL

One of the principal products of weathering is *soil,* which by definition is a collection of earth particles that are no more than 2 millimeters in diameter. Soil provides nutrients to plants, without which they simply would not thrive. The key to this is the process called *osmosis,* which is the transfer of nutrients from soil to plants through root membranes. When precipitation seeps into soil, it mixes with mineral and organic matter and becomes a kind of a "nutrient soup" that coats soil particles. Plants come into contact with soil by means of their roots, which take in the "soup" by osmosis, and thus receive nutrients from the very substance in which they grow.

A *fertile* soil is one that makes lots of nutrients available to plants. An *infertile,* or poor, soil is one that does the opposite. Soil fertility varies geographically and is one of several important elements that explains the distribution of different agricultural products. There's a clever little mnemonic device I learned in graduate school to remember the factors contributing to soil formation: CLORPT. The "word" is not that sexy, and in fact sounds like a bad skin disease. In any case, the factors that determine soil formation include the following: Climate, Organism, Relief, Parent Material, and Time.

Each of these is a geographic variable:

- *Climate.* Soil fertility tends to be best where the climate is not too hot, not too cold, not too wet, or not too dry. High heat speeds up organic decay, basically wasting a high volume of nutrients before soil can make them available to plants. Low temperatures slow down decay, and again render nutrients unavailable to plants. Very wet conditions flush away nutrients in soil (a process called *leaching*), while very dry climates produce very little "soup" to surround soil particles. In contrast, climatic moderation (such as occurs in the middle latitudes) encourages nutrient accumulation and retention in soil, and thereby enhances fertility.

- *Organism.* Organic matter contributes greatly to the quality of topsoil. Thus, plant cover is an important determinant of the geography of soil fertility. Generally, grasslands are best because their fine roots and root hairs readily decompose and are in the soil to begin with. Leaf-fall from trees has good nutrient potential, but these are deposited on top of the ground and therefore are dependent on climatic factors (rainfall and temperature) to mix with topsoil. Let's not forget other life, either. Earthworms, burrowing rodents, people: All are quite active in turning over soil.

- *Relief.* Shape, height, depth. These are the foundation of our physical landscape, it's topography. As you can imagine, a flatter landscape may drain more poorly while steep areas erode more quickly. North-facing slopes (in the Northern hemisphere) are generally wetter than those with a Southern aspect as the former receive less direct solar radiation. Soils are then influenced by different rates of evaporation and chemical reactions that vary by temperature.

(continued)

(continued)

- *Parent material.* This is the bedrock from which soil is derived. It can be extremely hard, like granite; or fairly soft, like limestone; or in between. Generally, soft rock is best because it weathers easily and produces much more soil than harder parent material. Thus, the nature of the bedrock that underlies an area may have much to do with the fertility of the soil. Soil particles can be large (sand), tiny (clay), or in between (silt). Sandy soils tend to be infertile because the "nutrient soup" tends to seep through rather than be held and made available to plants. Clay soils tend to be infertile because particles are spaced tightly together, making it difficult for the "soup" to seep in or roots to penetrate. Silty soils, in contrast, tend to be very productive because they both admit and hold good amounts of "soup," as well as roots. The bottom line, therefore, is that moderate-size particles generally make the best soil.

- *Time.* How long does it take to create soil? Much, much longer than we have here reading this soil sidebar. Soils generally form more quickly in warmer environments, influenced greatly by the factors already considered. That said, it still may take several hundred years for an inch of soil creation. That will be something to tell your grandkids about! Losing topsoil — that nutrient-rich upper layer that helps plants grow faster and healthier — is a critical ecosystem health and food production concern. How do geographers play a role here? They use satellite imagery to detect where lighter in color, carbon-deficient soils are located — an indicator of soil degradation and erosion.

On a totally different matter, weathering is fundamental to the creation of soil, which is discussed in some detail in the "Getting down and dirty: Soil" sidebar. Suffice it to say here that if you think gradation and weathering are relevant only to earth-science geeks, then guess again. It's critically important to *you*; without it, there would be no soil, and therefore next-to-no vegetation, and therefore next-to-no food.

Wasting away

Mass wasting is the movement of particles that are products of weathering. Thus, it's the aspect of gradation that is most directly concerned with creation and alteration of landforms and has two discrete components:

>> **Erosion:** This is the removal of *particulate matter* (pieces of soil and rock) from a particular location. Thinking back to the Grand Canyon, running water removed — or *eroded* — material from the space now occupied by the canyon. Of course, those eroded pieces didn't just disappear. Instead, they went someplace else.

> **Deposition:** This is the putting down (or coming to rest) of eroded materials. Returning once more to the Grand Canyon, all the eroded material that was carried away by the Colorado River was eventually deposited either downriver or settled in the Gulf of California, into which the Colorado empties.

The difference between erosion and deposition is rather like that between pick-up and delivery. Material is taken from one place and put in a different place.

Changing the Landscape

Beauty, as they say, is in the eye of the beholder. You can find it in art museums throughout the land and, as far as geography is concerned, in the land itself. Sculptures are made by removing material that once surrounded the final forms. Compositions are achieved by bringing together materials that were formerly separate. In a similar manner, erosion and deposition (see previous section) are creative processes. Like standard works of art, the results may inspire us or, frankly, leave us unmoved. Either way, the creative power of gradation produces works that we live on and, in the case of soil, cannot live without.

In the real world, nature has four means, or *agents*, at its disposal to quite literally carry out erosion and deposition: gravity transfer, flowing water, glaciers, and wind. Although a glacier is a form of (slowly) flowing water, the gradational actions of solids (ice) and liquids is sufficiently different to merit separate treatment.

Staying grounded: Gravity transfer

Gravity is constantly "pulling down" on surface material. If not somehow restrained, therefore, particulate matter ranging in size from soil particles to boulders may move downslope in an act of *gravity transfer*. This process can be awesome if swift, as in the case of a landslide. Much more common, however, are the decidedly unspectacular minute movements (*soil creep*) of small particles, and the occasional pebble and rock that roll a bit downhill. Given enough time, however, the cumulative effect of gravity transfer may be really noticeable. Uplands are eroded and reduced, while deposition creates new landforms, as when rock materials accumulate at the bases of mountains or cliffs to form features called talus cones.

Going with the flow: Water

TECHNICAL STUFF

Flowing water is far and away the principal agent of erosion and deposition. The extent to which it can rearrange the landscape is largely dependent on three things that vary widely.

- » **The amount of water.** This is a no-brainer. The larger a river or wave, the greater its ability to move surface materials.

- » **Velocity.** The faster water travels, the greater its ability to pick up and move surface material. On land, *gradient* (the steepness of a slope) is a major determinant of speed. The steeper the inclination, the faster water travels. Similarly, storms at sea greatly increase the size and velocity of waves, and therefore greatly increase their impact on coasts.

- » **Surface cover.** Generally, the more open or bare a landscape, the greater is the likelihood of its alteration by flowing water. Vegetation inhibits erosion by slowing down flow speed and by generating root systems that hold soil in place. Bare ground, in contrast, is rather at the mercy of water and velocity. Thus, one of the great geographical ironies is that flowing water is the principal agent of landscape change in dry areas, where bare ground is very common. Rain and streams may be scanty in arid and semi-arid regions, but it doesn't take lots of flowing waters to make major impressions on the land in those areas. Witness the Grand Canyon, for instance.

Any volume of water that interacts with Earth's surface can produce mass wasting. That includes small-scale phenomena like raindrops and rivulets. Thus, runoff on exposed soil in an agricultural field or even a backyard garden may erode soil and result in *gullying* that produces very miniature versions (an inch or so deep) of the Grand Canyon. From the perspective of geography, however, it is large-scale phenomena, like rivers, waves, and ocean currents, that are of greatest interest.

Rivers and streams

The characteristics of individual rivers and streams vary greatly. In very general terms, however, one may think of a river system as originating in highlands, gathering the waters of tributaries as it snakes through foothills, and then flowing through a low-lying coastal plain as it approaches the sea (illustrated in Figure 7-3). The nature and effects of mass wasting tend to be very different in each of these settings. Here is what happens in each setting:

- » **Highlands setting:** By their nature, highlands are high above sea level. Steep gradients are most likely to occur there, resulting in rapidly flowing streams that erode their beds and carry away weathered material with comparative ease. This process creates valleys. In highlands, V-shaped valleys are fairly commonplace, and a sure sign that erosion rather than deposition is playing the largest (in fact, almost exclusive) role in changing the landscape.

- » **Foothills setting:** In foothills, river gradients tend to be much less steep, resulting in decreased velocity and, therefore, decreased erosion of streambeds. In fact, erosion of valley walls may exceed erosion of the riverbed. As a result, the profiles of foothills valleys may assume the shape of a somewhat flat-bottomed V with much of their floors being occupied by land rather than river.

    While the fringing slopes — that is, the lines of the V — provide evidence of the continued presence and power of erosion, the appearance of a flat and comparatively (versus the highlands) expansive valley floor is testimony that deposition has been an active player in shaping the landscape. Not uncommonly, the width of these valley floors is sufficient to accommodate modest-size towns and substantial agricultural activity.

- » **Coastal plain setting:** In the coastal plain, rivers are at their peak volume by virtue of so many tributaries having added their waters to the combined flow. Velocity is fairly slow, however, since by definition coastal plains are low-lying and flat. Here the individual rivers flow through broad *floodplains* bound by diminutive valley walls. The rivers themselves are bound by even less diminutive natural *levees* that consist of sediments deposited during past episodes when the river overtopped its banks.

    Indeed, "floodplains" aren't so-named for nothing. Because the landscape is so flat and expansive, flooding tends to affect a very wide area. Such events are beneficial to the extent that silt and other sediments in floodwaters are deposited and enrich the soils. On the other hand, flat and productive land attracts people and enterprise, and thus virtually guarantees that floods have a major impact on life, property, and infrastructure (see the nearby sidebar).

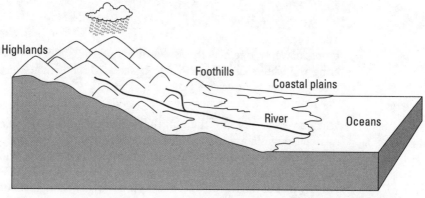

FIGURE 7-3: Rivers may flow through different landscapes on their journey to the ocean.

(© John Wiley & Sons Inc.)

THE FLOODPLAIN: LAND OF PROMISE AND PITFALLS

Floodplain refers to flat lands beside rivers that are prone to flood when the waters overflow their banks. Their extent may vary from a few feet to many miles on either side of the watercourse. The latter is common in the case of large rivers flowing through coastal plains.

Floodplains often are characterized by rich, fertile soils (*alluvium*) that have been deposited by past floods. For thousands of years, people have been attracted to these areas due to their favorable prospects for good harvests. Indeed, the attraction is so great that several alluvial regions — including the lower Nile, Ganges, Huang, and Yangtze (Chang) Rivers — have long supported very high human-population densities. And therein lies a major predicament: The alluvial lands that attract and nourish so many, may also be the scenes of terrible flooding and uncountable drownings. For that reason, the Huang (Yellow) River, which provides irrigation for millions of acres of cropland, is also called "China's Sorrow."

In modern times, engineering has sought to control (and ideally eliminate) flooding by building high artificial levees (banks) intended to make rivers stay put. While these defenses generally work, they occasionally fail in the face of record high water, as happened spectacularly and tragically during 2005's Hurricane Katrina when 80 percent of New Orleans was flooded. Indeed, critics claim that, for two reasons, these bulwarks merely guarantee that the inevitable flood will be exceptionally and unnecessarily disastrous. First, they say, the human earthworks (levees) create a false sense of security that encourages construction and settlement in patently hazardous areas. Second, artificially high levees negate the natural "sponge effect" of long, wide floodplains, and therefore magnify the effects of the inevitable flood.

Flood insurance is an interesting variable in relation to the floodplain. It clearly and compassionately helps to relieve the pain and suffering of people who "lose everything" in times of flooding. But critics claim its promise of compensating individual losses also encourages settlement of hazardous areas, and thus may serve to increase the very suffering it seeks to relieve. Far too many homeowners also find out after the fact that their insurance policies do not fully bail them out. There's always some fine print, and in this case, it is typically in learning that the house may have been covered, but none of its contents were.

When rivers empty into the sea, they disgorge eroded sediments that have been carried along. Where coastal waters are fairly calm, these particles may be deposited at the rivers' mouths, and progressively accumulate and form a delta, as shown in Figure 7-4. This landform is so-named because many of them — most notably that of the Nile River — assume a roughly triangular shape reminiscent of the Greek letter delta (Δ).

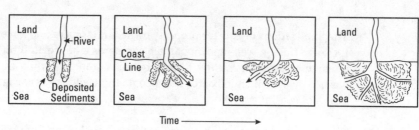

FIGURE 7-4: Deltas are formed from sediments deposited at river mouths.

(© John Wiley & Sons Inc.)

Waves and currents

Coastal areas bordering oceans and seas witness significant gradational activity. Every wave that strikes land potentially performs mechanical weathering, however minutely. Every drop of wave water that soaks land may contribute to chemical weathering. And, of course, every wave and coastal current, if strong enough, can erode sand and other coastal particles and deposit them somewhere else. As a result, coastal zones are among those parts of Earth that are most prone to change by natural means.

ERODING CLIFFS

Cliffs characterize many coasts, and are a sure sign of erosion. Each wave minutely helps to weather and erode the base of those landforms, undermining the entire cliff. Eventually, a portion of the cliff collapses and the shoreline *retreats,* that is, erodes inland. The rate of erosion varies from one cliff to the next depending on its composition. For example, a cliff composed of a mix of soil, gravel, and rock retreats far more rapidly than one composed of solid rock.

"Million-dollar views" may be had from cliffs, and that often results in prized pieces of property. The fact that a cliff is an erosional feature, however, virtually guarantees loss of adjacent real estate. Property owners who build houses close to drop-offs sometimes try to artificially stabilize the cliff to prevent further erosion. While such efforts usually are effective against minor storms, they can stand little chance against repeated major ones.

ERODING BEACHES

REMEMBER

Coastal areas that consist of sandy beaches and dunes are among the easiest landforms for nature to erode because they consist of small particles. The number of pieces of sand that make up a beach is beyond comprehension; but as far as mass wasting is concerned, particle sizes matter more than particle numbers. Take a nice sandy strand consisting of several giga-trillions of sand particles, let a severe storm pound away at it for several hours, and the result may be a greatly diminished beach.

CHAPTER 7 **A Nip and a Tuck: Giving Earth a Facelift** 113

The implications of this for property owners are severe and have been recognized for some time. A well-known parable compares a house built on rock to one built on sand. After a storm, the former remains standing while the latter has been washed away, which indicates that people thousands of years ago understood the basics of coastal erosion.

That knowledge has not, however, deterred people from building on sand. In the United States the number of people who live along the coast has soared in recent decades. Part of this is simply due to general population growth, as a result of which coastal cities have expanded up and down their respective shorelines. Probably of greater importance, however, in explaining the extent of coastal development is the American love affair with the beach, and the growing number of people who possess the financial wherewithal to purchase vacation or retirement homes by the water. Indeed, the combination of disposable income plus competitive bidding has served to make coastal real estate among the highest priced to be found.

Nature, however, is not impressed by price tags. In several instances severe storms have eroded dunes from under houses, literally leaving them high and dry. In response, several coastal communities have spent large sums of money to replenish beaches by dredging or pumping in offshore sand and building costly seawalls. At best, however, these are short-term cures with paltry prospects for long-term success, given the inevitability of future storms aided and abetted by rising sea levels (see Chapter 8). Moratoriums on building in these precarious environments, along with building set-back lines and purchasing insurance — with premiums that continue to climb and climb — are non-structural ways to deal with this issue. Another possibility also exists: moving people out of harm's way. Is it better to suffer the same repetitive losses in a place over and over again or should we buy out property owners and relocate them to less risky places? These are important questions economically for taxpayers and emotionally for communities. Towns are more than just a collection of buildings. Those places represent human memory and are called "home" for a reason.

COASTAL DEPOSITION

Of course, all the material that is eroded along beaches and cliff fronts goes somewhere. Much may be carried out to sea and come to rest on the ocean bottom. Sand, for example, may accumulate in the near-shore environment, forming shallow rises called *sand bars*. Other materials may create or add to *spits* (small points of land that extend outward into the water), or long narrow *barrier islands* that roughly parallel the coast. The latter are particularly important elements of the East Coast of the United States. New York's Fire Island, the Jersey shore, North Carolina's Outer Banks, South Carolina's Hilton Head, Cape Canaveral, Miami Beach — barrier islands are numerous and have seen significant development in recent decades. Each, however, represents a huge collection of small particles that can be eroded far more readily than they are deposited.

The chill factor: Glaciers

A *glacier* is a large, moving mass of ice on land. It originates when more snow falls in winter than melts in summer. If this is repeated for many years, then the annual surplus of snow compacts under its own weight and forms ice.

The massive weight of glaciers is capable of grinding up even the hardest rocks into soil particles. Since glaciers move, they are a combination earth crusher-bulldozer that perform weathering, erosion, and deposition all in one. The precise nature and results of these activities depends on whether the ice in question is a mountain glacier or a continental glacier.

Mountain glaciers

As the name suggests, *mountain* (or *alpine*) *glaciers* originate in snow that falls in mountainous areas. When large ice masses form, the slopes facilitate their downhill movement, and thus their power to erode and transform the landscape. Far up-slope, the erosional power of so much ice "gnaws away" at the mountain, turning rounded tops into pointed *horns* and nondescript ridges into jagged crests called *arêtes*.

Main valley floors are widened and deepened, changing from V-shaped valleys into larger U-shaped *glacial troughs.* Higher up, side valleys are also carved out. Less ice flows through them than the main valley, so less down-cutting occurs. Once the glaciers recede, the former side valleys are left "hanging above" the main valley, and are thus referred to as *hanging valleys.* In the mountainous coastlands of Norway, New Zealand, Chile, Western Canada, and Alaska, glacial troughs have filled with seawater, resulting in steep-sided ocean inlets called *fiords*.

Glaciated mountains have a spectacularly rugged look about them. Perhaps not surprisingly, therefore, such areas tend to be major magnets of travel and tourism. Also, one must remember that these landforms were created by erosion, which leads one to ask where all the removed material went. The answer is that it was deposited downslope, mixing in with soil or, in the case of fiords, on the ocean bottom.

But there's a danger here, too, in an era of rapid climate change. In Peru, for example, glaciers high in the Andes pushed forward in the past and left behind a *moraine*. Moraines are accumulated rock, a small ridge of sorts left along the front and side edges of a glacier. As the glacier recedes a moraine "dam" is left behind. Melted snow water can create a small lake — a *tarn* — behind the moraine. With enough climate change driven snow melt, that lake can become quite large and push against and through its weak natural dam. Disastrous flash flooding can result, flattening small Peruvian towns far below.

CHAPTER 7 **A Nip and a Tuck: Giving Earth a Facelift**   115

Continental glaciers

REMEMBER

Continental glaciers build up over large land masses in general, as opposed to mountains in particular. At the height of the last Ice Age, some 20,000 years ago, they covered substantial portions of North America, as well as fair portions of Eurasia. The present ice caps of Greenland and Antarctica (see Chapter 8), which are more than 2-miles thick in some places, offer insight into continental glacier scale and movement.

In North America, massive ice sheets (as seen in Figure 7-5) slowly built up over many years in what is now Canada. Eventually, the unimaginable weight and volume caused the ice to "ooze" outward in response to the pressure, at which point, the ice technically became a glacier. As the ice sheet advanced, its immense weight weathered the underlying earth and *gorged out* (eroded) low-lying landforms through which it passed. Thus were formed, for example, the beds of the Great Lakes, the Finger Lakes (New York State), and numerous other future water bodies. Much of Canada was scoured by that process, too.

FIGURE 7-5: Extent of continental glaciers in North America during the last Ice Age.

(© John Wiley & Sons Inc.)

Eventually, of course, all of the eroded material being carried or pushed along by the ice got deposited, particularly as climates warmed and the ice sheets waned.

In some places, these deposits of rock, sand, and other debris (called *glacial till*) merely coated the surface. In others, however, major accumulations occurred, resulting in landforms called moraines (see earlier in this chapter for more on moraines). Long Island's Harbor Hill Moraine is a noteworthy example. This landform was created by the bulldozing effect of the ice sheets and the transporting and deposition of debris beneath them. Today the feature serves as the island's drainage divide.

Generally, as the ice receded across the United States, it left behind a blanket of stones, gravel, and soil. Melt water issued from the retreating glaciers in innumerable streams that carried and deposited fine soil particles, which further transformed the post-glacial landscape. Today, thousands of years later, much of those materials underlie America's agricultural heartland and are important factors in explaining its productivity.

Making a deposit: Wind

Given sufficient velocity, wind can pick up soil particles and carry them long distances before they are deposited. *Aeolian* is the term used for this windborne process.

This was dramatically demonstrated in the 1930s when portions of the American High Plains endured the Dust Bowl, a phenomenon aided in large measure by poor land management practices. Wind-related erosion and deposition occurs most commonly in arid and semi-arid areas where ample bare ground is exposed to the atmosphere. In such venues, dust and sand "storms" of varying intensity are not unusual. *Sand dunes* are a classic landform that results from the inevitable deposition.

In a few non-arid parts of the world, substantial (that is, meters deep) deposits of wind-blown silt, called *loess*, are present. Because these are very fertile soils, loessal areas are among the most agriculturally productive to be found anywhere. Large areas of the American Midwest are covered by loess, as illustrated by Figure 7-6. Some of it probably blew in from arid areas. The common belief is that most loess, wherever they are found, originated long ago in dry glacial till that was exposed to wind after the ice sheet retreated.

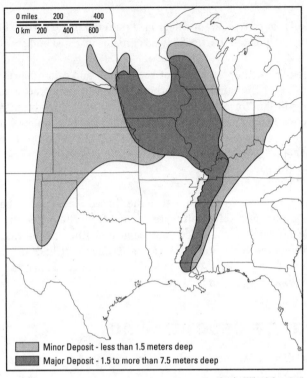

FIGURE 7-6: The geography of loess deposits in the United States.

(© John Wiley & Sons Inc.)

IN THIS CHAPTER

» Contemplating the uneven distribution of a precious resource

» Navigating the ups and downs of sea level

» Considering water rights (and water wrongs)

» Drinking water: Sometimes it's not a good idea

Chapter **8**
Making a Splash on Earth

If anyone in the universe needed to make a "splash," Earth is the place to do it. Around 70 percent of our planet's surface is covered by water, and the vast majority of that water is ocean. In addition, water is in the air you breathe and mixed in with the soil underfoot. Then you have lakes and rivers and icecaps. So, it really isn't a stretch to say, "Water, water everywhere."

The geography of water is vital to our lives. It determines where people live, the shape of our world, the content of the atmosphere, who can fish where, where travel is permitted, and how to supply the population with enough to drink. You may never look at a glass of water in the same way!

Many geography books don't include a chapter on water. Instead, the authors opt to sprinkle the subject throughout their books. Obviously, I've decided to go against the flow and put it all here in one big reservoir, with one exception: I'm going to hold off discussing ocean currents, which have a major influence on climate, until the next chapter.

Taking the Plunge: Global Water Supply

Although water is everywhere (see the "Where did all that water come from?" sidebar), it's very unevenly distributed. The oceans account for 96.5 percent of the global water supply. Ice caps account for another 1.75 percent, and together that makes 98.3 percent. Of the 1.7 percent that remains, nearly all is groundwater. Most of that is mixed in with soil, but a small portion is potentially available to humans as well water. All that remains (about 0.03 percent) can be found in all the world's lakes, all the world's rivers, and all the vapor in the atmosphere.

WHERE DID ALL THAT WATER COME FROM?

Actually, *all that water* isn't all that much considering Earth as a whole. Water may cover 70 percent of Earth's surface, but it accounts for only about 0.5 percent of the planet's weight. Oceans, on average, are about 2 to 3 miles deep, but the average distance from Earth's surface to its center is 3,960 miles. So the deepest of the deep blue seas is but a shallow veneer of surface material.

Apparently, all that water was here from the beginning. When Earth was a newborn fireball, its water was mixed together with other planetary matter. Indeed, probably much, much more water was on the inside of the planet than on the outside. Any substance that can exist as a solid, liquid, or gas is rather amazing. That's water. Because early-Earth was so hot everywhere, water was then in a gaseous state instead of liquid. Gases are lighter than solid matter, so they want to rise. In time, the gaseous water inside the Earth migrated (rose) to the outside and into the primordial atmosphere. This migration continued until the crust cooled (blocking the migration of additional internal water vapor to the surface), although subsequent volcanic activity did spew additional water into the atmosphere, along with a lot of other stuff.

Eventually the water vapor in the atmosphere condensed and formed rain. Some scientists believe that downpours began very early on. Others believe that didn't happen until eons later, after the crust cooled. But most agree that the vast majority of Earth's surface waters originated in rain that fell from the sky over a very, very long period of time.

REMEMBER

What these data tell us is that the vast majority of Earth's water is unavailable for human use. In some countries, there are "desalination" plants that — as the name implies — remove the salt from seawater and produce potable fresh water (take a look at Figure 8-1 and the sidebar "Why are the oceans salty?"). The process is rather expensive and very energy intensive, however, and helps satisfy the needs of only a few localities — coastal cities, mainly. And if the amount of potable water isn't worrisome enough, the geography of supply is often way out of whack with the geography of need. Fresh water is often abundant in areas where human need for it is scarce. And fresh water is often scarce where human need is abundant. The following sections show you just how the global water supply is broken down.

FIGURE 8-1: The desert island of Aruba, home to more than 100,000 residents and highly visited by tourists, uses desalination as it has no substantive natural fresh water supply.

(© Jerry T. Mitchell)

Those ice caps are really cool!

Earth's "ice caps" are actually continental glaciers, whose origins are described in Chapter 7. Cool? They sure are. Ice caps account for just less than 2 percent of Earth's water. That may not seem like much, but as a portion of all the water that exists on this planet, 2 percent turns out to be a lot of wet stuff — or actually hard stuff, because it's water in a solid state.

You want data? Antarctica's ice cap is as deep as 15,760 feet. If you stood at that spot, you would not literally be *on top of* Antarctica. Instead, you would be on top of a 3-mile-thick piece of ice that is on Antarctica. Indeed, the total volume of that ice cap is estimated to be about 30 million cubic kilometers. That's a lot of margaritas.

Want more? Well, there's Greenland, which is one of the great misnomers in the history of real estate. It really ought to be named Whiteland, because close to 99 percent of it is covered by ice cap. Actually, Greenland's ice cap is almost 10,000 feet deep at its thickest, so it's practically minor league compared to Antarctica.

REMEMBER

What's not cool? How much of these ice caps have we been losing! Estimates in the early 2020s for both Antarctica and Greenland together are more than 450 million tons of ice *each year*. And each melted drop means that inch by inch our ocean levels are rising.

Getting out: Oceans, seas, gulfs, and bays

Most of Earth's water is oceans, including seas, gulfs, and bays. The difference between these terms is basically a matter of size and location. *Ocean* comes from the *Okeanos* of Greek mythology, which was a river thought to encircle Earth. And indeed, *the ocean* is a continuous body of water that encircles the land, but it also consists of a handful of divisions that are also referred to as oceans — Atlantic Ocean, Pacific Ocean, Indian Ocean, and Arctic Ocean. Some geographers like to denote a fifth ocean, the Southern Ocean. This would be all the water south of 60 degrees South Latitude and encircling Antarctica.

Seas, gulfs, and *bays* are parts of the ocean that adjoin land bodies. Generally, seas are ranked second (after oceans) with respect to size. They may be relatively enclosed by land, as is the Mediterranean Sea, or they may be "open" to the ocean, as is the Arabian Sea. A *gulf* is a part of an ocean or sea that extends into the land. Dictionaries suggest that gulfs are smaller than seas but bigger than bays. A bay may be defined as an inlet of a sea or gulf. For example, Tampa Bay is an inlet of the Gulf of Mexico.

REMEMBER

That's by the book, of course. In reality, things aren't so neat and tidy. Considering the Indian Ocean and parts thereof (see Figure 8-2), for example, you can see that the Bay of Bengal is bigger than the Gulf of Oman. When it comes to these place names, it's kind of like being told "size matters" and then finding out it doesn't. Basically, what happened is that, way back when, some explorer or mapmaker simply labeled a water body and the name stuck. Maybe that person didn't appreciate the true size of the feature being named or didn't appreciate the nuances of vocabulary. Whatever the case, the result is a dictionary that suggests a definite rank order with respect to size of seas, gulfs, and bays, and a world map that says it just isn't so.

WHY ARE THE OCEANS SALTY?

Dissolved mineral salt is the key. As water flows to the seas, it comes into contact with lots of rock and rock particles. These contain mineral salts, minute quantities of which are dissolved and carried along by running water. It's probably in the stuff you drink, but because the salt content is so low, you don't notice anything peculiar.

Ultimately, that water with its low salt content joins the sea. The sun continually evaporates sea water, producing vapor that becomes future rainfall. But here's the punchline: It's the ocean water that gets evaporated and not its salt content. So during evaporation, the salt gets left behind, only to have "new salt" added to it as freshwater eternally runs to the sea. Give this process a couple million years (which it did), and the result is an ocean of water too salty for human consumption.

How about de-salting the waters? That certainly is possible. *Desalination* (distilling sea water to remove the salt) is a fairly simple process, and goodness knows the oceans contain more water than humans will ever need. But desalting seawater in the copious quantities that cities require is very, very expensive. Few can afford it, so desalination is a viable option only in few places at present.

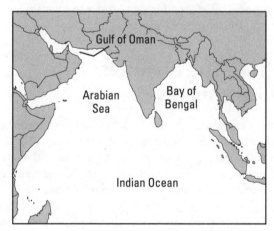

FIGURE 8-2: The Indian Ocean and parts thereof.

(© John Wiley & Sons Inc.)

Coming inland: Lakes

A similar brand of confusion reigns with respect to inland bodies of water. A *lake* is a body of water completely surrounded by land. But when you open an atlas and browse the lakes, you find some are called seas. For example, the Caspian Sea and Aral Sea are in Asia. The Salton Sea is in California, and the Dead Sea is between

Israel and Jordan. Logic would suggest that a "sea" should be bigger than a "lake." In reality, every one of the Great Lakes is larger than the Dead Sea, and so are a bunch of not-so-great lakes.

The explanation is that the previously named seas (which are really lakes) are salty. Each occupies a *basin* — a depression in the landscape. Fresh water flows in, but nothing flows out. So you end up having an ocean in miniature. That is, the rivers bring minute quantities of dissolved salts to the lake/sea, which is essentially a dead-end repository. When the sun evaporates the surface waters, the salt gets left behind. Keep this up for many, many years, and what you have is a lake whose waters are not only salty, but are even saltier than your average ocean water. That's the way it is with the Great Salt Lake, which apparently has every right to be called a sea but is not. Again, somebody way back when gave a name to a water body, and the label stuck regardless of the logic.

Shaping Our World: Oceans

Being the largest supplier of water on the planet, oceans have a big responsibility. This is where major transportation takes place between countries, where wars are sometimes fought, and where food is caught, not to mention the physical effects that oceans have on the land they wash up upon.

Going where the action is: The continental shelves

Off the coasts of continents are relatively flat expanses of ocean bottom that average about 300 feet in depth — shallow by oceanic standards. Such a feature is called a continental shelf (see Figure 8-3). The distances that the shelves extend from the shore vary. While the continental shelf off California extends 2 to 3 miles, the one off Newfoundland extends seaward some 200 miles.

Altogether, the continental shelves define what is arguably the most economically important part of the ocean. A growing human population translates into a need for more food (especially stuff that is high in protein) and more mineral resources. Increasingly, oceans are being looked to as sources of supply. And in that regard, the continental shelf is where the action is.

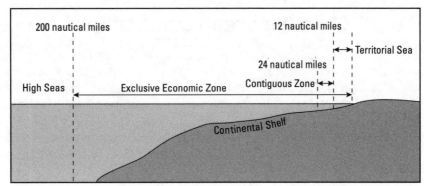

FIGURE 8-3: Profile of the continental shelf.

(© John Wiley & Sons Inc.)

Something very fishy going on

If you were a fish and lived in the ocean, then chances are you would hang out over the continental shelf, just like almost every other fish. Schools are in session there for good reason — lots of food is available. Here's how the food gets there:

- » **Plant life:** Because shelf water is shallow (compared to mid-ocean depths), some sunshine hits bottom and gives rise to plant life that serves as food for small fish that serve as food for bigger fish, and so forth.

- » **River flow:** Rivers empty onto the continental shelves. Their flow typically contains a lot of organic matter (especially dead, decaying, or dissolved plant parts), which adds to the abundance of fish food.

- » **Vertical mixing:** Wave action and turbulence produce a considerable amount of vertical mixing over the continental shelves. As a result, organic matter gets distributed over the depths, making for a very robust feeding environment.

Fuel for thought

Petroleum and natural gas underlie some areas of the ocean bottom, just as they are located under some areas of dry land. In the oceanic setting, these resources are found exclusively under continental shelves because they alone in the marine environment are composed of the geological features in which oil and gas are found. So, when you see photos of oil rigs offshore in places like California or Louisiana, you are looking at an important economic activity that coincides with a continental shelf. Again, that's where the action is.

Claiming ocean ownership

The existence of all the marine goodies leads to the rather important issue of ocean ownership. That is, who owns them? How far offshore does a nation's sovereign territory extend — if at all? For countries that have coastlines — and for ships at sea — having an answer is important for the following reasons:

>> **National security:** Countries have the right to defend themselves from attack or intrusion. It's crucial to be able to determine, therefore, the point at which a ship "crosses the line."

>> **Police power:** Criminal activities can take place on water as well as land. Drug smuggling is a key example. Legally, however, the police and Coast Guard can only board vessels at sea within their jurisdictions. Thus, it is of some importance to clearly define how far out to sea those jurisdictions extend.

>> **Trade and commerce:** Thousands of freighters and tankers ply the ocean. The captains and navigators who set ships' courses need to know where they can "sail as they please" and where they need permission by virtue of having entered a foreign country's territorial waters. Only by defining the extent of ocean ownership can that be determined.

>> **Resource ownership:** Disagreement between countries concerning ownership of or access to resources can lead to conflict. "Marine goodies" mentioned earlier are potential sources of contention. Therefore, clear definition of the extent of ocean ownership may prevent conflict over oceanic resources.

>> **Dire straits:** The oceans contain several *straits* — narrow waterways that separate land bodies. About a dozen of them are major bottlenecks as far as international shipping is concerned. It is important to determine whether these are international waterways through which ships of all nations may freely pass, or if they belong to the countries that border them — and which would therefore have the right to deny passage or charge tolls.

REMEMBER

For decades the United Nations (and the League of Nations before it) sponsored conferences on "The Law of the Sea" to determine the nature and extent of offshore jurisdiction as well as other issues related to ocean use. In 1982, those efforts resulted in a draft treaty entitled the *United Nations Convention on the Law of the Sea* (UNCLOS), which became international law in 1994. More than 160 United Nations members are treaty signatories, showing you just how important drawing an invisible line in the water can be. The jurisdictional provisions are depicted in Figure 8-3. If you think of yourself as the coastal country in that diagram, then the following list details what the different zones mean to you.

- **Territorial Sea:** Adjoining your coastline is a zone known as territorial sea that extends 12 nautical miles (nm). (A *nautical mile* equals 1.15 "regular" miles.) This area is all yours to do with as you please. Vessels that might engage in fishing or mineral extraction in this area are subject to your licensing and permission. Vessels that just want to pass through have the *right of innocent passage,* meaning they are free to travel provided they don't threaten your security or laws. Warships of other countries have the right to be here, but in normal times that is considered unnecessarily provocative.

- **Contiguous Zone:** The Contiguous Zone extends another 12 nautical miles past the border of the territorial sea. You don't own this area in all respects, but you can enforce your customs, immigration, and sanitation laws within this area. *Hot pursuit,* chasing and boarding vessels suspected of involvement in illegal acts, is also allowed. Imagine the Coast Guard as a county sheriff chasing bootleggers only having to give up pursuit not at the next county line but 12 nautical miles offshore.

- **Exclusive Economic Zone:** Exclusive Economic Zone (EEZ) extends as far as 200 nautical miles from the coast. All fishing and mineral exploration in this area is under your control. The presumption, of course, is that you exercise this power wisely for the management of marine resources. Also, it's no accident that the dimension of EEZ places almost all of the world's continental shelf under somebody's official jurisdiction.

- **The High Seas (International Waters):** *The High Seas* (also known as *International Waters*) extends beyond the EEZ. They are open to vessels of all nations for whatever purpose. Thus, everybody enjoys the right of innocent passage in this area. In theory, deep ocean mineral resources are for the benefit of all the peoples of the world. In reality, the minerals down there are at the disposal of a handful of countries that possess the technological wherewithal to go get them.

REMEMBER

The various offshore zones may appear straightforward, but in the real world they often don't work. Consider the United States and Cuba, two countries that aren't exactly bosom buddies. Each country is theoretically entitled to their EEZ and its contents, but Key West, Florida, is only about 90 miles from Cuba. Adjustments had to be made, therefore, beginning with a zone of High Seas in between to guarantee the right of innocent passage to ships of all countries. After that, the U.S. and Cuba were each allotted equal, but reduced, amounts of the other zones shown on the diagram.

When you give the world map a major look-see, you find virtually dozens upon dozens of other watery expanses where the Law of the Sea "doesn't work." Each was reconciled on a case-by-case basis. Straits were given special attention to guarantee the right of innocent passage.

The Law of the Sea as currently constructed promotes the past and enables creative future thinking. The United Kingdom maintains ownership of the Falkland Islands in the South Atlantic not only because the people there want to be citizens of the UK. By being UK territory, the island provides them with an EEZ and all the resource rights that come with it. Argentina interprets the question of island ownership differently (calling them Las Islas Malvinas), declaring sovereignty on maps and postage stamps despite losing a war with the UK over the islands in 1982.

China, in contrast, is not only squabbling with neighbors over rock specks in the South China Sea — they are also building islands of their own. The Spratlys — a series of rocks, atolls, and reefs — lie near Vietnam, the Philippines, and Malaysia. Each country wants to raise a flag there for fishing rights *and* natural gas reserves. China has occupied several of the islands, expanding them for airfields and other infrastructure to extend their military and economic might far beyond what the Law of the Sea would allow from their traditional land border.

Getting a rise out of oceans

Sea levels are rising. The ice caps on Greenland and Antarctica, plus a majority of the world's glaciers, are slowly shrinking. The shrinking is a by-product of *global warming*, which you can read about in Chapter 18. As the glaciers slowly recede, their melt water returns to the oceans, which rise. Perhaps the most important word there is "slowly." Unless you live very close to sea level, you have no cause to have a nightmare about the ocean rising around your ankles. I would caution you, though, to think very carefully about beachfront real estate as a long-term investment.

How high the oceans will rise is open to debate because nobody knows for certain the future course and severity of global warming. But consider the past as a portent of possibilities. About 18,000 years ago, at the peak of the last Ice Age, sea levels were about 350 feet lower than they are today — because so much water was "locked up" on land in the great glaciers. About 400 years ago, the English founded Jamestown, Virginia. Today, sea levels around and the water table under that site are rising, giving us a glimpse about the historic and present losses we may be facing.

REMEMBER

Geographically, this process is very important. Perhaps the most basic element of geography is the familiar outline of the continents and other major land bodies that you see on a world map. For the last 18,000 or so years, the world map has been changing, with the ocean on the up-and-up. And this trend will continue for the foreseeable future. Thus, you may legitimately think of today's world map as a single frame of a very long movie. You know the plot: Oceans are rising. As for the ending . . . well, several outcomes are possible. All involve higher sea levels.

Even if the change is slight, the impact will be significant. If the rise is just a couple of feet in the next century, millions of people will feel the effects. The highest elevations of several island countries (the Maldives in the Indian Ocean, for example) are "just a couple of feet" above sea level. Much of the Bahamas is "just a couple of feet" above sea level. There's a reason that in 1990 a new organization was born — the Alliance of Small Island States (AOSIS). The nearly 40 members of AOSIS work collaboratively to give voice to these places so vulnerable to a changing climate. Other low-lying coastlines, such as large areas of Bangladesh — one of the most densely populated countries in the world — are similarly at risk. And literally dozens of populous port cities around the world have millions of people and trillions of dollars of infrastructure "just a couple of feet" above sea level (Figure 8-4). Significant change is on way. The map will look different. The world's geography will be different.

FIGURE 8-4: Singapore, a coastal city with much infrastructure and population at risk from rising sea levels.

(© Jerry T. Mitchell)

Getting Fresh with Water

Water, water everywhere. While the oceans may be the biggest water bodies out there, they're not necessarily the most important — it's time to talk about some drops you can actually drink. And that entails the *water cycle* (or *hydrological cycle*), which, as far as physical processes are concerned, is complicated, amazing, and elegant at the same time. See Figure 8-5 for a look at the water cycle. You simply cannot live without it, and neither could any other living thing that requires fresh water because the water cycle is the world's one and only producer of fresh water — no water cycle, no fresh water — it's as simple as that.

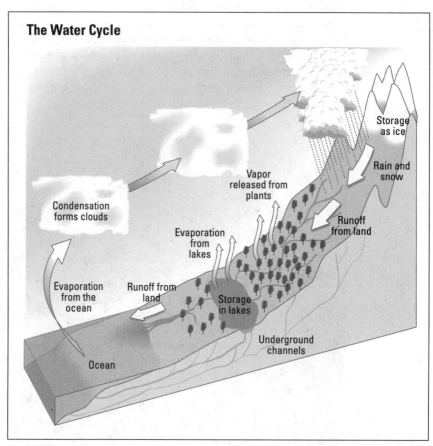

FIGURE 8-5: The water cycle.

(© John Wiley & Sons Inc.)

In addition to the stages described in the next section, the water cycle has two overriding characteristics that are good to keep in mind. Here they are:

» **It really is a cycle.** The water really does go round and round just like the diagram suggests. "Cycle," therefore, very appropriately describes what's going on.

- » **The cycle is a closed system.** By closed system, I mean that nothing gets added to it or subtracted from it. For all intents and purposes, the amount of water on Earth is fixed — closed to change. Thus, if a lake or reservoir dries up, the water is not really "gone." Instead, it has relocated elsewhere within the system.

The stages of the water cycle

Here are the principal components of the water cycle:

- » **Solar energy:** Technically, sunshine isn't part of the water cycle for the simple reason that sunlight isn't water. But it's the sun that sets the cycle in motion. The sun is the pump that gets things moving.

- » **Evapotranspiration:** Whew! That's a mouthful. And it's a combination of two words: evaporation and transpiration. With regards to the first word, some water evaporates when it receives solar energy. That is, it changes from a liquid state to a gaseous state: vapor. With salty seawater, only the water evaporates, and not the salt. So salty seawater becomes freshwater vapor in the atmosphere.

   With regards to the second word, you perspire, plants transpire. Plants sweat when the sun heats them up — only it's called transpiration. Plant transpiration is a key input into the vapor in the atmosphere, especially in lush, tropical areas.

- » **Condensation:** Water vapor is so small that it's invisible. But when lots of individual bits of vapor cluster and combine, they become visible miniature droplets of water. This is condensation, and if it happens by the gazillions, you get a cloud. As to exactly what happens to produce condensation, well, I'm holding that until Chapter 9.

- » **Precipitation:** If condensation continues, then the droplets continue to get bigger and put on weight until they are too heavy to remain suspended in the atmosphere. They then fall to Earth as precipitation. This phenomenon may take various forms, including snow, sleet, hail, and most commonly, of course, rain.

- » **Run-off.**

- » **Infiltration.**

No, I didn't forget to write something. Instead, each deserves a separate heading because, no matter where you live, one or both is responsible for your water supply.

Run-off: Going with the flow

Some of the water that falls to Earth collects on the surface and begins a downslope journey to the sea (or bay, or gulf) — there to complete the water cycle. Trickles join to form babbling brooks (why do brooks babble, anyway?), which join to form rivers. Basically, it's fresh water on the move, and every last drop of it is potentially available to people.

Rivers have long served as water supply systems for cities and towns. But rivers that bring drinkable water can also take away sewage and waste. That's great unless you happen to live downstream — there to discover your drinking water is no longer drinkable thanks to your upstream neighbors. And most people live downstream from somebody else, so problems afloat.

REMEMBER

To solve the problem, many towns and municipalities have gotten into the business of capturing drinkable run-off by building dams that create reservoirs that store water that can be transferred to where people need it. And if those dams can create power in the form of electricity, then all the better. These public works tend to get located in not-yet-contaminated headwaters. Sometimes, however, these headwaters are far from the people who will consume them, and therefore require construction of aqueducts to resolve the geographic difference between supply and demand. Water from Yosemite in the Sierra Nevada snakes its way west through the Hetch Hetchy Aqueduct to greater San Francisco. Similarly, some of the water that supplies New York City comes from a reservoir more than a hundred miles away.

Infiltration: Out of sight, not out of mind

Some of the water that falls to Earth *infiltrates* the soil. That is, it seeps into the ground. Because nature has been at this for a long, long time, some lands are covered by substantial *aquifers* — subterranean accumulations of water. Better to think of these not as underground lakes but as areas of super-saturated soil or porous rock (yes, rock!) with a high water content. If your water supply comes from a well, then you live on water that has infiltrated. Well water can be suitable to drink as is, or it may require minimal treatment. But infiltrated water has two potential problems — contamination and depletion.

Contamination

Along with rainwater, chemical fertilizers and industrial wastes can seep into the soil and contaminate aquifers. Agriculture has used pesticides, herbicides, fungicides... with a key word stem –cide ("to kill") being a clue to their danger in our water supply. From industry, we have an examples like mercury from mining that can harm one's kidneys and nervous system and volatile organic compounds (such as varnishes, dyes) that have been linked to cancer.

Depletion

Quite often, people consume water from aquifers much faster than nature replaces it. Nature was putting water in aquifers long before people came along. But the amount of water in the bank can diminish quickly when people start drawing from that watery account in quantity.

Consider, for example, the Ogallala Aquifer, which underlies a considerable portion of the American High Plains (see Figure 8-6). If you reside in the U.S., then that aquifer is vitally important to you even if you live nowhere near it because a host of crops that feed people and fatten livestock are grown in great quantities in the High Plains with the use of irrigation water from the Ogallala Aquifer. And the water within it is being consumed much faster than is being replenished. Wells must be dug deeper and deeper. Theoretically, one day, they may all dry up.

FIGURE 8-6: The Ogallala Aquifer.

(© John Wiley & Sons Inc.)

Doom and gloom? Not necessarily. Advances in agriculture are making it possible to produce good harvests with less water. Greater conservation also is possible, as is greater use of regional rivers for irrigation. And if worse comes to worse, well, portions of the High Plains could revert to something akin to the lush natural grazing lands (the prairies) that were done away with to make way for farms.

Good to the very last drop

REMEMBER

The bottom line with respect to run-off and infiltration is that you can't take more water out of the system than nature puts into it. Reservoirs and aquifers *can* dry up. In that event, water will still be everywhere, albeit in forms that are not readily accessible (such as vapor in the air or veneer surrounding soil particles) or in amounts that satisfy local needs. Humans number around eight billion and counting. More people mean more direct consumption, more irrigation, and more industrial use. How these needs will be met remains to be seen, but they will clearly reflect the geography of a precious resource.

APPLIED GEOGRAPHY: DRIP IRRIGATION AND CLOUD CATCHING

In arid and semi-arid lands, water for irrigation has long been applied to fields by means of open-air ditches between rows of crops. Although this method has helped satisfy the food needs of unknown numbers of people over the millennia, it has three significant drawbacks. First, a substantial volume of water may evaporate before it reaches the plants. Second, the amount of water that seeps into the soil is usually far more than the plants actually need. Third, the mineral salts that build up in the soil as a consequence of evaporation may ultimately undermine the usefulness of the farmland and lead to its abandonment.

In some countries, however, these environmental effects have been largely nullified by introduction of drip irrigation. Though practiced in some form a thousand years before the present in China, it should be no surprise that advances in the twentieth century came via pioneers in dry places in Australia and Israel. In drip irrigation, water reaches the fields and is distributed up and down crop rows by means of thin plastic tubing that is perforated by tiny holes every few inches. Water is forced through the tubing at very low pressure; so instead of squirting out of the holes like so many tiny fountains, the water slowly drips out. Because of the tubing, very little water is lost to evaporation. Moreover, drop by drop application results in little water wasted and minimal salt accumulation. Greater efficiency is being achieved by using computers to monitor conditions to distribute the water at a time when even less could evaporate. All in all, therefore, drip irrigation is proving to be an effective and fairly low cost means of making maximum use of a scarce arid-land resource.

Yet another novel form of water capture is *cloud catching*, a still-developing process to catch water droplets from fog onto tightly woven netting. Practiced in limited form along the dry west coasts of Peru and Chile, this fog — *camanchaca* — passes through the nets and deposits water into tubes for consumption. The environmental impact is negligible and the water costs little to transport as the nets sit atop hillsides; gravity does the work to bring the captured water to the people.

The photo here shows the lush landscape in the midst of semi-arid land in Chile's Jorge National Park. The fog deposits scare moisture here — a nod toward the potential for cloud catching.

© Jerry T. Mitchell

> **IN THIS CHAPTER**
>
> » Feeling the heat
>
> » Tilting along
>
> » Changing seasons
>
> » Observing current affairs

Chapter **9**

Warming Up and Chilling Out: Why Climates Happen

Several years ago, I happened to be nearly atop the equator, our planet's imaginary line that separates the Northern from the Southern Hemisphere. At 1.4 degrees North latitude, Singapore is pretty close to that 0-degree line. And it was hot. I was miserable pedaling around the muggy city on my bike. Few people were on the streets. They were below the street in the air-conditioned subway or the mall that burrowed six stories into the ground. Everyone knows that it is hot at the equator, but still it was pretty sticky.

Not long after, say a few months, I was milling about Nairobi, Kenya. I was located at about 1.3 degrees South latitude and I was ready for the equator to serves up its best heat. But it wasn't hot at all. The temperature was fairly mild, probably in the mid-70s Fahrenheit. What explained this variation from the expected? Elevation. The city sits at almost 6,000 feet — more than one mile — above sea level, and the air is cooler there. Local men strolled around in pants and long-sleeved shirts that could be rolled up or down as the temperature changed. I was going on safari in the eastern Kenyan lowlands later that week and soon had to think more carefully about how I might dress myself comfortably in an environment where temperatures could change more than I anticipated in my trip preparations.

My experiences, and ones that you may have experienced as well, shows that you can travel the world and encounter all kinds of climates. Together they constitute a vast array of atmospheric characteristics that concern temperature, precipitation, and seasonal change. But no climate "just happens." While all are products, directly or indirectly, of the same Sun, factors are afoot that give each of them characteristics that really do make all the difference in the world.

The factors that determine climate in different parts of the world are of central interest to geography and are the subjects of this chapter. The characteristics, locations, and consequences of climate occupy the next chapter.

Getting a Grip on Climate

REMEMBER

Like climate, weather is concerned with atmospheric conditions. The difference between the two is a matter of time. *Weather* refers to day-to-day conditions and changes in Earth's atmosphere. *Climate* refers to the average of weather conditions at a location over a long period of time — 30 years, as far U.S. government climatologists are concerned. Climate, therefore, is the more appropriate topic for this book because it concerns general characteristics of a location or region. Accordingly, among the things you will not read about here are tornadoes, hurricanes, thunder, lightning, hail, and other forms of short-term atmospheric mayhem that fall within the purview of weather.

So what factors cause the different kinds of climates to occur? (Drum roll, please.) Six determinants, which may act singly, in combination, or in opposition to each other, make climate occur:

- The angle at which solar energy strikes Earth
- The tilt of Earth on its axis
- Altitude with respect to sea level
- Solar absorption properties of land and water
- Ocean currents
- High and low atmospheric pressure belts

The following sections show how each bullet point creates climate.

Playing the Angles

REMEMBER

Parts of Earth receive different amounts of *solar energy* — heat from the Sun. The dosages are greatest in the equatorial realms and progressively diminish as one approaches the Poles. For this reason, rather warm climates generally dominate the low latitudes and give way to cooler and cooler climates in the mid-latitudes and polar regions.

These differences are due to the angles at which solar energy strikes Earth at different latitudes. Figure 9-1 shows three "bundles" of sunshine whose widths are the same, so it can be assumed they contain equal amounts of solar energy. But if you examine the amount of Earth that each impacts, a key difference appears. Bundle A illuminates a much smaller area than Bundle B, which in turn illuminates a much smaller area than Bundle C.

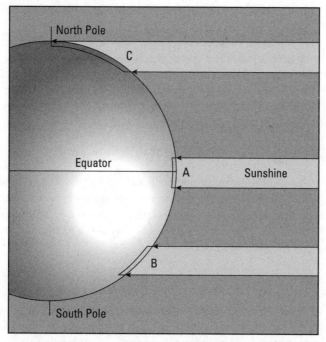

FIGURE 9-1: Solar energy strikes the Earth at three latitudes.

(© John Wiley & Sons Inc.)

Making hot and cold

REMEMBER

The differences among the Bundles in Figure 9-1 are determined by the curvature of Earth. Bundle A contains *vertical rays*, which strike Earth perpendicularly. Due to curvature, however, Bundle B strikes Earth at a sharper angle. As a result, its solar energy is spread over a larger area than is Bundle A's. Sharper still, thanks to

CHAPTER 9 **Warming Up and Chilling Out: Why Climates Happen** 139

curvature, is the angle at which Bundle C strikes Earth. And as a result, its heat is spread over the largest area of the three examples.

Absent other factors that affect temperature, Area A has the warmest climate because it has the greatest concentration of solar energy. That is, the heat in Bundle A is brought to bear on a relatively small area. In contrast, Area C has the coldest climate because the heat it receives is spread over the largest of the three areas. Intermediate conditions are present in Area B.

Let's look at some real locales and compare the climates of Manaus, Brazil, and Churchill, Canada. Manaus is located at about Latitude 3° South and therefore exemplifies Area A on the diagram. Churchill is located at Latitude 58° North and therefore has the characteristics of Area C.

In Manaus, the annual average temperature is 79° F. (That means that if you recorded the temperature every hour of every day for a year, the average would be 79° F.) In Churchill, it's 19° F. The difference is 60° F. Manaus is much warmer.

REMEMBER

Beware the reason! Is it because Manaus is closer to the equator? Nope. Proximity to the equator *per se* is not the explanation. Instead, the answer lies in the angle. The Sun is more directly overhead at Manaus than at Churchill throughout the year. Thus, Manaus experiences a greater concentration of solar energy throughout the year and is therefore a warmer climate.

Making rain and snow

Climate is about precipitation as well as temperature. Regarding wet stuff, Manaus receives 82 inches of precipitation on average each year. Churchill, in contrast, receives 15 inches. Thus, the precipitation difference in inches is even greater than the temperature difference in degrees Fahrenheit. Again, sun angle plays a major role in these particular cases.

In Chapter 8, I discuss how solar energy causes water to evaporate and plants to *transpire* (sweat), producing atmospheric vapor, the building blocks of raindrops and of other forms of precipitation. Generally, the greater the solar energy, the greater the amount of evaporation and transpiration, which result in vapor in the atmosphere. And the greater the amount of atmospheric vapor, the greater the likelihood of precipitation. Given that Manaus experiences much more intense concentrations of solar energy than does Churchill, its atmosphere is much more humid and therefore has a higher rainfall potential. Indeed, the very heat that produces that vapor may also create an atmospheric upwelling, or *convection current*, which carries vapor to a high altitude where it cools, condenses, and falls as rain.

Tilt-a-World: The Reasons for the Seasons

REMEMBER

Earth's axis is not, in reality, "straight up and down" as indicated by Figure 9-1. Instead, it's tilted by 23½° from the perpendicular. As a result, the vertical rays do not "stay put" at the equator as Earth orbits the Sun. Instead, they "migrate" north and south of the equator at different times of year, bringing with them patterns of seasonal change (as illustrated by Figure 9-2). Take, for instance, the United States. During summer, the vertical rays move into the Northern Hemisphere, increasing its "dosage" of solar energy and producing the warm temperatures that are associated with that time of year. During winter, however, the vertical rays move into the Southern Hemisphere, greatly increasing that region's solar dosage but at the same time greatly decreasing the dosage that hits the United States.

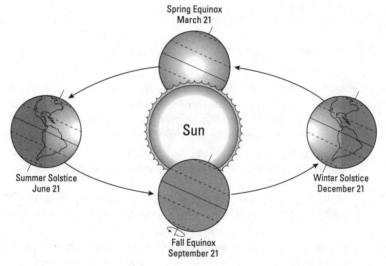

FIGURE 9-2: Earth's inclination at different points in its orbit of the Sun.

(© John Wiley & Sons Inc.)

Special lines of latitude

The inclination of Earth on its axis accounts for four special lines of latitude that appear on most world maps and globes (see Figure 9-3). Two of them are the *Tropic of Cancer* (23½° N) and the *Tropic of Capricorn* (23½° S). The area between them may properly be called *The Tropics*.

TIP

What's an easy way to remember which tropic is which? People get corns on their feet, so the Tropic of Capricorn is the one "below" or to the south.

CHAPTER 9 **Warming Up and Chilling Out: Why Climates Happen**   141

The other two lines are the *Arctic Circle* (66½° N), and the *Antarctic Circle* (66½° S). The area north of the Arctic Circle may properly be called *The Arctic* or *The Northern Polar Region*. Similarly, the area south of the Antarctic Circle may properly be called *The Antarctic* or *The Southern Polar Region*. But why are these lines there in the first place? For now we will concern ourselves with the Tropic lines because they have much to do with defining seasonal change.

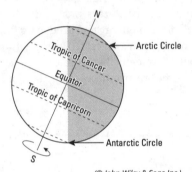

FIGURE 9-3: The tilted Earth showing special lines of latitude.

(© John Wiley & Sons Inc.)

Because of the angle of Earth's tilt, during summer in the Northern Hemisphere, the vertical rays move north of the equator as far as Latitude 23½° North. Conversely, during winter, the vertical rays migrate south of the equator as far as Latitude 23½° South. Thus, the Tropic of Cancer marks the most northerly latitude that is struck by the Sun's vertical rays at some point during the year. Conversely, the Tropic of Capricorn marks the most southerly occurrence.

Parenthetically, "Tropic" comes from the Greek *tropos,* to turn. The ancient Greeks observed that during their summer the Sun's vertical rays moved northerly until they reached Latitude 23½° North, whence they "turned" back to the south. Cancer and Capricorn refer to stellar constellations that were prominent in the Greek sky when the vertical rays struck one or the other tropic lines.

Defining the seasons

REMEMBER

Four days each year, vertical rays strike either the equator or one of the Tropic lines. On two of those dates, called *equinoxes,* the vertical rays strike the equator. On the other two dates, called *solstices,* the vertical rays strike one of the Tropics. The significance of these dates is that they mark the beginnings of the four seasons of the year. The following sections show the annual cycle as it relates to the Northern Hemisphere.

Keep in mind that seasons are relative. Summer never happens everywhere at once. Ditto fall, winter, and spring. Instead, summer occurs in one hemisphere while winter happens in the other, and vice versa. Likewise, spring occurs in one hemisphere while autumn falls in the other, and vice versa. Thus, when the "Summer Olympics" were held in Sydney, Australia (Southern Hemisphere), it wasn't summer as far as the locals were concerned, but instead late winter.

Spring

Sometime around March 21, the vertical rays strike the equator, marking the spring equinox. This is the first day of spring in the Northern Hemisphere and the period of daylight and darkness are the same. Every day for about the next three months, the vertical rays strike the Earth at progressively more northerly latitudes. The daylight hours get longer while night gets shorter.

Summer

On or about June 21, the vertical rays strike the Tropic of Cancer, marking the summer solstice. This is the first day of summer as well as the date that has the longest period of daylight and the shortest nighttime. From this point, the vertical rays then "turn south." Daylight hours lessen while nighttime hours increase.

Fall

On or about September 21, the vertical rays again strike the equator, marking the fall, or autumnal, equinox. This is the first day of fall and again the period of daylight and darkness are equal. The vertical rays then move into the Southern Hemisphere, striking ever more southerly latitudes each day for about the next three months. In the Northern Hemisphere, nighttime now exceeds daytime by a margin that increases each day.

Winter

December 21 is the approximate date of the winter solstice. The vertical rays then strike the Tropic of Capricorn, marking the first day of winter. Also on that date, the Northern Hemisphere experiences its longest period of night, not shortest period of night. The vertical rays then "turn northward." For about the next three months, nighttime periods continue to exceed daytime periods in the Northern Hemisphere, but by a difference that diminishes daily. Finally, on or about March 21, the vertical rays are back at the equator, marking the spring equinox. Day and night are again equal and the seasonal cycle is complete.

Special lines of latitude revisited

Before leaving this section, discussion is in order of those other two "special lines of latitude" — the Arctic and Antarctic Circles. Each demarcates parts of the world where something peculiar happens. Specifically, every location north of the Arctic Circle and south of the Antarctic Circle experiences at least one continuous 24-hour period of daylight, and at least one continuous 24-hour period of darkness during each year. Moreover, the farther north and south one goes with respect to those two lines, the greater is the number of days of complete daylight and darkness. The extreme cases occur at the two poles, where the year is divided into one six-month-long period of daylight followed by one six-month-long period of darkness.

TECHNICAL STUFF

To help understand this, look back at Figure 9-3. The North Pole is 90 degrees' worth of latitude from the equator. On the first day of fall, the Sun is directly overhead at the equator, but will appear to an observer at the North Pole to be on the horizon (90° from overhead). Every day for the next six months, the vertical rays strike Earth below the equator. From the perspective of the North Pole, the Sun is below the horizon all the while, so darkness ensues. Every other latitude between the North Pole and the Arctic Circle, Latitude 66½° North, also experiences at least one continuous 24-hour period of darkness. The significance of Latitude 66½° North is that it's exactly 90° from the Tropic of Capricorn, which marks the farthest southerly point that feels the Sun's vertical rays.

The opposite occurs during spring and summer, when the North Pole has continuous daylight. As the summer solstice approaches, more and more latitudes south of the North Pole experience similar conditions. Finally, on the summer solstice, the experience of a continuous day of sunlight reaches its most southerly locale — the Arctic Circle.

TECHNICAL STUFF

One last technical exception to everything just written: these lines, seasons, and associated periods of light and dark assume that Earth's tilt remains constant. It doesn't. For the record, Earth's tilt does stay stationary for a long time from a human perspective. But wobble it does between 22.1 and 24.5 degrees. This shift takes place over a 41,000-year time period to be exact. *Milankovitch Cycles* describe orbital movements that include not only Earth's tilt but its orbit around the Sun. Changes in each over time influence the amount of incoming solar radiation received by one place over another. The meaning? In the future, there will be a bit more fuzziness associated with seasonal activity and its timing at the Tropics of Cancer and Capricorn.

Hot or Cold? Adjust Your Altitude

Altitude has an important impact on climate. The rule of thumb is that temperature and elevation are inversely related. Or in every day speech, "the higher you go, the cooler it gets and vice versa." Thus, highland areas have cooler climates than lowland areas.

Consider this example: Fewer than 200 miles separate Guayaquil and Quito, the two largest cities in Ecuador. But the annual average temperature is 77° F in Guayaquil and only 56° F in Quito. The explanation is that Guayaquil is virtually at sea level while Quito is up in the Andes at an elevation of about 9,250 feet. Despite being nearly on the equator, the city does not experience the warm climate one would normally expect at that latitude, due to the altitude factor. How does that work? Glad you asked.

Warming the atmosphere

Part of the answer concerns how the atmosphere obtains heat. A portion of the solar energy that reaches Earth (18 percent) is absorbed directly by the atmosphere, while an even larger percentage (32 percent) reflects back into space. The largest portion by far (50 percent), however, is absorbed by Earth, which then re-radiates that heat into the atmosphere. Thus, solar energy turns Earth's surface into something like a giant frying pan that heats the atmosphere above it. Generally, therefore, air that is at or near Earth's surface is relatively warm, while increasing elevations above "the frying pan" brings progressively cooler temperatures.

Weighty matter

Another reason why "the higher you go, the cooler it gets" is the fact that the atmosphere has weight. Gravity is constantly "pulling down" on it. Thus, as altitude increases, the amount of air decreases. And because the atmosphere holds heat, less air means colder temperatures. Therefore — and to return to our opening examples — Quito, which is way up in the Andes, has a colder climate than Guayaquil, which is down by sea level.

Seeing (and feeling) is believing

Because relatively cool temperatures characterize high elevations, precipitation in highlands and mountainous regions is apt to be snow rather than rain for a good portion of the year — if not for all of it in the case of really high mountains. That results in what many regard as one of nature's most aesthetically pleasing sights — a snow-capped mountain (as you can see in Figure 9-4). But the effects of elevation may be felt as well as seen. Go up a high mountain, and not only does it get colder as you go higher, but also breathing becomes increasingly difficult. It makes sense. Air has weight. It wants to sink towards sea level. So the higher you go, the less air is available to help warm things up and to help you breathe easier.

FIGURE 9-4: From Pau, France, one can see the snow-capped Pyrenees Mountains, demonstrating the relationship between altitude and temperature.

(© Jerry T. Mitchell)

The lapse rate

TECHNICAL STUFF

The numerical relationship between temperature change and elevation change is called the *lapse rate*. It works out to about 3.5° F per 1,000 feet, or 6.4° C per 1,000 meters. That is, if you have two hikers on a mountain separated by 1,000 feet of vertical distance, then the person higher up is experiencing a temperature that is about 3.5° F colder than the person down slope. Parenthetically, I say "about" because *humidity* — the amount of vapor in the air — can and does tamper with these formulae. Therefore, think of the above numbers as average figures.

Windward slope, leeward slope

REMEMBER

In addition to temperature, altitude can have a profound effect on patterns of precipitation. An air current is forced to rise when it meets a mountain (see Figure 9-5). As it does so, its vapor cools and condenses, forming raindrops (or snowflakes). Precipitation that is produced in this manner is called *orographic* (from the Greek *oros*, mountain), meaning that it is mountain-related. The heaviest rains and snows tend to occur on the *windward* slope, which is the side of the mountain from which the wind is blowing.

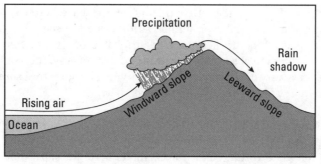

FIGURE 9-5: Given a dominant wind direction, mountains and mountain ranges may exhibit a wet windward slope and dry leeward slope.

(© John Wiley & Sons Inc.)

After cresting the summit, the air descends. As it does so, it warms, which is the opposite of what air needs to do in order to condense and form rain. But the air also has little vapor remaining, so the prospects for precipitation are just about zilch. The result is a dry *leeward* slope, whose droughty environs are said to be in a *rainshadow*.

Instead of a lone mountain, as in Figure 9-5, imagine a lengthy mountain range. The windward side has a rather wet climate. The leeward side, being in the moisture-deprived rainshadow, is a desert or semi-desert. For example, the Himalayan Mountains lie perpendicular to seasonal rain-bearing winds that come from the Indian Ocean. The result is a virtual tropical rainforest climate on the southern windward side, and a vast expanse of arid and semi-arid conditions (including the Gobi and Takla Makan Deserts) on the northern leeward side.

Similarly, the Coast and Cascade Ranges in Northern California, Oregon, and Washington State, intercept moisture-bearing winds that enter the mainland from the Pacific Ocean. The result is a very moist coastal fringe (noted for its tall trees — the redwoods — and lush forests) on the windward side. But the lands to leeward are arid and semi-arid. Picture in your mind the dry landscape of Nevada. Now consider that the state lies on the leeward side of the Sierra Nevada Mountains. Starting to make sense, isn't it?

This phenomenon is also present on a smaller scale as many islands have distinctive windward and leeward sides. On the windward coast the wind comes off the ocean, often resulting in rough surf conditions. If there are highlands inland, there may be more rain here too as air is forced to rise over the higher elevations. The leeward side will have less rain and is generally drier.

Gaining Heat, Losing Heat

REMEMBER

Locations in the middle of continents tend to have hotter summers and colder winters than do locales at similar latitudes by the sea. This condition is called *continentality*. It occurs because land and water have very different characteristics when it comes to absorbing and retaining solar energy. It occurs because land and water have very different characteristics when it comes to absorbing and retaining solar energy as the next local examples show.

Earth and sand are not transparent, so most of the solar energy that strikes them is absorbed by and concentrated in the top-most inch or half-inch of surface material. As a result, sand, for example, becomes super-hot. If you have ever experienced scorched feet while walking barefoot on dry sand on a sunny summer day, then you know exactly what I am talking about.

In contrast, because water has a certain transparency, solar energy penetrates the surface and, depending on the clearness of the water, spreads itself out over the depths. Also, wave action and other flow mix the upper layers of water, and thereby transport the absorbed heat away from the surface. As a result, a much greater volume of water, compared to the sand, absorbs heat. Therefore, the temperature of the water on a sunny summer day tends to be somewhat cooler than the temperature of the sand.

This difference has significant implications for an outing at the beach, as explained below. The same is true regarding the tendency for mid-continent locales to have warmer summers and cooler winters than locations by the coast.

Afternoon versus evening

Assume it's a boiling hot summer afternoon. You and some friends are on a beach blanket on the lake shoreline. You decide to take a swim. The sand is terribly hot against your feet as you run for the water and take the plunge. That water sure feels cool!

Now assume it's 9 p.m. and you are still hanging out by the lake. The Sun has gone down. The temperature has cooled considerably, and the sand is now somewhat cool against your feet. Someone suggests you all go for a swim and you decide to join in, however reluctantly. So you run for the water and take the plunge. That water sure feels . . . warm!?

How can the same body of water that cools you off at midday warm you up at 9 p.m.? Well, the beach gets super-hot on a sunny summer because solar energy is absorbed and concentrated in the top veneer of surface material as I explained earlier. In contrast, the lake heats up more slowly because the solar energy it absorbs is spread out over a much larger volume of matter. Thus, you experience a cooling sensation when you jump into the lake in the middle of the afternoon.

But it's a different matter at 9 p.m. Because nearly all of the land's heat is contained within a thin veneer of surface sand, it tends to radiate back into the atmosphere rather fast after the Sun goes down. But the lake is different. Only a small portion of its solar inventory is at the surface — exposed to the atmosphere. After the Sun goes down, the lake's heat is radiated back into the atmosphere at a slower rate. Thus, it retains its absorbed energy longer than land, resulting in a warming evening swim.

Summer versus winter

The fact that land and water heat up and cool down at different rates has significant implications for climate. Picture now instead of a beach by a lake a continent next to an ocean. Also assume that there is a city in the middle of that continent and that there is a coastal city a thousand or so miles away. Timewise, consider summer versus winter as opposed to mid-afternoon versus 9 p.m.

During the summer, the city in the middle of the continent is likely to be warmer. That is because, just like in the beach example, solar energy will concentrate at Earth's surface and heat the atmosphere overhead. In contrast, the coastal city is likely to be less warm during summer. That is because its atmosphere will be warmed by heat that radiates off both the land around it and the water offshore. But because the surface of the water contains so much less heat than the land, the total amount of heat that is radiated into the atmosphere is far less than occurs in the middle of the continent. In the parlance of climatology, the water body has a *modifying* or *mitigating effect.* That is, it results in the atmosphere being less warm than would be the case if there were land all around. The net result, whatever the vocabulary, is that the city in the middle of the continent will experience a warmer summer than the one by the sea.

Winter is another matter, however. Regarding the mid-continent city, the heat that was absorbed by the surrounding land during the summer — being concentrated at the surface and now exposed to long winter nights — radiates into the atmosphere rather rapidly, contributing to cold temperatures. These land-related conditions also apply to the coastal city. But something else of significance also affects the temperature of the atmosphere in the latter locale. Because the heat absorbed by the water body is distributed over a certain depth — and is therefore not concentrated at the surface and exposed to long winter nights — it is radiated back into the atmosphere at a much slower rate. In effect, it may serve as a source of atmospheric warmth for a significant portion of the winter, and result in a warmer winter for the coastal city.

Consider a comparison of Pierre, South Dakota and Portland, Maine. The former is in the middle of a continent, the latter is on the coast, and their latitudes are nearly the same. In Pierre, the coldest month of the year averages 17° F and the warmest month of the year averages 75° F. That makes a 58-degree *annual temperature range* (the difference between the coldest and warmest month). In Portland, the coldest month averages 23° F and the warmest month averages 68° F, making its annual average temperature range 45° F. So winters are harsher and summers are hotter in the mid-continent city.

Oh, How the Wind Blows

The heat differential between land and water also has influenced how people live, and how they have designed their architecture. Consider hot and muggy Charleston, South Carolina. The hot warmed land radiates heat into the air above during the day and due to air pressure differences onshore breezes develop as "new" air fills the void. This brings cooler air — a sea breeze — inland. Architects designed long open porches facing the direction of the wind to take advantage of this "air conditioning" and they built rooms with high ceilings to better circulate the air inside.

Going with the Flow: Ocean Currents

REMEMBER

The oceans have warm and cold surface currents that act like a global heating and air-conditioning system. They bring significant warmth to high latitude areas that would otherwise be much cooler, and significant coolness to low latitude areas that would otherwise be much warmer. Remember that the Equator is 0 degrees, thus the middle of Earth is the "low" latitudes and one goes to the "high" latitudes even if you traveled south!

Ocean currents also play a major role in determining the global geography of precipitation. The Sun can more easily evaporate warm water than cold water, and thereby produce the atmospheric vapor that results in rain. Therefore, lands that get *sideswiped* or impacted by warm currents tend to have abundant precipitation in addition to a comparatively warm climate. Conversely, lands impacted by cold currents tend to receive very little precipitation in addition to a comparatively cool climate.

Generally, surface currents exhibit circular movements (see Figure 9-6). North of the equator, the flow is usually clockwise. South of the equator, the flow tends to be counterclockwise. These movements are principally products of prevailing winds that "push" the ocean's surface. On the map you can see occasional exceptions to the general rules of circulation. They are the results of deflections caused by the angle at which a current strikes a land mass or the continental shelf, or by the direction of prevailing sea level winds at particular latitudes.

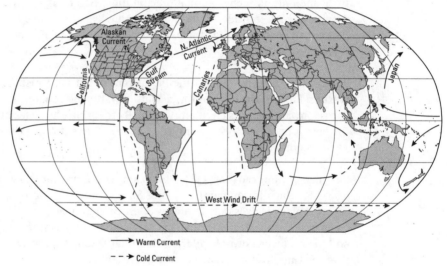

FIGURE 9-6: A generalized geography of ocean surface currents.

(© John Wiley & Sons Inc.)

Warm currents, cold currents

The warm and cold portions of these circulatory systems have rather predictable geographies. As ocean currents move westward along the equator, they absorb lots of solar energy, heat up, and become warm currents. As they turn away from the equator, they generally continue to absorb about as much heat as they dissipate, at least while they remain in the Tropics — that is, the region between the Tropic of Cancer and the Tropic of Capricorn.

After leaving the Tropics, the reverse starts to happen: the currents radiate more heat than they gain — but slowly for the reasons you read about in the previous section concerning the ability of water to store and retain heat. Thus, the currents remain comparatively warm longer after they have left the tropics. The *Gulf Stream*, for example, is a warm-water current that moves up the Eastern coast of the United States and then becomes the North Atlantic Current (see Figure 9-6). Although it loses a fair amount of heat as it moves eastward across the mid-Atlantic, the North Atlantic Current reaches Europe with a considerable amount of stored heat remaining. As it continues to radiate that heat, it contributes to the climate of Northwestern Europe a degree of warmth that is unusual for those latitudes, and also abundant rainfall.

Take a look at an example of an area that is affected by the North Atlantic Current. Bergen, Norway (Lat. 61°N) has an annual temperature of 45° F and receives 77 inches of precipitation per year. Compare that to Churchill, Manitoba (Lat. 58° N), which, as mentioned earlier in the chapter, has an annual average temperature of 19° F and only 15 inches annual precipitation. Bergen is significantly warmer — despite its high latitude — and much wetter. The difference is partly a matter of Churchill's continentality, and partly a matter of the relatively warm current that sideswipes Bergen.

But the Gulf Stream-North Atlantic Current is not yet finished. After impacting Western Europe, the current turns south towards the equator, now as the *Canaries Current*, to complete its circulatory cycle. By that time, however, it has lost most of the heat it once had. As a result, the Canaries Current that sideswipes Northwest Africa is quite cool.

Casablanca, Morocco (Lat. 33° N), for example, has an annual average temperature of 63° F, which is comparatively cool for a country on the fringe of the Saharan realm. It also receives only 17 inches of precipitation per year. That is a paltry sum compared to the 77 inches that the same circulatory systems dumps on Bergen, and the 46 inches of precipitation that Charleston, SC, receives by being located near the Gulf Stream directly across the Atlantic.

Going against the norm: El Niño and La Niña

You should remember that climate is an average of yearly conditions, but that in any given year very "un-average-like" events can occur. El Niño and La Niña, which happen every so many years, provide good examples. (Niño and niña mean boy and girl in Spanish.) As you can see in Figure 9-7, during an El Niño, the surface waters become unusually warm in the tropical portion of the Pacific. The reasons for this are not fully understood but are primarily related to changes in the tradewinds (generally the surface winds across the Pacific have become weaker). Because the conditions occur around Christmas in the waters off western South America, the local populace call it El Niño, referring to the Christ child. During La Niña, the opposite happens ("girl" being the opposite of "boy") — the water is unusually cold.

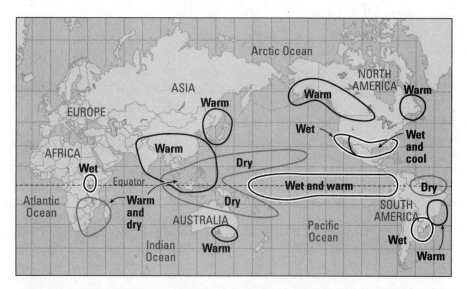

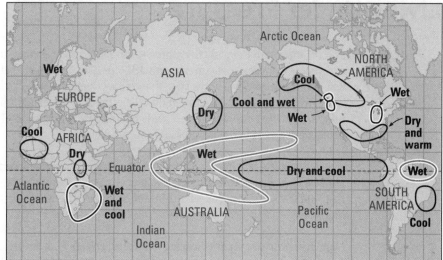

FIGURE 9-7: Conditions associated with El Niño (top) and La Niña (bottom).

(© John Wiley & Sons Inc.)

Because the affected ocean water circulates, and also influences the behavior of atmospheric pressure belts (which you can read about in the next section), the impact can be substantial and widespread. Just what that means varies from place to place and year to year. Sometimes, for example, rainy seasons become extremely stormy and dry seasons become prolonged droughts. On the other hand, the effects are not always bad, as may be evidenced perhaps by a normally harsh winter that turns up mild.

CHAPTER 9 **Warming Up and Chilling Out: Why Climates Happen**   153

> ## COASTAL DESERTS
>
> Casablanca, Morocco, highlights one of the world's most provocative geographic juxtapositions: places where oceans border deserts. Indeed, a couple of coastal deserts exist. What most have in common is a neighboring cold-water current that makes evaporation difficult and rainfall unlikely.
>
> Another example is found in southern Peru and northern Chile. Here we have the Atacama desert, the world's driest. A number of culprits are at play. One is the moisture blocking power of the towering Andes to the east, shielding the area from moisture in the continent's interior. Another is the location under a high-pressure belt with drier air. Finally, there's a very cold ocean current working its way north from Antarctica. Over time this has been called the Humboldt or Peru Current.

Living Under Pressure

REMEMBER

You're under pressure all the time — atmospheric pressure, that is. Just about everybody has seen a weather map with big "Hs" and "Ls" here and there. And just about everybody knows that they respectively stand for: high pressure and low pressure. But just about nobody understands what exactly they mean, except maybe that lows are associated with cloudy, rainy (or snowy) days, and highs usually are associated with pleasant, sunny days.

A *low-pressure system* is an area of relatively warm, moist ascending air. A *high-pressure system* is an area of relatively cool, dry descending air (see Figure 9-8). In general, therefore, you can think of low pressure as being a rainmaker, and high pressure as a drought-maker.

High pressure is so-named because the atmosphere is pressing down on Earth. In contrast, low pressure is so-named because, due to its upward-moving air, the pressure (or weight) of the atmosphere against Earth is comparatively low. Both are linked in a three-dimensional pattern of atmospheric circulation as shown in Figure 9-8.

Solar energy sets this circulatory system in motion. Some parts of Earth heat up more rapidly than others. Over those areas that do, the air tends to warm, expand, and rise. The vapor in the air cools as it rises in a *convection current*, causing condensation and (in all probability) precipitation to occur. Thus, low pressure is associated with cloudy, rainy (or snowy) conditions.

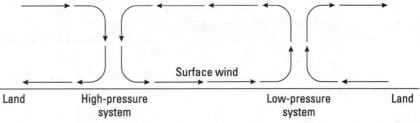

FIGURE 9-8: A cross-section of high- and low-atmospheric pressure-systems.

(© John Wiley & Sons Inc.)

After precipitating, air at the top of a low-pressure system is cool, dry (having "lost" its moisture) and heavy. It wants to sink back down to Earth, but can't because of other air coming up from underneath. Air in the upper atmosphere therefore moves laterally until it finds a place where it can descend as a high-pressure system composed of comparatively cool, clear, and dry (low humidity) air.

Pressure belts

Because the equatorial latitudes receive a greater degree of solar energy than anyplace else on Earth, a global "belt" of low-atmospheric pressure characterizes them (see Figure 9-9). This phenomenon is called the *inter-tropical convergence zone* (ITCZ), because air from the tropics north and south of the equator is drawn into (converges on) this zone before it rises in a convection current.

The result is a warm, humid "rainmaker" that produces the tropical climates presented in Chapter 10. As implied by Figure 9-9, the air that rises in this low-pressure belt must fall to Earth elsewhere. Generally, this occurs in two subtropical high-pressure belts that roughly correspond to Latitudes 25-30 North and South. Given these belts of "drought-makers," it's not surprising to see desert and semi-desert conditions over much of these latitudes (see Figure 9-10). Remember the Atacama Desert mentioned in a sidebar earlier in this chapter (pay attention at the back!)? It's located at roughly 24° South latitude. How about the Kalahari Desert in Africa? The Great Victoria in Australia? All at the same latitude or pretty darn close. There's a distribution for all Earth phenomena, from people to climates. Sometimes they make patterns that are clear to see. The location of deserts is among the patterns most readily observed.

Monsoons

Now you know how and why the Sun's vertical rays migrate north and south of the equator during the year. (You may want to refer back to "Tilt-a-World: The Reasons for the Seasons," earlier in this chapter.) Because the equatorial low-pressure belt is a product of those very same rays, it migrates as well, and so do the related subtropical highs. During the year, therefore, tropical latitudes may be alternately dominated by the low-pressure belt, which brings a rainy season, and one of the high-pressure belts, which brings a dry season.

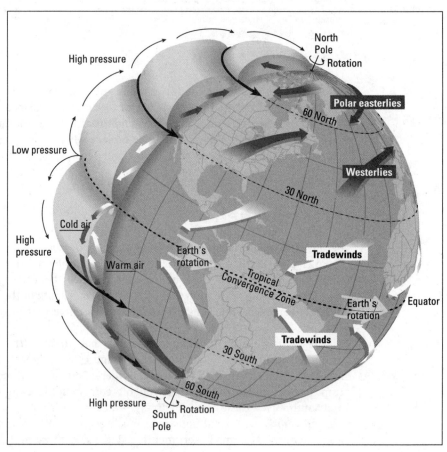

FIGURE 9-9: Inter-tropical convergence zone and sub-tropical high-pressure belts.

(© John Wiley & Sons Inc.)

Wet season-dry season transitions happen in many parts of the tropical world, the most well-known example being the *monsoons* of South Asia. As the vertical rays move northward during summer, the low-pressure belt moves with it, drawing moisture-laden air from the Indian Ocean and producing a pronounced rainy season. During winter, the ITCZ moves south of the equator. At that time the subtropical high-pressure belt moves over South Asia and surface winds blow from the interior of the continent to the Indian Ocean, resulting in a relatively rainless period (see Figure 9-11). To see the impact of monsoons on a local place, see the sidebar "The wettest place on Earth?".

FIGURE 9-10: The sand dunes of the Sahara? No, these travelers are wandering coastal Peru – an area located under a high-pressure belt.

(© Jerry T. Mitchell)

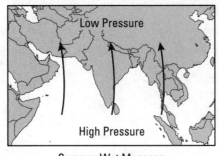 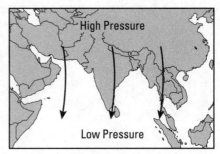

FIGURE 9-11: Pressure belt locations during the wet and dry monsoons.

(© John Wiley & Sons Inc.)

For example, Mumbai, India (Lat. 19° N) receives 72 inches of rain during June–September. In contrast, it receives less than half an inch during December–April. For some reason, many people associate "monsoon" strictly with the rainy season. In reality, there are two monsoons, wet and dry, and each significantly impacts the region, albeit in entirely different ways.

THE WETTEST PLACE ON EARTH?

Cherrapunji, India, claims the wettest recorded climate of any settlement on Earth. This place holds the world record for the greatest amount of rainfall in a year — 1,042 inches — and also in a single month — 370 inches! I include the question mark with the title because there may be wetter locales that go unrecorded. In any event, it may be near impossible to beat an average of 450 inches of precipitation per year. That's Cherrapunji — more than an inch of rain per day on average. But even that doesn't tell the whole story. Check out the following table, and pay attention to the substantial monthly variation.

Cherrapuinji: Average Monthly Temperature and Precipitation

	J	F	M	A	M	J	J	A	S	O	N	D
Temp °F	53	56	61	65	67	69	68	69	69	68	62	56
Rain (in.)	1	2	13	33	54	102	122	72	45	18	3	1

Cherrapunji has a dry monsoon season that runs from November through February, during which it receives about 7 inches of rain. But then things change rather dramatically. In an average June and July, the town receives more than 3 inches of rain *per day*, before things taper off to a mere 2 inches per day in August, and an inch-plus in September. Cherrapunji exemplifies the extreme conditions that can occur when two climatic determinants "pull together." In this case, a fortuitously located low-pressure system and the effects of altitude combine. The town is 4,300 feet above sea. Thus, when the wet monsoonal winds are forced to rise in and around Cherrapunji, the results are very wet indeed.

IN THIS CHAPTER

» Classifying climates

» Getting sticky with humid tropical climates

» Drying out with dry climates

» Staying comfortable with humid mesothermal climates

» Chilling out with humid microthermal climates

» Freezing with polar climates

Chapter **10**
Connecting Climates and Vegetation

The ancient Greeks divided the world climatically into a tropical *torrid zone*, two mid-latitude *temperate zones* (one in each hemisphere), and two high-latitude *frigid zones*. The Greeks lived in the temperate zone of the Northern Hemisphere and never sojourned to the frigid zone to their north or torrid zone to their south. That suited them just fine, for what they knew — or rather believed to be true — about those areas was fearsome.

The torrid zone in particular inspired dread. It was believed that the sun could literally burn people to death or set fire to a ship. The Greeks' direct experience with Saharan temperatures reinforced that perception. These ideas persisted for nearly two thousand years, until 1434, when the Portuguese captain Gil Eannes rather clandestinely navigated into the area without ill effect to any of his crew.

REMEMBER

Nowadays the study of *climate*, the average temperature and precipitation conditions that occur at a location over a long period of time, is looked upon more as a source of useful information than of fear. Knowledge of climates and their distribution help us to understand, for example, why particular patterns of agriculture are practiced in particular parts of the world. This may prove very useful in devising development scenarios aimed at increased food production. In addition, knowledge of climate helps us to understand why people live where they do (as well as where they could live), the problems and potentials of various regions, and the geographies of architecture and dress.

Obviously, therefore, climate is an immensely important geographic variable. Accordingly, *climatology* is a major sub-field of study and research within geography, and the subject of this chapter.

TIP

I'd like to throw in my own two cents here. Although each chapter is designed to stand on its own, I advise you to read Chapter 9, which recounts the reasons why particular climates occur in particular regions, prior to this one in order to grasp the full meaning of weather and climates. Global warming and climate change are important and timely topics. While this is a logical place to discuss them, I'm going to mainly hold off until Chapter 18, which focuses on current issues of human-environment interaction. A sneak into how warming can impact people is in the sidebar "Malaria: A case study in climate and disease" later in this chapter.

Giving Class to Climates

One thing that has not changed since ancient Greece is the need to classify climates. Because no two Earth areas have exactly the same average temperature and rainfall regimes, various categories — climate types — have been defined on the basis of maximum and/or minimum temperature and precipitation data.

In 1898, Vladimir Köppen, a German geographer and climatologist, developed the climate classification system that is most in use today. He identified about 25 specific climates and used a rather arcane letter code (using codes such as BWh, Dfb) to identify and define them. If this book were *Climate For Dummies*, then it would be appropriate to discuss the Köppen system *ad nauseum*. But because this book's title is *Geography For Dummies*, and because I'm a nice guy and don't feel it necessary to bog you down with unnecessary information, I'm going to forego the letter code and several of Köppen's climate-types and aim for just enough descriptive treatments of just enough climates to provide a global overview consistent with the goals of this book.

REMEMBER

Before getting down to the nitty-gritty of classifying climates, we must keep in mind a common thread among climates. Each climate type has an associated assemblage of *natural vegetation* that is likely to occur provided human beings and natural catastrophes do not interfere. Thus, if one could journey overland from the equator to the North Pole, different natural vegetation assemblages would be encountered with every change in climate (and altitude! – see later sidebar "Vertical zonation and 'highlands climate'"). This is illustrated in Figure 10-1. In reality, of course, people have modified or eliminated natural vegetation in many areas, as by converting grasslands and rainforests to farms. When one visits a particular climatic realm, therefore, purely natural vegetation may or may not dominate the landscape, or be present at all. In any event, the climate-natural vegetation connection is so strong that some climates are named for the plant cover that is associated with them (including the tropical rainforest, steppe, and tundra.)

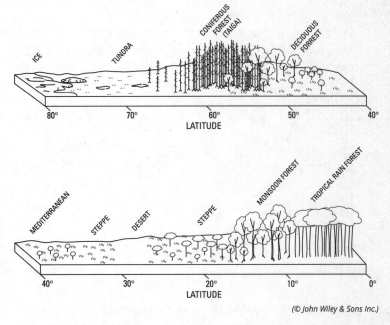

FIGURE 10-1: The hypothetical sequence of natural vegetation from the equator to the North Pole.

(© John Wiley & Sons Inc.)

In general terms, the world's climates may be grouped into five classes. They are humid tropical climates, dry climates, humid mesothermal climates, humid microthermal climates, and polar climates. Each is discussed in the following sections.

CHAPTER 10 **Connecting Climates and Vegetation** 161

Mixing Sun and Rain: Humid Tropical Climates

REMEMBER

In humid tropical climates, the average temperature of each month is 64° F or higher. The warmth is a function of vertical rays (see Chapter 9) and near-vertical rays that strike the tropical latitudes pretty much throughout the year. All that sunshine, in turn, generates high evapotranspiration (for more on this, see Chapter 8), producing a moisture-laden atmosphere, and also creates convection currents (see Chapter 9 for more details) that cause the air to rise, cool, condense, and cause rain. In consequence, annual precipitation is abundant and may occur year-round or in distinctive wet seasons that vary in intensity and duration. This variation in precipitation distinguishes the three principal climates in the Humid Tropical category (see Figure 10-2): tropical rainforest, tropical monsoon, and savanna — the latter is also known as tropical wet and dry.

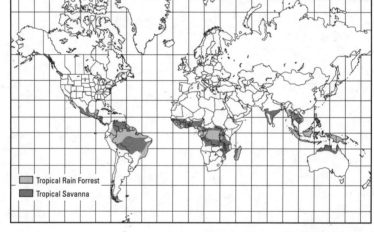

FIGURE 10-2: The geography of humid tropical climates.

(© John Wiley & Sons Inc.)

TECHNICAL STUFF

As "tropical" suggests, these climates generally occur between Latitudes 23½° North and South. Figure 10-2 shows, however, a few decidedly non-equatorial areas where a tropical humid climate prevails due to warm water currents, orographic rainfall (see Chapter 9 for more details), or some other mechanism. Conversely, "non-tropical" climates are occasionally seen between the Tropics of Cancer and Capricorn thanks to the cooling effects of elevation, cold-ocean-surface currents, or predominant wind directions.

Tropical rainforest

REMEMBER

In areas of Tropical Rainforest climate all months average above 64° F and the driest month of the year averages above 2.4 inches (6 cm) of rain. For all intents and purposes, therefore, this climate may be described as warm and wet year-round. Although the equatorial low-pressure belt shifts north and south with the seasons, it never wanders far enough afield to result in a genuine dry period.

Incredible forests!

Tropical rainforest vegetation is the definitive natural-landscape feature. This plant assemblage is dominated by broadleaf evergreen species that grow to about 150 feet in height. Their adjoining tree tops create a "closed canopy" that in turn give rise to vertically arranged ecological zones between the ground and tree tops, each comprised of different plant species. Add to that the following:

» A year-round growing season (which accommodates a wide range of species)

» The lack of frost and drought (which also accommodates a wide range of species)

» The great age of the rainforest (which has encouraged mutation and genetic drift)

The result is the greatest concentration of living things (especially as regards to plants) to be found anywhere on Earth. How great, you ask? Well, in a square mile of forest in Vermont, you may find 12 to 15 different species of plants. In tropical rainforests, 300 to 400 different species are not unknown in comparable-size areas.

Not only is the variety of plants found here great. So, too, is their potential as sources of food and medicine. The latter is particularly important because a high percentage of medicinal drugs — including anti-carcinogens — utilize chemical compounds derived from tropical plants. Thus far, however, only a relatively small percentage of rainforest plants have been fully analyzed for their food and medicinal values. Also, because plants take in carbon dioxide and give off oxygen, the tropical forests are important to maintaining global atmospheric balance.

Endangered forests

Despite their real and potential benefits, tropical rainforests have been disappearing at an alarming rate. Reasons include:

» Rapidly rising populations in rainforest countries, which encourage conversion of forests to farmlands

» Global demand for timber coupled with technological developments that make rainforests more harvestable than ever before

>> Government programs that encourage settlement of rainforests either to assert ownership of remote areas or to relieve population pressure in other parts of the country

Fortunately, countries that contain rainforests have created national parks and preserves that will save millions of acres for posterity. But millions of acres more stand to be lost unless preservation efforts are greatly expanded.

Poor soils

TECHNICAL STUFF

Despite the lush forest cover, the tropical rainforest realm is underlain by infertile soils called *latosols*. These are products of warm temperatures and high rainfall, which respectively encourage high microbial activity that breaks down topsoil nutrients, and wash them away (a process called *leaching*) by means of runoff or downward percolation of water through the soil. Either way, the effect on soil nutrients is much like what happens to the contents of a tea bag after it has been used a couple of times — it becomes weak.

If the soil is so bad, then why (you may ask) is the natural plant cover so lush? The answer is found in root systems that tend to fan out laterally from the bases of plants rather than dig vertically into the soil. This allows trees to effectively absorb nutrients in the topsoil before leaching does its thing.

Shifting cultivation

A large percentage of the people who live in rainforest countries farm for a living, many of whom practice *shifting cultivation* (see Figure 10-3), which has a particularly devastating effect on rainforests. In this form of agriculture, farmers (and their extended families) clear an area of forest, grow crops on the plot for a year or two, and then abandon it, only to move on (hence, shift) to a new area of forest and repeat the process.

Soil infertility explains this practice. When farmers remove the trees, they also remove the sources of leaf-fall that contribute to productive topsoil. With the nutrient source literally cut off and the soil exposed to direct sunlight and rainfall, leaching is swift and sure.

When a plot of land has been abandoned, the forest reclaims it and the fertility of the soil gradually improves. After lying *fallow* (plowed land that's not being farmed) for a number of years, it may be used again. But population growth in most rainforest countries is so high that "recycled" land alone is insufficient to meet local food needs. As a result, new areas of virgin rainforest must be annually cut down and the acreage added to the inventory of land that is used for occasional shifting cultivation.

FIGURE 10-3: Fire is used to clear land for a type of agriculture — shifting cultivation — in Peru.

(© Jerry T. Mitchell)

Tropical monsoon

Unlike the rainforest realm, tropical monsoon regions experience a distinct dry season. That is because tropical monsoon regions generally lie in the area located between Latitudes 5° and 10° North and South, where the drying effect of the subtropical high-pressure belts are felt for part of the year. Despite high annual rainfall, the periodic dryness is sufficient to prevent the presence of many plant species that grow under tropical rainforest climatic conditions. Other than that, the characteristics and problems facing the world's monsoon lands are rather similar to the rainforest realm. In fact, except for the monsoon's wet and dry seasons, the characteristics are so similar that many climate maps include tropical monsoon regions within the tropical rainforest category — as is done in Figure 10-2.

Savanna (tropical wet and dry)

Savanna climate is distinguished from its tropical climatic kin by pronounced wet and dry seasons. The rather even duration and importance of these seasons have given rise to a self-explanatory climatic alias, tropical wet and dry. Most of the savanna realm is located between latitudes 5° and 20° North and South. These regions are alternately affected by passing low- and high-atmospheric pressure belts, which bring with them the wet and dry seasons respectively.

MALARIA: A CASE STUDY IN CLIMATE AND DISEASE

The most recent statistics from the World Health Organization indicate that about 230 million cases of malaria occur globally and some 500,000 people — mainly residents of sub-Saharan Africa (over 90 percent of deaths) — die from the disease or related complications. The disease is caused by a parasite that consumes red blood cells, causing high fever and other side effects. Species of mosquitoes that belong to the genus *Anopheles* are responsible for malaria in humans.

When a mosquito "bites," it actually sticks a syringe-like appendage into its victim and sucks up some blood. (Here's a fun fact for you: mosquitoes have no teeth, so they literally couldn't bite if their lives depended on it.) This rather antisocial behavior is necessary for mosquito reproduction, blood being required to produce eggs. Accordingly, and with no disrespect intended, it is only the females that bite. When a mosquito sucks up blood from a person or animal that has malaria, it may also suck up the malarial parasite. Hundreds of species of mosquitoes exist; and fortunately for humans, in just about every case, sucked-up malarial parasites die soon after entering the insect's body. For whatever reason, however, in the body of an *Anopheles* mosquito, the parasite remains viable, and thus is capable of being spread to the next human or animal that the insect "bites." The geography of malaria, therefore, is largely determined by the geography of the *Anopheles* mosquito, which is in turn is determined by the geography of the environmental conditions that the mosquito requires in order to live and reproduce. The principal criteria are temperatures that stay above about 75° F, and standing, shaded fresh water in which the insect can lay its eggs. In other words, it requires conditions that exist in abundance in areas that experience humid tropical climates.

People used to think that tropical air was unhealthy (hence, *mal aria* — "bad air"), and in many circles, belief persists that tropical climates are dangerous to humans. In truth, nothing is innately harmful about a warm and humid atmosphere. But what does occasionally happen is that climate gives rise to environmental conditions that are ideal for the proliferation of an insect or critter that is instrumental to the spread of a particular disease. The connection between humid tropical climates and malaria is a case in point.

One point to ponder further: If climates warm further, do previously mosquito-free areas now become hospitable to malaria-carrying mosquitoes? Climate change has implications far beyond rising sea levels.

"Savanna" refers to the natural vegetation that occurs under these conditions: a mix of trees and grasses (see Figure 10-4). The relative abundance of these elements generally varies, however, with annual precipitation. Accordingly, trees dominate where rainy seasons are relatively long. Grasses dominate under opposing conditions. In some parts of the world — Africa in particular — the grasses attract large herds of grazing animals (herbivores) and the meat-eating animals (carnivores) that prey upon them. But the grasses and relatively fertile soil that underlie them also attract herdsmen and farmers. The result, as described in Chapter 2, has been a steady decline in wild-animal habitat.

FIGURE 10-4: A savanna landscape surrounds Mount Kilimanjaro in Kenya and Tanzania.

(© Jerry T. Mitchell)

Going to Extremes: Dry Climates

"Dry climate" would seem to be a pretty straightforward concept. Wrong. Technically, it occurs where warm temperatures cause potential evaporation to exceed rainfall. Don't worry if that leaves you scratching your head.

REMEMBER

So for the sake of convenience, let's settle for the notion that a dry climate is characterized by no more than 20 inches of precipitation during the course of a year. Climatologists, true and exacting scientists that they are, may scream upon reading this, but I think they will agree that the 20-inch threshold is pretty close to accurate, and entirely appropriate for the purposes of this book. The geography of dry climates (see Figure 10-5) is made up of two areas: desert and semi-desert (or steppe).

CHAPTER 10 **Connecting Climates and Vegetation** 167

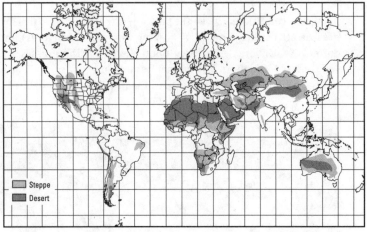

FIGURE 10-5: The geography of dry climates.

(© John Wiley & Sons Inc.)

Desert

Desert climate pertains to areas that average less than 10 inches of precipitation per year (we usually think of them as hot, too; see sidebar "Hot times on Planet Earth"). As noted in Chapter 9, cold ocean currents, persistent high-atmospheric pressure, and mountain ranges that produce *rain shadows* (leeward slopes that lack rain) create dry, desert conditions. These causative factors occur over a wide latitudinal range, which explains why deserts are rather widely distributed.

TECHNICAL STUFF

Hollywood movies have a penchant for depicting deserts as seemingly never-ending seas of sand dunes. People who live in desert areas or who have traveled through them know different, however. Most deserts are covered mainly by gravel, with enough sand and soil mixed in to support plant life (take a look at Figure 10-6). The descriptive term for this is *reg*, a word that English-speaking students of desert environments have borrowed from Arabic, a language which was born in desert surroundings and therefore has a much richer desert-related vocabulary than does English. Contrary to what Hollywood might have you believe, about 65 percent of the Sahara is *reg*. Another 30 percent is *erg*, the classic sand dune landscape. The remaining 5 percent is *hammada*, or rock-covered.

The natural vegetation of deserts consists of *xerophytes* ("dry-loving" species), which are plants that have adapted to dry conditions. To help conserve internal moisture, and thus live in lands where they would otherwise transpire to death, most xerophytes have defense mechanisms. These may include tough (even waxy) exteriors often complemented by thorns that ward off birds and other animals that might peck away at their exteriors and expose fleshy innards to the hot dry air. Because it takes a rather specialized plant to thrive in desert conditions and usually a very long time to grow to maturity in these regions of low-moisture availability, some xerophytes of the American Southwest (such as the giant saguaro cactus) are now protected by law.

HOT TIMES ON PLANET EARTH

Earth's all-time recorded high temperature is subject to debate. Al-Azizia, Libya recorded 136° F on September 13, 1922, but that record has been tossed out. Many are also suspicious of the 134° F, recorded in Death Valley, California, on July 10, 1913. A new front runner is the 130° F recorded at Death Valley National Park's Furnace Creek Visitor Center on July 9, 2021. In all likelihood, higher temperatures have occurred on Earth but have gone unrecorded. In any event, these three numbers are testimony to the fact that by far the world's highest temperatures occur in sub-tropical deserts.

Given the discussion of sun angles in Chapter 9, this may surprise you. The equator receives higher concentrations of solar energy than these two places in Libya and California, so you'd figure the equator would be warmer. However, in the equatorial realm the atmosphere tends to be somewhat cloudy and contain lots of water vapor. These respectively reflect a good portion of incoming solar energy back into space and directly absorb solar energy, both before the sunshine touches Earth. Moreover, much of this area is covered by vegetation instead of bare ground, so the "Earth as frying pan" analogy simply does not work to anything near maximum efficiency at the equator.

However, desert areas are a different matter. The lack of cloud cover and scarcity of vapor in the clear desert air means that a very high percentage of the solar energy that strikes this area reaches the surface, much of which is bare ground. Thus the frying pan analogy works to near perfection. The bottom line is that even though the place receives less intense "dosages" of solar energy than does a point on the equator, it heats up to a much greater extent.

Semi-desert (steppe)

Areas with semi-desert climate receive between 10 and 20 inches of precipitation per year. They normally are located between deserts and humid climate-types of either the tropical or middle latitudes. Semi-desert owns the record for the greatest latitudinal range. Instances of it are found on the equator in East Africa and in Western Canada at about Latitude 52° North. The same climatic determinants that explain the rather broad distribution of deserts also generally explain the geography of semi-desert. The natural vegetation of this climate is *steppe* — short grasses that grow in clumps with bare earth in between. Steppe is Russian in origin and describes what one sees in the vast, treeless, semi-arid plains of south-central Eurasia.

FIGURE 10-6: The desert landscape of central New Mexico.

(© Jerry T. Mitchell)

Crop-growing without irrigation in semi-desert areas is an iffy proposition (check out Figure 10-7). During those years when precipitation is average to above average, some production is possible. Below-average years, however, bring with them the high likelihood of crop failure and famine (see sidebar "Applied Geography: Drought mitigation"). Agriculturally, the steppe realm is *marginal land*, meaning that it's on the fringe (or margin) of that portion of Earth that is suitable for crops.

In contrast, the raising of livestock on the natural grasses has long been a principal activity. Accordingly, traditional pastoral nomads are associated with steppe environments, as are cowboys and cattle drives of the United States and *gauchos* (cowboy-like herdsmen) of South America. Nowadays, each of these is much more in the realm of lore than life, thanks to economic and political forces that have turned stock-raising into a rather sedentary endeavor. But raising livestock on steppe is risky. Wise resource management is essential, lest too many animals feed on the grasses, resulting in overgrazing and potential *desertification* — the conversion of non-desert lands to desert.

170 PART 2 **Let's Get Physical: Land, Water, and Air**

FIGURE 10-7: Steppe agricultural production can have disastrous outcomes as shown in this ship graveyard in a declining Aral Sea (Kazakhstan/Uzbekistan).

(© meiram / Adobe Stock)

APPLIED GEOGRAPHY: DROUGHT MITIGATION

Droughts have long been a source of human misery and death. The most horrendous ones typically occur in regions of steppe climate when the characteristic dry season is drier than usual and just won't quit. Areas to the south of the Sahara Desert have been particularly prone to these occurrences in recent decades. Adding to their devastation is the relative remoteness of this region, which hampers awareness of the drought in the outside world, thus inhibiting the ability of relief agencies to mount an effective response.

Thanks to *remote sensing,* the use of satellite imagery to monitor Earth's surface, it's now possible to monitor the onset of drought as it happens. This is made possible by satellite-based infrared imaging, which is somewhat like picture taking. Lush, healthy vegetation has lots of chlorophyll, which is an excellent reflector of the infrared energy that is a part of sunshine. In color infrared imaging, such vegetation registers as bright red. In contrast, dry vegetation (which is low on chlorophyll) appears as brown. Thus, when the dry season ends and the rainy season begins, the landscape rather immediately changes from brown to red, at least as far as an infrared sensor is concerned. If however, brown persists, then that means that the rains have not arrived, possibly indicating the onset of drought.

Enjoying the In-between: Humid Mesothermal Climates

REMEMBER

Humid *mesothermal* (moderate temperature) climates are located in the low-middle latitudes. They typically receive more than 20 inches of precipitation per year and therefore are not "dry." Also, the coldest month is less than 64° F but above 27° F, which places these climates between the tropical and polar temperature thresholds (see Figure 10-8). Mild winters and natural vegetation that is dominated by deciduous trees (which shed their leaves annually, as opposed to evergreens) are definitive characteristics of these areas, which include the Humid Subtropical, Mediterranean, and Marine West Coast climates.

Humid subtropical

This climate type is characterized by year-round precipitation, warm summers, and cool winters. Much of the Southeastern United States is in this category, which helps explain why balmier parts of that region are favored nowadays by so many people as a desirable place to live and retire. Other parts of the world that experience this climate include Southern China and the sub-Himalayan lands, Southeastern South America, and parts of eastern Australia.

The short winter of this climate-type makes for a long *growing season*, the average number of days between the last frost of spring and the first frost of fall. As a result, agriculture tends to be an important activity. In the lower latitudes, where freezing temperatures are rare, citrus and other frost-sensitive tree crops may be plentiful. But most anything can grow here to good effect. That includes a majority of Earth's rice crop (mainly in Asia), arguably the world's most important staple food.

Mediterranean

The principal characteristic of this climate is its dry summer. Though found on all continents, Mediterranean climate is most associated with — you guessed it — the land around the Mediterranean Sea. The natural vegetation consists of grasses and scrubs of various sorts, which may become a tinderbox during the dry season. Much of California (save the mountains, deserts, and northern coast) experiences this climate, which explains why wildfires are a common dry-season hazard in that state. Parts of Chile, South Africa, and Australia fit the Mediterranean climate profile, too.

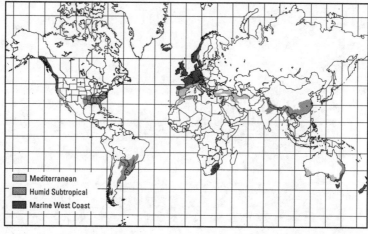

FIGURE 10-8: The geography of humid mesothermal climates.

(© John Wiley & Sons Inc.)

Various fruits prized for their sweetness and/or juices (grapes in particular) dominate agricultural land use. Lack of rain during summer, when the fruits ripen, deprives plants of moisture that would be taken up by root systems and dilute the natural juices. This is a very big deal for the wine industry, for which valuable vintages have everything to do with lack of rain while grapes mature (see Figure 10-9 while thinking of the wine!). As to why this kind of climate happens in the first place, predominant wind systems (associated with shifting atmospheric pressure belts) blowing from the land to the sea during summer is the culprit.

Marine west coast

This climate is characterized by mild to cool summers and cool winters. Its name pretty much tells you where you'll find it — on the west coast of continents in the middle to high-middle latitudes. For virtually every occurrence, the principal determining factor is a relatively warm ocean surface current immediately offshore. Coastal mountain ranges prevent this climate from spreading inland and occupying large areas in North America, South America, and Australia. Conversely, lack of Atlantic coastal mountains in much of Europe allows the marine atmosphere to uninterruptedly waft eastward and characterize most of that region.

The natural vegetation of this climate is mixed coniferous (needle-leaf) and deciduous forests, the two types respectively dominating in the higher and lower latitudes. Over the centuries humans have deforested much of that portion of Europe where this climate is found and converted the land to agricultural use — the warm temperatures attendant to the warm currents serving as a boon for agriculture. The same is generally true of the marine west coast areas of Australia, New Zealand, and South Africa.

FIGURE 10-9: Chile produces a variety of (mostly) red and white wines in its Valle Central near Santiago.

(© Jerry T. Mitchell)

In North America, relatively little conversion of forests to farms occurred in areas of marine west coast climate because of the prevalence of mountainous terrain. As a result, the natural vegetation serves as the preferred crop — which is to say that forestry is a major endeavor. And what forests they are! The relatively high temperatures and abundant precipitation wrought by the warm-water current offshore created lush and majestic stands, including the redwoods of Northern California. As demand for timber — much of it emanating from Asian markets — has grown, so has controversy between loggers and environmentalists regarding the extent and nature of cutting.

Cooling Off: Humid Microthermal Climates

REMEMBER

Humid *microthermal* (low temperature) climates are found in the high-middle latitudes. Sun angles are rather low in these areas. As a result, the average temperature of the warmest month only surpasses 50° F while the average temperature of the coldest month is 27° F or less. The two principal climates in this group are humid continental and subarctic (see Figure 10-10). Summertime differences distinguish the two climates. In *humid continental* regions at least four months of the year average above 50° F. In *subarctic* climatic regions, in contrast, fewer than four months average above 50° F. Forest is the dominant natural vegetation in both areas. Lack of land in the high latitudes explains the absence of these climates in the Southern Hemisphere.

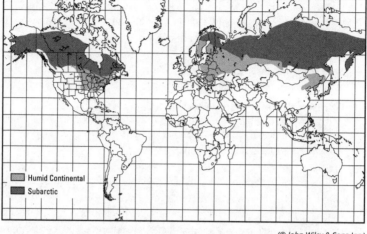

FIGURE 10-10: The geography of humid microthermal climates.

(© John Wiley & Sons Inc.)

Humid continental

This climate is found principally in Northeastern China, Eastern Europe, and, in North America, the Northeastern and Upper Midwestern parts of the United States and adjacent areas of Canada. Over much of these lands, the natural vegetation has given way to farmland. In the United States, dairy farming and the corn-soy complex (popularly called *the corn belt*) dominate the more humid east, while wheat and other hardy grains dominate the drier west. Much of the more northerly part of this realm is a bit too cool for agriculture, so forestry is intact. Coniferous softwoods, highly prized sources of pulp, dominate and support locally important logging economies.

Subarctic

This climate is generally found immediately north of the humid continental realm. Temperatures are too cold for too long for deciduous trees to thrive, and therefore coniferous forest (called *taiga*, a word of Russian origin) dominates the natural vegetation. Indeed, for the most part these forests are intact because the same chilly climes that discourage deciduous tree-growth also preclude agriculture. As a result, the broad belt of subarctic climate that extends all across the northerly portions of North America and Eurasia represents the largest expanse of forest on Earth. Generally, however, these forests are well removed from markets and mills and are therefore relatively untapped. On the whole, the subarctic realm is lightly populated. Mining is locally important and accounts for most towns' economies.

> ### VERTICAL ZONATION AND "HIGHLANDS CLIMATE"
>
> *Vertical zonation* refers to the changes in climatic conditions and their associated vegetation that are observed between the base of a high mountain and its summit. To take an example from East Africa, the base of Mount Kilimanjaro lies in tropical savanna climate. Its summit, however, which is 19,340 feet above sea level, is covered partly by snow and ice. For all intents and purposes, therefore, a hike from the base to the summit is tantamount to traveling from the tropics to the poles, experiencing enroute climate and vegetation change that would normally require a journey of several thousand miles. On some world climate maps, mountain ranges are shown as having *highlands climate*. This refers to the presence of the multiple climates of vertical zonation, instead of a singular climate type that is unique to mountains.
>
> It is common in Latin America to describe this vertical zonation as *altitudinal life zones*. *Tierra caliente*, the lowest level (sea level to 3000 feet), is hot and home to tropical crops and cities such as Colombia's Cartagena. *Tierra templada*, the "eternal spring", sees coffee and vegetable crops planted between 3,000 and 6,000 feet along with people in population clusters such as San Jose, Costa Rica. In *tierra fria* (6,000 to 12,000 feet), you will find potatoes among other crops; Mexico City and Bogota are found in this zone. *Tierra helada* (12,000 to 15,000 feet) is sparsely populated, primarily with people and grazing animals.

Dropping Below Freezing: Polar Climates

REMEMBER

Cold temperatures are the dominant characteristic of polar climates. The average temperature of the warmest month is less than 50° F, and most months typically average below freezing. The very small doses of solar energy that occur at these polar latitudes, despite the long daylight hours of summer, explain the frigidity. The resulting natural vegetation (if any) consists of short grasses, mosses, lichens, and an occasional stunted tree or shrub. The two climates that make up the geography of polar climates (see Figure 10-11) are tundra and ice cap.

Tundra

In areas with *tundra* climate, at least one month of the year averages above freezing (32° F), but not above 50° F. Like the humid microthermal climates, and for the same reason, tundra is almost exclusively a Northern Hemisphere phenomenon (small parts of Antarctica experience it). *Tundra* is a Russian word that refers to the vast, nearly treeless landscapes that are characteristic of this climatic region.

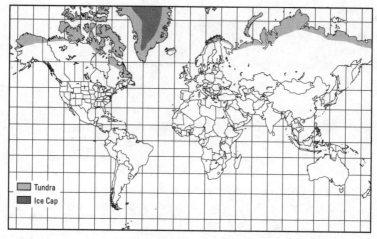

FIGURE 10-11: The geography of polar climates.

(© John Wiley & Sons Inc.)

Lack of forest is not a function of cold air temperatures *per se*, but rather the frozen soil that persists for nearly the entire year, and which prohibits tree roots from taking in sufficient nutrients (learn more about this frozen soil in the "Keeping permafrost frozen" sidebar). Thus, grasses that grow in abundance during the long daylight hours of the short and chilly summer dominate the natural landscape. This vegetation, in turn, attracts huge herds of caribou that annually migrate to the tundra to feast and fatten up for the long winter ahead. This relationship between plant and animal is the principal reason why, in Alaska at least, large portions of the tundra region have been designated as National Parks or National Wildlife Refuges.

Because the growing season is so short, agriculture is virtually unknown. Thus, the livelihood of the traditional societies who have long inhabited this realm — the Inuit and neighboring Native Americans in the Western Hemisphere, and the Lapps and neighboring peoples of the Russian Arctic, have relied on hunting or herding. In the last several of decades, however, the economic importance of the tundra has increased dramatically as a result of the discovery of significant quantities of petroleum and natural gas. Exploration continues in Alaska and the far northern reaches of Russia with American and Russian commercial and governmental interests hoping to lay political and thus economic claims to territory harboring these resources. Serious and sometimes acrimonious debate has resulted, pitting proponents of resource exploitation and pristine wilderness.

CHAPTER 10 **Connecting Climates and Vegetation**

KEEPING PERMAFROST FROZEN

A peculiar environmental phenomenon that has a powerful and direct bearing on debate between drilling for natural resources and providing for wildlife is *permafrost*, permanently frozen soil that underlies the tundra. If you drill for oil in the tundra and send it through a pipeline to wherever, then the oil needs to be warm. That is because crude oil, which is thick and viscous, flows very haltingly through cold pipe. And cold pipe is something that Alaska's climate virtually guarantees for most of the year. Fortunately, crude oil is hot as it comes out of the ground, and the warmth helps make it less viscous as it flows through the pipe. But the warm oil warms the pipe, which can melt the permafrost underneath, causing the pipe to sink into the ground and break, resulting in oil spills. For that reason, and at great expense, the Trans-Alaska Pipeline is elevated on stanchions for much of its route. This and other permafrost-related problems are at the heart of debate concerning possible future oil exploration and drilling in the tundra.

Ice cap

In areas of ice cap climate, every month averages below freezing. As the name suggests, its distribution generally coincides with the ice caps that overlie Greenland and Antarctica. At low temperature air can contain very little water. Also, due to the cold temperatures, relatively little evaporation takes place in the polar realm. On both accounts, therefore, the air has a rather low supply of vapor, which in turn depresses the possibility of precipitation. Thus, ice cap climate technically qualifies as desert because it receives less than 10 inches of precipitation per year. The astonishing and uniquely low temperatures, however, result in its being granted its separate climatic status.

As noted in Chapter 8, the ice caps in Greenland and Antarctica are a couple of miles thick, and all of it is the result of precipitation, which would seem to contradict what I just wrote. But all of that ice, however, is the product of thousands of years' worth of small annual accumulations of snow, which, due to the year-round cold temperatures, tend not to melt, but instead accumulate, compact, and add to the ice cap.

These facts mean that boring down into the ice cap is rather like going back in time. The deeper you bore, the older the ice — and thus the older the precipitation (snow) of which the ice is made. Moreover, because precipitation captures minute but measurable amounts of atmospheric gases, it constitutes a record of the nature of the atmosphere at the time that it fell. Thus, the ice caps are important sources of information about the atmosphere and climates long ago — like tree rings can be —, which figures prominently in contemporary inquiry concerning our changing climates worldwide.

3 Peopling the Planet

IN THIS PART . . .

Where have you gone, Adam and Eve? Life was so simple then. Two people. One culture. Nice garden. Plenty of room for growth. If somebody else had been around to write about population geography, I'm guessing three paragraphs would have sufficed. Not anymore.

Today nearly 8 billion people are divided into who knows how many thousands of cultures. And these folks just won't stay put. Ever since the original twosome got their eviction notice, people have been moving and migrating, rendering population geography into something akin to a restless tide. In the midst of change, however, discernible patterns (constants, if you will) emerge that concern the geographies of population, culture, migration, and control of the planet.

In this part, you'll find out about some of the key concepts and concerns of human geography. And yes, it takes more than three paragraphs. Indeed, it consumes four chapters that address the topics just mentioned. Even that doesn't complete the story, for we still have the matter of how people use and misuse the planet. Stay tuned.

> **IN THIS CHAPTER**
>
> » Living in crowded spaces, but not empty quarters
>
> » Studying a major league curve
>
> » Charting the stages of change
>
> » Grappling with the question of overpopulation

Chapter 11
Nobody Here But A Few Billion Friends

Not long from now, the global population — the number of people worldwide — will pass the 8 billion mark. That number has little meaning by itself. But if you consider that 200 years ago the global population was "only" 1 billion, then today's total gets your attention pretty quickly. A logical first reaction is that the birds and the bees have been working overtime. Indeed, those little critters have a certain way about them. But global population trends involve more than what happens in the privacy of a nest or hive.

The pages ahead focus on *population geography,* which analyzes the distribution of people and their characteristics over the face of Earth. Of necessity, this involves a smattering of *demography,* the science of vital statistics. "Vital" refers here to life, as when medical equipment is used to monitor a patient's "vital signs." Thus, demography involves birth rates, death rates, life expectancy, and other numerical indicators of the human condition.

For people who love to calculate statistics, demography is a dream come true. Chances are good, however, that you are not one of those people. So I forego the arithmetic and focus on generalizations and implications that result from it. Most of all, I focus on how humans and some of their vital attributes vary geographically. If you are wondering how we know how many people exist on Earth, take a look at the final chapter sidebar "Applied Geography: Census-taking from above."

Migration is an important factor in population change both internationally and within individual countries. Indeed, I am going to hold off on that subject for now and instead devote the entire next chapter to it because of its importance.

Going by the Numbers

REMEMBER

The world's 7.9 billion people are spread very unevenly across the planet's surface. Virtual empty quarters — large and totally uninhabited realms — correspond with the ice caps and tundra of Antarctica, Greenland, and the very high latitudes in general. Similarly, large desert areas often are low on people. Indeed, if you read the chapter on climate (Chapter 10), then it should come as no surprise that the Sahara, Gobi, Arabian, and other desert realms are fairly devoid of people. Also, most of the world's rainforest realms have low population densities, as the Brazilian interior and central Congo indicate.

But for every desolate area, you must consider the likes of Hong Kong, with some 17,000 people per square mile, or Singapore with its 20,000 people per square mile. Those are small dots on the world map that complement large areas of comparatively high density: the northeastern U.S. and adjoining areas of Canada; much of Western and Central Europe; the Nile valley; north central India; eastern China, and Japan and Java. Figure 11-1 gives you an idea of the world's population density unevenness from one country to another.

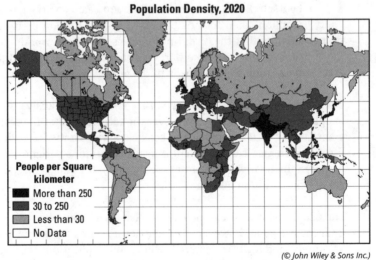

FIGURE 11-1: Global population density by country in 2020.

(© John Wiley & Sons Inc.)

So people largely avoid some places and are heavily clustered in others (see the two sidebars "Dispersion versus clustering" and "OK, everybody into Rhode Island!" for more on this point).

Table 11-1, showing the world's most populous countries, highlights the dominance of China and India, which respectively are home to 17 percent and 16 percent of all the people on this planet. Given those two population powerhouses, Asia (stretching all the way from Israel to Japan) contains some 60 percent of the world's population — the largest continental percentage by far. This would be even more if we counted Russia, but that country is typically lumped into the European total. The United States is now the third most populous country on Earth, but North America as a whole contains only about 6 percent of the human population. All told, the 10 most populous countries account for fully 56 percent of humanity.

TABLE 11-1 The 10 Most Populous Countries in 2021

Rank	Country	Population (in millions)
1	China	1,398
2	India	1,339
3	United States	332
4	Indonesia	275
5	Pakistan	238
6	Nigeria	219
7	Brazil	213
8	Bangladesh	164
9	Russia	142
10	Mexico	130

Source: U.S. Census Bureau

Opportunity for livelihood

Trying to fully explain global population geography in a one-liner is impossible, but perhaps "opportunity for livelihood" is a good start. Human population densities tend to be high where opportunity for livelihood is favorable and low where the opposite is true. Opportunity for livelihood takes different forms and therefore, so does the characteristics of regions that support high densities.

Agricultural land in the Nile, Ganges, and Indus river valleys, plus the valleys and coastal plains of eastern China support large populations due to their rich alluvial soils. How rich? Well, rich enough that since the beginning of recorded time, people who possess even the most modest agricultural technology have been able to realize sizeable harvests on relatively small acreage. In complete contrast, high densities are also found in countries where industrial and post-industrial economies dominate. Examples include the Northeastern U.S., Western Europe, and Japan.

Populations cluster near water — like the rivers noted earlier — for other reasons, too. Navigation and trade, drinking water, and the potential for hydropower have all been draws. The vertical dimension is also important. Higher elevations are generally avoided as being too cool and deficient in something we humans like to have: oxygen.

DISPERSION VERSUS CLUSTERING

Two areas can contain the same number of people, yet have a totally different look and feel because of the ways their populations are distributed. For example, the United States is among the minority of countries in which farmers typically live on their farms. That statement may cause you to ask, "Where else would a farmer live?" The answer is, in a village or town, and therein lies a significant difference in the way people are distributed. Rural population geography in the United States generally exhibits *dispersion,* which entails considerable open space between individual farmsteads. In contrast, the pattern in much of the rest of the world exhibits clustering. That is, farming families tend to live in a compact village, from which they walk or otherwise "commute" to the land that they tend. The two patterns are depicted graphically in the following figures, both of which contain 21 dots that represent homesteads.

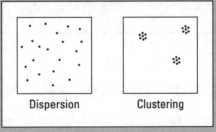

(© John Wiley & Sons Inc.)

Urban growth

Because people round the world view cities as centers of opportunity for livelihood, one of the most significant population trends today is urban growth. I talk about that more fully in Chapters 12 and 17. But for now, you can clearly see its effects on the world population map in the likes of Greater Mexico City, the Sao Paulo–Rio de Janeiro complex in Brazil, and the major cities on the East and West Coasts of the United States.

Globally, about 56 percent of the human race is categorized as urban, but the figures vary sharply from one region to the next. The populations of North America, South America, and Europe are each at least 75 percent urbanized. In contrast, the urban population percentages of South Asia and Sub-Saharan Africa are 35 percent and 40 percent respectively, which largely reflect continued heavy reliance on manual labor in the agricultural sector of countries' economies, plus relative lack of employment opportunities in the service and manufacturing sectors, which globally tend to be more urban based.

Going Ballistic: Population Growth

REMEMBER

For the vast majority of human history, total population was much, much lower than it is today. The United Nations estimates that it was only about the year 1650 that, for the first time, as many as 500 million people were alive at any one time (as illustrated by Figure 11-2). Until then, population growth could be fairly characterized as a gently rising straight line. But around 1650 something began to happen. The line started to curve upward — gently at first, but then ballistically.

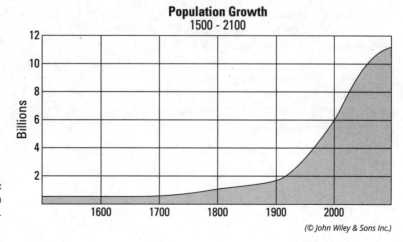

FIGURE 11-2: Global population growth.

(© John Wiley & Sons Inc.)

Total population passed 1 billion around 1800. Thus, while it took untold eons for human numbers to reach 500 million, a mere 150 years were required to double that number. By about 1925, the figure had doubled again to about 2 billion. In the next 50 years, it doubled again, reaching 4 billion sometime around 1975. And here we are today at slightly less than 8 billion.

Global population growth has not stopped. Instead, it will continue to rise in the present century before leveling off near 11 billion in the year 2100. Naturally, that may cause you to ask, "How can demographers be so certain about the future course of the world's population?" In fact, the experts are not certain. Instead, the future projections are (highly) educated guesses based upon reasonable assumptions concerning the global courses of birth and death.

OK, EVERYBODY INTO RHODE ISLAND!

Rhode Island is the smallest state by area in the United States. Imagine if every person on Earth went there to participate in a meeting. Would they all fit? And what would be the population density?

Rhode Island contains 1,212 square miles, and about 8 billion people currently live on Earth. If all of them went to Rhode Island and stood evenly spaced apart, that would work out to 6,600,660 people per square mile. One square mile (5,280 feet x 5,280 feet) equals 27,878,400 square feet. Dividing that many square feet by 6,600,660 people gives each person 4.22 square feet in which to stand. That works out to a square that is just a tiny bit longer than 2 feet on each side.

Get out a foot ruler, measure an area that size on the floor, and stand in it. Now imagine being surrounded for miles and miles by people who are allotted an equal amount of space. Would you feel cramped? Probably so! But with 4.22 square feet per person, 8 billion people could stand (wedged?) in Rhode Island and all of the rest of the world would be completely empty of humans.

This mathematical exercise is just that. It's not meant to play down the global impact of 8 billion people who need to be housed, fed, and otherwise sustained — all of which requires considerably more space than exists in Rhode Island.

Checking Behind the Curve: Population Change

Population change is a matter of birth, death, and/or migration. That is, in a given year in a given country, some people are born, some people die, some people move in, and some people leave. Demographers have developed a statistically based vocabulary that addresses these issues. Three terms in particular are worth passing along to you at the present time because they appear frequently in the following pages:

> » **Birth rate:** The annual number of births per 1,000 population.
>
> » **Death rate:** The annual number of deaths per 1,000 population.
>
> » **Natural increase:** The annual rate of population change as calculated by subtracting the death rate from the birth rate. (Typically, the birth rate exceeds the death rate, so population rises. But occasional short-term calamity such as a plague, war, or economic turmoil may produce the opposite effect.)

Dealing with births and deaths: Natural increase

Just as humans are unevenly spread across Earth's surface, so, too, is population growth. Indeed, perhaps the single most important demographic reality of our times is that the rate of population increase differs dramatically in different countries and regions of the world. The highest rates tend to be found in Africa and Southwest Asia. The lowest rates occur in North America, Europe, and northern Asia.

A "high" rate of increase is considered to be in excess of 2 percent. That may not seem like a lot, but a 2 percent rate of natural increase will double a country's population in 35 years. Therefore, every country in the "high" category could double its citizens in some time less than that. Indeed, approximately half of the "high" countries have rates of natural increase in excess of 2.5 percent (a doubling time of 28 years) and about half of those have rates of 3.0 percent (23 years) or greater.

This assumes of course that the rate of increase remains constant. Be careful about another issue: countries with large population numbers can have a lower rate of increase but still add a whole lot of people!

ON THE DOUBLE!

How long does it take a country to double its population given a particular rate of natural increase? The following table provides some answers. To take two examples, if a country has a 1.5 percent rate of natural increase, then it will take 48 years to double its population, assuming the rate of natural increase does not change in the interim. In contrast, a 3.0 percent rate of natural increase is sufficient to double the population in only 23 years. Remember that natural increase equals birth rate minus death rate. It does not consider either in-migration or out-migration, both of which could be key factors in a country's population change.

Rate of Natural Increase (percent)	Doubling Time (years)
0.5	141
1.0	70
1.5	48
2.0	35
2.5	28
3.0	23
3.5	20

Rapid growth, poor country

REMEMBER

Rates of natural increase and their associated "doubling times" (see the "On the double!" sidebar) have incredible implications for economic development and social well-being. A 2.5 percent rate of natural increase means there will be about twice as many mouths to feed in 28 years; twice as many people in the schools; twice as many people seeking health care, energy, employment, and transportation. Now, it would be one thing if this pertained to affluent countries that could adequately meet the needs of their citizens. But what about a poor country?

And indeed, that's the rub. Generally, countries that exhibit the highest rates of natural increase are relatively poor. Stated differently, the highest rates of population growth are generally taking place in countries with the least amount of financial resources to address the needs of their growing population.

Slow growth, affluent country (with some notable exceptions)

The "low" category for the rate of natural increase is under 1.0 percent. That corresponds to a population "doubling time" of 70 years, so every country in this category takes at least that long to double its present citizenry (migration not considered). Often, low rates of natural increase coincide with relatively affluent countries. A principal reason is that the economies and livelihoods of well-to-do countries do not require large amounts of manual labor, so families have comparatively few children, and that depresses the rate of natural increase. (Additional reasons are discussed in the next section.)

However, many countries have doubling times far longer than 70 years, and some even have negative rates of natural increase. Germany, for example, has a natural increase of -.2 percent. Russia's figure is the same. In fact, Eastern Europe as a whole is at −0.3 percent. That means the death rate is actually exceeding the birth rate in those countries, whose populations may begin to decline should these conditions hold for the foreseeable future. Part of the explanation in this case is that many former Communist countries experienced economic difficulties as they transitioned to market economies in the 1990s. One way to economize in tough times is not to have children, which is why those countries' birth rates dropped below their death rates — negative growth.

Increasing for a reason: The demographic transition model

The geography of natural increase doesn't "just happen." Instead, particular rates of natural increase are occurring in particular countries for particular reasons. To help explain these circumstances, demographers have developed a widely applicable set of generalizations (based on the experiences of many countries) called the *demographic transition model,* which is shown in Figure 11-3. Because the topic is people, *demographic* makes complete sense. Also, the model begins and ends with nominal population change. But in between a period of *transition,* characterized by substantial growth, occurs.

The demographic transition model considers the relationship between birth rates and death rates over time, and consists of four stages:

>> **Stage 1:** High stationary

>> **Stage 2:** Early expanding

>> **Stage 3:** Late expanding

>> **Stage 4:** Low stationary

FIGURE 11-3: The demographic transition model shows the relationship between birth rates and death rates over time.

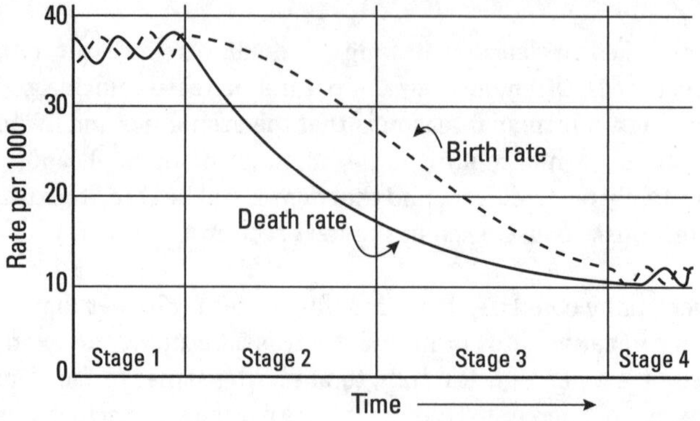

(© John Wiley & Sons Inc.)

I am going to make a big deal of them for two reasons. First, if you understand the model, then the factors that gave rise to the historical population growth curve illustrated in Figure 11-2 become crystal clear. Second, different countries of the world are in different stages of the demographic transition. Therefore, if you understand the model, then you can understand why a particular country is experiencing its particular rate of natural increase and its attendant social characteristics. Parenthetically, several developed and developing countries have already completed the transition. Demographers have used the experiences of those countries to predict the course of global population growth into the next century.

Stage 1: High stationary

REMEMBER

In this stage, birth rates and death rates are *high* and about equal (see Figure 11-3). Thus, population growth is *stationary* (exhibiting little or no change) because the numbers of births and deaths are "canceling out" each other. These conditions were characteristic of global population prior to 1650. The high death rates of those times were products of poor sanitation, tainted water supply systems, faulty food storage, lack of education, and absence of medicines and vaccines. The results included the following:

» *Infant mortality* (the incidence of death before a child's first birthday) was astonishingly high, meaning that it claimed about 25 percent or more of newborns even in some "advanced" societies.

» *Average life expectancy* (the number of years a newborn could expect to live) was low. How low? Well, today the average citizen of France can expect to live about 80 years. But church and cemetery records suggest that in the 1600s, French life expectancy was about 35 years.

On average, therefore, people died young. Many never reached reproductive age, and those who did tended not to live that many years during their fertile time of life.

Human societies have typically responded to high death rates by having high birth rates, and the time prior to 1650 was no exception. Factors that contributed to high birth rates included the following:

» Most families farmed for a living, so more children meant more hands to do the manual labor that was necessary in those days before widespread use of machinery.

» Retirement plans and 401(k)s, life insurance, and social security checks were unknown. Having children (and the more, the better) guaranteed there would be somebody to look after you if you were fortunate enough to reach old age.

» Given short life expectancy and need for children, people — especially females — married young. Most women were wed by their mid-to-late teens and, not withstanding the often-fatal rigors of childbirth, had been through a couple of pregnancies by age 20.

Virtually no country currently experiences this range of conditions in its entirety. Nevertheless, an understanding of these circumstances is very important because they set the stage for other phases that are very real in the present age.

Stage 2: Early expanding

REMEMBER

Birth rates exceed death rates by a widening margin in this stage (see Figure 11-3). When that happens, population does more than simply grow: It increases dramatically. Look again at Figure 11-2 and note the S-shaped curve of population growth. The conditions just described correspond to the bottom — that is, *early* — half of the curve, when population was *expanding* at faster and faster rates after years of being stationary.

Hence the name of this stage is *early expanding*, which nevertheless perplexes many people because, as you can see on Figure 11-3, birth rates and death rates are declining throughout it. The key thing to focus on in that diagram is the widening gap between birth rates and death rates that is characteristic of this stage. Even though both rates are dropping, the gap between them is widening, birth rates being the higher of the two. Thus, population grows (expands).

But why are the rates dropping and the gap widening? Basically, birth rates drop because of a tempering of the last bulleted items of Stage 1, but death rates are declining much faster because of the following:

» Basic and widespread improvement in water supply and sanitation are having a very positive effect on public health.

» Medicines and vaccines are becoming accessible to more and more people.

>> Infant mortality is dropping and life expectancy is rising. More people are reaching reproductive age and are reproducing. People are living longer in their reproductive years and are reproducing more.

>> Death rates are dropping faster than birth rates, so population grows — slowly at first, and more dramatically more recently.

The widening gap between birth rates and death rates results in growing rates of natural increase.

Stage 3: Late expanding

REMEMBER

Birth rates exceed death rates by a narrowing margin in this stage (see Figure 11-3). When that happens, population grows but at rates that are progressively slowing. Look once at Figure 11-2 and its S-shaped curve of population growth. The conditions just described correspond to the top — that is, *late* — half of the curve, when population was *expanding* at slower and slower rates after years of skyrocketing. Hence the name of this stage is *late expanding*. Because birth rates exceed death rates by a *decreasing* margin, the result is a lowering of the rate of natural increase. Countries experiencing this stage tend to have the following conditions:

>> Improvements continue to be made in public health, resulting in lower infant mortality and longer life spans. Thus, the death rate continues to decline.

>> As the economy develops, machines perform increasing amounts of work that used to be done manually. Thus, the incentive to produce children strictly for their labor potential drops.

>> More people gain work in jobs that provide retirement options. This lessens another historic incentive to produce children.

>> Increasing numbers of women encounter career and educational opportunities that have the effect of delaying marriage and childbearing. Contraceptive availability is also more widespread.

>> Increasingly, couples consider the costs of raising and educating children and opt to limit the size of their families.

As a result of the last four factors, the birth rate begins to decline — slowly at first, but then more rapidly as the modern economy encompasses more and more families. As the gap between death rate and the birth rate diminishes, the rate of population increase begins to slow, and the curve exhibits signs of leveling off.

Stage 4: Low stationary

REMEMBER

In this final stage birth rates and death rates are *low* and about equal (see Figure 11-3). Thus, population growth is *stationary* (exhibiting little or no change) because the numbers of births and deaths are "canceling out" each other. (Global population as a whole, as per Figure 11-2, will probably not experience this stage until early in the next century.) The characteristics of the low stationary stage are as follows:

» Continued improvement and increased availability of health care results in continued lowering of the death rate.

» The economy is overwhelmingly industrial or post-industrial, resulting in diminished need for manual labor except in those remaining occupations that require a high degree of craftsmanship.

» Institutional retirement systems and benefits are widespread, nullifying the need to have children for the sake of social security.

» More women defer marriage and motherhood (or opt out entirely) as educational and career opportunities become more widely available and socially acceptable. The effects lower the birth rate.

» Average family size continues to decrease as more families factor in the costs of child raising and educating children.

Making connections

In each stage of the demographic transition model (see the previous section), natural increase is closely related to other demographic variables, each of which can be mapped and analyzed, and thus reveal a broader appreciation of the geography of the human condition. The following sections offer maps and brief discussions of three variables that illustrate the possibilities.

Wealth (Gross National Income [GNI] per capita)

A map of global wealth reveals that the world's most affluent countries are found in North America, Western Europe, and selected "outlying" places such as Japan, Australia, and New Zealand (illustrated in Figure 11-4). At the other extreme are numerous countries in Africa and Asia. If the overall pattern of rich countries and poor countries on this map seems familiar in your head, it ought to. As suggested earlier and confirmed here, generally we witness an inverse relationship between higher population increase and wealth. That is, countries that have a high rate of natural increase generally have low average wealth, and vice versa. And again, the most significant implication is this: The highest rates of population increase are occurring in countries that have the least financial means to see to the needs of their rapidly expanding populations.

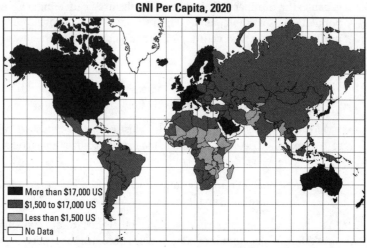

FIGURE 11-4: Wealth (Gross National Income per capita).

(© John Wiley & Sons Inc.)

Percent of population ages 0–14

Figure 11-5 — a world map showing percent of population under 15 years of age (by country) — reveals a familiar pattern. The highest category on the map pertains to countries in which more than 40 percent of the population is in that age category. And basically, those countries are found in exclusively in Africa, except for Afghanistan. The lowest rates, in contrast, tend to be found in North America, Europe, and East Asia, plus Australia and New Zealand (Singapore and Japan at 12 percent are the lowest, by the way). Thus, the highest percentages of young people tend to be found in countries that have rapidly expanding populations and the least financial wherewithal that can be brought to bear on the education, health, and nourishment of the next generation. You can understand these age distributions better after reading the sidebar "Population pyramids."

Infant mortality

Infant mortality is a sensitive indicator of public health and education. Figure 11-6 shows the highest rates of infant mortality are occurring in countries that have the highest rates of natural increase, which also tend to be poor. That lack of wealth is largely responsible for the poor states of health care and sanitation that produce high rates of infant death. Co-occurrence of high rates of infant mortality with high rates of people under 15 years of age suggests a societal preference for high birth rates to offset high death rates. It likewise suggests existence of characteristics described in the high stationary and early expanding stages of the demographic transition model.

POPULATION PYRAMIDS

Population pyramids provide a graphic means of depicting and comparing the populations of different countries. In the diagrams below, the vertical axis shows age groups, while the horizontal axis indicates the percent of a country's population that is in each of those groups. Most countries have more young people than old people, so the graph typically has a wide base that tapers upward — rather like the shape of a pyramid. But the width of the base may vary substantially. Pyramids of developing countries, which typically have high rates of natural increase and therefore large percentages of their populations in the younger age groups, tend to have wide bases. In contrast, the pyramids of developed countries, with their low rates of natural increase, tend to have narrower bases.

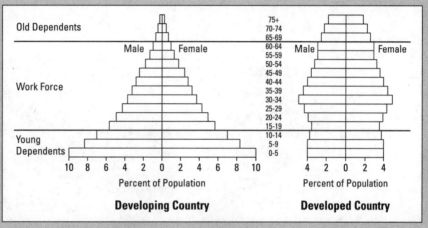

(© John Wiley & Sons Inc.)

Population pyramids are particularly useful for contrasting *dependents* and *non-dependents*. These are terms demographers use to characterize people in the under-15 and 65-and-over age groups (*dependents*) and the middle-aged people *(non-dependents)* on whom they must generally rely for their sustenance. This is sometimes also referred to as the *dependency ratio*. Typically, a low percentage of dependents is desirable because it means a high percentage of non-dependents is available to see to the needs of the young and the old. And in fact, that tends to be the case in well-off countries. In developing countries, in contrast, a high percentage of dependents relative to non-dependents is the norm. The population pyramids shown here make these differences very apparent.

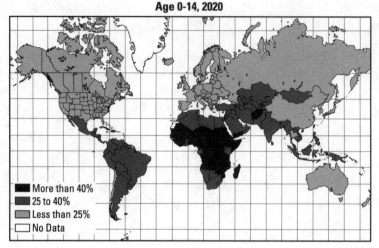

FIGURE 11-5: Percent of population ages 0–14.

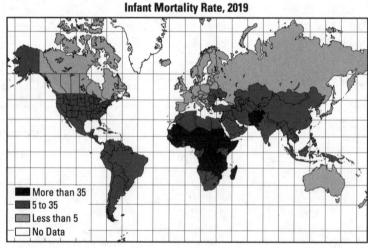

FIGURE 11-6: Infant mortality rate.

Considering "Overpopulation"

REMEMBER

Given the billions who live in poverty, poor health, and crowded conditions, people sometimes suggest that Earth is overpopulated. This controversial subject is made all the more problematical because it has proven difficult, if not impossible, to define. The clear implication of "overpopulation" is "too many people." But how many is too many, and is that number the same everywhere, or is it dependent on local conditions?

TIP

Reputable demographers agree there is no "magic number" of people or of people per square mile beyond which a country or region is overpopulated. Being statistically inclined, however, they do look to numerical data and analyses to gain perspective. Perhaps their most intriguing concept is *carrying capacity* — the number of people that a country or region can sustain at an acceptable level of well-being given its prevailing technology.

As you will see in the "Regarding overpopulation and carrying capacity" sidebar, one can argue that technologically advanced societies have higher carrying capacities than developing nations that lack similar expertise. But differences in culture, life experience, and personal preference have generally rendered inconclusive mathematical attempts to precisely determine carrying capacities. At best, and in response to the questions above, demographers' numerical exercises suggest it is impossible to determine how many is too many, and that thresholds — if and as near as they can be determined — do vary from place to place depending on local conditions.

Lack of conclusive definition and indicators of overpopulation — statistical and otherwise — has not prevented people from taking sides on this issue. Some argue passionately that Earth or parts thereof are overpopulated and espouse policies to rectify the perceived problem. Others argue just as passionately that overpopulation does not exist and espouse policies aimed at relieving malnourishment, poor health, and its other would-be symptoms. Proponents of these viewpoints occasionally and respectively are referred to as *neo-Malthusians* and *cornucopians*.

Neo-Malthusians

If you hail from the camp of neo-Malthusians, your camp leader is Thomas R. Malthus (1766–1834), an English political economist and theologian, who believed population increase is a prelude to disaster. In his famous "Essay on the Principle of Population" (1798) he stated that "population increases in a geometric ratio, while the means of subsistence increases in an arithmetic ratio." In other words, human population is increasing and at a rate much faster than food supply.

During Malthus's time, agricultural technology offered limited prospects for improved harvests. Thus, the only real option, as Malthus saw it, was for humans to have fewer babies. Artificial means of birth control were available, but Malthus opposed them on theological grounds. He preached restraint, but conceded that human passions were not likely to be held in check by "Just say no." Thus, he saw no solution save the grim reaper. Population would continue to increase faster than food supply until, ultimately, large-scale famine and starvation rectified the imbalance.

REGARDING OVERPOPULATION AND CARRYING CAPACITY

The following are selected demographic data for two honest-to-goodness countries. If I showed only the first two rows of data and then asked which country could be considered overpopulated, I suspect most people would say Country Y. Although it has the same number of people as Country X, it has a population density (people per square mile) that is almost 40-times greater than its counterpart's. That might suggest a lot of stress on a small amount of land or the resources available there. But when you throw in the last three rows of data, a different picture emerges.

	Country X	Country Y
Population (millions)	17	17
People per Square Mile	34	1,316
Female Life Expectancy	59	83
Infant Mortality (per 1,000)	69	4
Per Capita GNI ($US)	1,580	59,700

Women live considerably longer in Country Y, and infant mortality is overwhelmingly *less* prevalent. Also, per capita GNI data suggest personal income is astronomically higher in Country Y. The implication, therefore, is that citizens of Country Y are generally well off while those of Country X are not. This leads to consideration of carrying capacity — the number of people a country can sustain (carry) at an acceptable level of well-being. Review of the data suggest Country Y has a high carrying capacity. Though densely populated, it clearly has the capacity to carry its relatively large population at a high level of well-being. The opposite might be said of Country X. Even though the country is lightly populated, it appears not to have sufficient resources to sustain its population at a high level of well-being. Therefore, one might argue that Country X has exceeded its carrying capacity. What is certain, however, is that overpopulation cannot be determined simply on the basis of a country's population or its density.

By the way, Country X is Chad, and Country Y is The Netherlands.

The world has changed a lot since Malthus. Human numbers have exceeded his wildest dreams, but so has agricultural productivity. Also, in Malthus's time, transportation technology was such that food had to be produced fairly close to consumers. Nowadays, however, food travels hundreds — even thousands — of miles to get to your supermarket. Thus, a local crop failure or poor harvest need not have the devastating effects of yesteryear because food can be brought in from somewhere else.

Today few reputable scholars espouse the literal word of Malthus. But lots of neo-Malthusians believe the old bloke basically got it right. Namely, they say that many countries suffer from too many people (see Figure 11-7). And while technology may someday improve the average welfare, population reduction is the most effective and reliable way to achieve a better balance between people and resources.

FIGURE 11-7: A photo of a slum in a developing country. The low quality of life in places such as this is central to neo-Malthusian arguments to limit population.

(© Dmitry Rukhlenko / Adobe Stock)

Cornucopians

This term *cornucopia* recalls "the horn of plenty," that curly-cued overflowing basket of food one tends to associate with Thanksgiving decorations. If you camp with *cornucopians,* the viewpoint is that the world has a food supply problem, not a people supply problem.

And in their view, the solutions are not futuristic. Rather, means are available now to greatly increase global food supply and therefore improve carrying capacity throughout much of the world. They include:

> » **Greater use of green revolution know-how.** *Green revolution* refers to a number of agricultural innovations designed to increase food production in developing countries. Chief among them are varieties of rice (Figure 11-8) and wheat that have been genetically engineered to increase their yields as well

as their resistance to crop diseases. Increasing access to these relatively inexpensive strains could greatly help developing countries to increase their carrying capacities.

» **Improved grain storage.** In several countries a significant portion of grain harvests are lost to vermin due to poor storage. Modest expenditures on secure storage could substantially increase available food.

» **Improved transportation.** Modest investment in the most basic forms of infrastructure could greatly enhance food supply and well-being. Roads in parts of many developing countries are little more than dirt tracks that may be nearly impassable during a rainy season or other time of year. This diminishes access to markets and to goods that might improve agriculture. Modest investment in road-building could lead to a much improved picture.

» **Greater distribution of food surpluses.** Some developed countries have massive food surpluses that are merely stored, as well as policies that pay farmers not to grow crops (in order to maintain decent price levels). Cornucopians regard these issues as ethically unconscionable and believe that the surpluses should be readily available sources of food.

Cornucopians, in short, believe that a number of means are available that can significantly improve global carrying capacity. What is lacking, in their view, is the will to do the right thing.

FIGURE 11-8: A photo of a high-yield rice paddy. High-yield rice is one reason Cornucopians are optimistic about sustaining the world's people at an acceptable level of well-being.

(© Thirawatana / Adobe Stock)

APPLIED GEOGRAPHY: CENSUS-TAKING FROM ABOVE

In some places census-taking is sometimes inhibited by the remoteness of towns and the reluctance of their inhabitants to be enumerated. Intent on conducting the best possible head count, some nations — including the United States — have successfully overcome these impediments through careful use of two geographic techniques: spatial sampling and aerial photography. Specifically, inhabitants in a number of accessible and representative towns throughout the country are surveyed (spatial sampling) with special emphasis on determining the average number of people per house. Afterwards, aircraft fly over remote areas and photograph them. Photo interpreters then examine the pictures, taking special care to identify and count the houses. That number is then multiplied by the average number of people per household, the result being the estimated population of the photographed areas. Geographic variation in house-types and social structure may complicate matters, but adequate sampling coupled with skillful photo interpretation may result in a reasonably accurate census.

- » Studying early human dispersal
- » Deciding where to go
- » Marking migration magnets
- » Marketing and mental mapping

Chapter 12
Shift Happens: Migration

New Britain and New Holland are towns in Pennsylvania. New Prague and New Ulm are towns in Minnesota. And who could forget New Lisbon, Wisconsin; New Leipzig, North Dakota; and New Hamburg, New York? One can't help but wonder what was going on in Old Prague, Old Lisbon, and Old Leipzig that caused people to up and embark on the journey of their lives. Goodness knows, shift happens. And thank goodness it does, because geography would be pretty dull without it.

REMEMBER

At issue in this chapter is migration, which is key to understanding

- » The distribution of people at the global, regional, and local scales
- » Differences in population growth
- » Patterns of ethnicity and culture
- » Environmental issues related to population growth

Migration is travel that involves a change in residential location. Together with birth rates and death rates (described in Chapter 11) it's a central component of population geography. More specifically, migration is fundamental to understanding population shifts present and past, including ones that ran their courses long before the dawn of recorded history.

Populating the Planet

Perhaps the greatest migration story of them all involves the populating of the planet. Scientific evidence suggests humans originated in East Africa, originally occupying one very small part of planet Earth.

That didn't last. When Columbus and other Europeans reached the New World, they discovered that other humans had gotten there long before they did. Throughout the Age of Discovery, other explorers also found that, time and again, other humans had beaten them to their newly found lands. Throughout the Americas, the Pacific Islands, the far Northlands, Eurasia, Africa, Australia, and New Zealand, you name it. Humans were just about everywhere except Antarctica. How had they done it? That is, assuming humans originated in a single region (and no credible evidence has been found to the contrary), how had they managed by 1492 to assume a near-global distribution? The answer is land bridges and ocean voyages.

Bridging the oceans

Once upon a time it was possible for humans to walk between certain continents and other land bodies that are today separated by straits and shallow seas. The key word in that sentence is walk — and on dry land, too. That was made possible by something rather peculiar that happened during the last ice age.

Earth has experienced several ice ages during the past million years. During those times climates then were definitely cooler on average than they are today. The most recent ice age began about 120,000 years ago and ended about 10,000 years ago. During that period, humongous amounts of seawater evaporated, condensed in the atmosphere, fell to Earth as snow, and compacted to form glaciers instead of returning to the sea as runoff. Thus, as the glaciers grew, sea level dropped. The exact extent of the decline is unknown, but for several thousands of years sea level was as much as 475 to 500 feet lower than today.

Thus, a world map at the height of the ice age would have looked a lot different from today's (see Figure 12-1). Substantial areas that had been ocean bottom became dry land, so the continents and other land bodies grew while the oceans shrank. Most importantly, several land bodies that had been separated by water became connected by *land bridges* — dry land in places that had been straits or shallow seas.

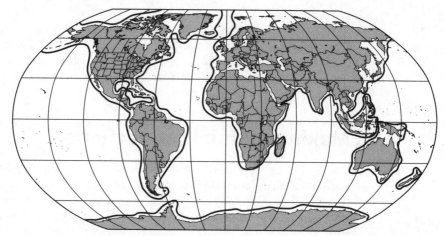

FIGURE 12-1: The world at the height of the last ice age. The space between the bolded lines depicts what used to be solid land.

(© John Wiley & Sons Inc.)

The land bridges lasted for thousands of years. Formerly ocean bottom, they became grasslands and woodlands that provided habitat for animals. Most importantly, of course, the land bridges provided firmament that allowed humans (over many generations) to migrate and occupy lands that had been unknown or out of reach. For example, people could have walked from present-day France to Ireland (see Figure 12-1). Similarly, of course, the ancestors of Native Americans migrated over dry land from Siberia to Alaska and a land bridge at Gibraltar allowed movement between Africa and Europe. Other connections are also evident on Figure 12-1, though some are controversial. We're not certain, for example, as to the location of ice age coastlines in the Indonesian and Philippine archipelagos.

Eventually, the last ice age came to an end. Climates warmed, and as glacial ice receded their melt waters flowed to the oceans, which rose and inundated the land bridges. Thus, continents and land bodies became disconnected — just as they had been before the ice age started. But something significant had happened between the appearance and disappearance of the land bridges. Humans had migrated across them, giving rise to native populations that, many generations later, would greet European and other explorers.

Voyaging afar

Sometimes people were separated from their nearest neighbors by thousands of miles of ocean far too deep (14,000 feet or so) for any land bridge to explain their arrival. Therefore, ancient voyages of substantial scale must have occurred. Research suggests the requisite navigational skills were based on knowledge and application of star positions, bird behavior, cloud types, ocean swells, and wave refraction. For example, when Captain James Cook reached the Hawaiian Islands in 1778, he found natives of Polynesian ancestry. Thousands of miles of ocean separated these people from any neighbors, proving that voyages had occurred. Other islands were found to have similar populations, and not just in the Pacific. Thus, the Merina people of Madagascar, off the southeast African coast, speak a language with Austronesian roots.

Making colonial connections

The Age of Discovery foreshadowed an era of expansionism and colonial acquisition on the part of several European powers that would have major consequences for population geography. Specifically, exploration led to knowledge of and interest in distant lands that were perceived to have strategic or economic value — or both. These interests led to claims on territories that became colonies.

Establishment of colonies led to creation of *migration fields,* which are countries or regions that generate (such as the European powers) or receive (such as Australia, New Zealand, and parts of the Western Hemisphere) major migration flows. These migrations, in turn, led to the creation of regionally distinctive population characteristics that endure to this day. Thus, for example, what were once sparsely populated "native" lands in Australia, New Zealand, and parts of the Western Hemisphere now bear the unmistakable imprint of European settlement. In addition to free men and women, large numbers of indentured people — including non-Europeans — also relocated. Many Indians (by which I mean South Asians instead of Native Americans), for example, migrated within the British Empire and added significantly to population geography as far away as the Caribbean. The effects of all migration on native populations varied from elimination, to relegation onto reservations, and absorption or intermarriage.

Reciprocal flows of goods and people between European countries and their colonial possessions became commonplace. Low-cost raw materials were sent from colonies to colonizer, where they were made into higher-cost manufactured items and sent back to the colonies for sale. Thus, began business and trade relationships that persist to the present, though not necessarily in the same, one-sided manner.

Likewise, in matters of migration, disproportionately strong migration fields exist among the countries of former empires. Thus, a visit to virtually any large city in England reveals ethnic neighborhoods dominated by West Indians, Pakistanis, and other ex-colonials. Similarly, a review of immigration in Canada reveals a migration field in which the countries of the former British Empire are disproportionately evident.

Forcing involuntary migration

Migration is not always a matter of personal choice. Sometimes it's involuntary — forced upon certain populations. Examples include the expulsion of Jews from parts of Europe, Native Americans from their homelands, Syrians escaping civil war, or Chinese towns flooded when constructing the Three Gorges Dam. But undoubtedly the most terrible case of them all, and the worst chapter of colonial history, is the Trans-Atlantic slave trade.

Beginning in the 1400s and continuing for nearly four centuries, millions of Africans were kidnapped and forcibly shipped off to the Caribbean islands, South America, and North America. They worked sugar plantations, silver mines, and all forms of agriculture. The impact is, of course, clearly seen today in the demographics of receiving areas.

Choosing to Migrate

In a majority of cases nowadays, people migrate because they choose to do so. The decision to migrate varies from simple to complex. Sometimes the key element is a *push factor*. This is a characteristic of a region that causes dissatisfaction among residents and encourages them to emigrate.

REMEMBER

The most common push factors are war, political unrest, famine, persecution, dislike of the physical environment, and economic hardship. One can certainly argue whether some of these push factors really involve choice. *Voluntary* migration when the choice to stay can mean death might not really be a choice. Push factors have been responsible for some of the greatest migrations in history. Emigration from Ireland due to the potato famine is a case in point. Another — and perhaps the greatest migration in recorded history — is the relocation of Muslims and Hindus within the Indian subcontinent after its partition into India and Pakistan.

In other instances, the decisive element in the decision to migrate is a *pull factor*. This is a characteristic of an area that exerts an attractive force that draws people from other regions. Being the opposite of push factors, pull factors include peace and harmony, lack of persecution, a pleasant environment, and the perception of economic opportunity. Many times, of course, both push factors and pull factors play a role in decision-making. The Great Migration of 6 million African Americans from the southern states of the U.S. to the northern states in the early and mid-20th century makes that case.

While push and pull factors encourage migration, potential *barriers to migration* have the opposite effect. These deterrents may include

- **Physical barriers** such as oceans, mountain, and deserts
- **Economic barriers** such as the costs of migration and of establishing a new home
- **Cultural barriers** that involve the sobering prospect of leaving a familiar religious, linguistic, and relational environment for an unfamiliar one
- **Political barriers** that may include policies of one's own country that discourage emigration, as well as those of potential receiving countries that discourage immigrants of one sort or another

Numerous countries offer numerous examples of these and other migration concepts. Possibly none, however, surpass the United States.

Coming to America

The United States is often and justly referred to as a nation of immigrants. Prior to 1875, anybody from any foreign land could legally and freely enter the country and become a resident. Thereafter, Congress began passing a series of laws (which exemplify political barriers) that restricted immigration on such criteria as morality, race (starting with the Chinese Exclusion Act of 1882), and national origin. Laws in recent years have been characterized by annual caps on the number of newcomers and a system of preferences for family members, skilled workers, and people from under-represented countries (see sidebar "The brain drain").

Review of sources of immigrants reveals a significant change in migration fields. Considering U.S. immigration history from 1820 (the beginning of record keeping) to 2009, Europe was the principal source area, accounting for seven of the top ten countries overall (see Table 12-1). In various ways these Europeans satisfied the cost of migration (an economic barrier) and made passage across the Atlantic Ocean (a physical barrier). Once in America, of course, they often found that their Old World culture traits were obstacles (cultural barriers) to integration into an existing social fabric. But migrate they did, ultimately because negative conditions in the Old World (push factors) and the attractions of the New World (pull factors) proved more powerful than barriers to migration.

TABLE 12-1

Top 10 Sources of U.S. Immigrants Historically and Recently

1820–2009: Country – Number	2010–2019: Country – Number
Mexico – 7,640,159	Mexico – 1,506,738
Germany – 7,282,458	China – 707,314
Italy – 5,463,516	India – 623,919
U.K. – 5,424,352	Philippines – 508,391
Ireland – 4,793,973	Dominican Republic – 503,978
Canada – 4,707,449	Cuba – 479,818
Russia – 3,962,201	Vietnam – 327,997
Philippines – 2,012,673	Haiti – 201,329
Austria – 1,866,331	Jamaica – 199,928
Hungary – 1,687,401	El Salvador – 198,974

Source: U.S. Department of Homeland Security, 2019 (Lawful Permanent Resident Status).

For a long time, Germany had the distinction of being the number one donor nation of all-time, but Mexico closed the gap at a torrid pace and surged into first place in the early 21st century. The fact that about 75 percent of all Mexicans who have ever immigrated to the U.S. have done so since 1980 is astonishing. While future Americans still come from Europe, many, many more now come from Latin America and Asia. Indeed, countries in the latter areas accounted for all of the top ten sources of immigrants between 2010 and 2019.

Just as source areas have changed, so, too, have the final destinations of immigrants. Cities have always attracted large numbers of them, either as points of arrival or potential employment. But in the first half of America's history, an abundance of rural land (due to displaced Native Americans) was also available to pioneer settlers.

In many instances, immigrant groups of like origin from different parts of Europe settled large contiguous tracts, basically transforming the frontier into an ethnic quilt. Often, of course, settlements thrived, populations grew, and towns arose, resulting in the New Pragues, New Lisbons, and New Leipzigs that dot America. In these rural (or formerly rural) areas, land tends to be owned rather than rented, and therefore it gets passed down over the generations. As a result, a persistent ethnic geography that dates from pioneer days remains over much of America.

By the second half of America's history, the frontier was largely gone and with it the opportunity for new immigrant groups to settle large rural tracts. Accordingly, immigration assumed an increasingly urban focus (where jobs were more plentiful, too), and so it remains. In cities, more people tend to rent their residences rather than own them. Thus, the propensity is for ethnic turnover to occur in certain neighborhoods with each new wave of immigrants. As a result, if you live in New Prague, New Lisbon, or New Leipzig, then the countries in the right-hand column on Table 12-1 may surprise you. On the other hand, if you live in New York, Chicago, or Los Angeles, then this information may be yesterday's news.

TECHNICAL STUFF

Channelized migration, which links geographically specific points of origin and destination, characterizes the growth of several urban immigrant enclaves. For example, a sizable Dominican neighborhood exists in Upper Manhattan in the area of Washington Heights. Detailed examination reveals that it is far too simplistic to describe the migration as people from the Dominican Republic moving to New York. Rather, people from particular towns or regions of the Dominican Republic are settling in particular parts of the ethnic enclave in Manhattan. Thus, specific channels are found within the overall flow of immigrants from the Dominican Republic to the United States. Similar examples appear in other parts of the United States regarding this and other immigrant groups. Similar examples have also been documented in other countries all over the world.

Neither channelized migration nor migration in general is strictly international. Many countries provide cases of internal (domestic) migration whose characteristics and impacts are no less significant. Again, the United States provides excellent examples.

Migrating at home

Domestic migration involves residential relocation within a given country. While some moves may involve crossing the street, others may be cross-country or inter-regional. The latter are of particular interest to geography for two reasons. First, and as we shall shortly see, they may be symptomatic of waxing and waning economies of different areas. Second, they may have significant political ramifications. In the United States, Canada, Britain, and other democracies, the number of legislative representatives allocated to a state, province, or region is based on the number of residents. In the United States, for example, the population of each state determines the number of U.S. Representatives that its citizens elect to Congress. As state populations rise and fall because of migration and other factors, the allocation of delegates to Congress changes, and therefore so does the geography of political clout.

Once again, it is useful to focus on a single country to examine the causes and consequences of domestic migration. And once again we will consider the United States.

Relocating within America

Americans are a people on the move. They commute, shop, take their kids here and there, and go away on vacation. But they also move in the sense of changing their residential locations. In fact, data suggest that about 10 to 20 percent of Americans move each year (that percent has been trending downward). About 14 percent changed their state of residence in 2019 alone. The latter figure may not seem like a lot. But given some 332 million Americans, that means about 46 million people were hiring movers or loading up all their belongings themselves for a new life across state lines.

REMEMBER

Principally because of this domestic (that is, interstate) migration and international additions, the populations of states and regions of the United States are growing at different rates (see Figure 12-2). Nevada, for example, grew by an astounding 288 percent from 1980 to 2020, while the population of West Virginia declined (a loss of 8 percent). Overall, we can pinpoint discernible regions of growth. The highest rates of increase form an arc sweeping south from Washington state towards Texas and then north again towards Virginia. The lowest rates of growth are occurring in the Northeast and Upper Midwest.

"THE BRAIN DRAIN"

Brain drain refers to the global tendency for highly educated and skilled citizens of developing countries to migrate to the more developed nations (*brain gain*). This population shift saps (or drains) the developing world of intellectual resources as it adds them to the already more well-to-do countries. People the world over seek education that prepares them for modern highly skilled, good-paying jobs. The economies of developing countries typically do not, however, generate large-scale employment of this type. In contrast and by their very nature, the economies of developed countries do. Indeed, in some developed nations the number of such jobs exceeds the local supply of qualified applicants. One result is immigration laws in developed countries that give preference to highly trained foreigners who are prepared to fill the vacancies. In the United States, for example, laws have reserved some slots annually for foreigners with valuable skills. The years since have witnessed pressure to increase that number, especially in tech-dominated industries. The effect of these and similar policies is to increase the brain drain. While this policy clearly benefits the receiving developed countries, it also exemplifies the saying that "the rich get richer, and the poor get poorer." Changing scales, a brain drain is also evident within countries. Young, college-educated migrants, for example, have gravitated away from the Rust Belt and Midwest to the Southeast and West Coast regions.

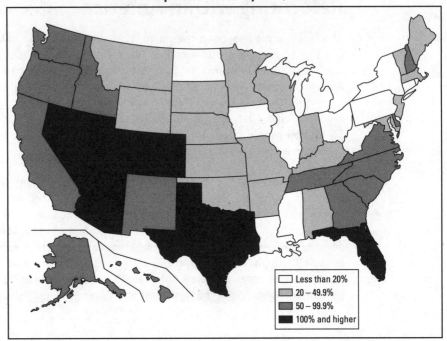

FIGURE 12-2: U.S. population growth by state, 1980–2020.

This relationship has been in high gear for the past 50 years. Reasons for it include

» A surge in retirees migrating from the northern Snowbelt (a push factor) to the warmer climes of the Sunbelt (a pull factor)

» A rise in employment opportunities (pull factors) in the southern tier (and the southeast in particular) as businesses moved in to take advantage of the region's comparatively low wages, low taxes, and low energy costs (which also attract good numbers of immigrants)

» A related rise in unemployment (a push factor) in the Northeast and Upper Midwest (due to the closure of numerous heavy industrial plants in the traditional manufacturing belt) and the desire and willingness of the jobless to relocate where jobs are available

» The attractive outdoor amenities and recreational potential (pull factors) associated with the climate, beaches, landscapes, and open spaces of parts of the South and West

The slower growth in some states reflects the reverse (i.e., the population losses incurred by these four factors), as well as others. Mississippi, for example, saw a brief period of population decline after 2005's Hurricane Katrina sent people scrambling to other states.

Normally, one thinks of growth as a good thing. But some high population growth areas are now victims of their success at attracting migrants. Virtually every one of the 15 fastest growing cities in the U.S. are in the Sunbelt or West (9 of the 15 are in Texas!), where the imagery of open space and the great outdoors has met with traffic jam and urban sprawl (see Table 12-2.) Other areas have witnessed a growing disparity between the need for water and the supply of it. Primary examples are California, Arizona, and Florida, where thirsty cities and even thirstier agriculture are increasingly pitted against each other.

TABLE 12-2: **The 15 fastest growing cities in the United States, 2000–2020 (100,000 or more residents).**

City	Percent Increase 2010–2020
Frisco, Texas	79.5%
Meridian, Idaho	61.4%
McKinney, Texas	58.8%
Sugar Land, Texas	49.6%
Kent, Washington	41.4%
Murfreesboro, Tennessee	38.6%
Round Rock, Texas	37.7%
Pearland, Texas	35.4%
Irvine, California	33.6%
Bend, Oregon	32.9%
Edinburg, Texas	32.8%
Midland, Texas	32.3%
Cape Coral, Florida	30.2%
Denton, Texas	30.1%
League City, Texas	29.9%

In addition to the fastest growing cities, we have those that have added the greatest number of people. From 2010 to 2020, the top five include Phoenix, San Antonio, Houston, Austin, and Fort Worth (in that order, highest first). That Sunbelt or Western distribution remains evident, with only Columbus, Ohio (11) and Washington DC (13) breaking the pattern.

Giving a Good Impression

Decisions to move to or vacation in particular destinations tend to be based largely on the knowledge and impressions people have about them. Different people have different attitudes toward different places. Some locales may impress you as desirable places to live, some may be undesirable, and some may make no impression one way or the other. Sometimes these attitudes are based on personal experience and verifiable facts (objective reality), and sometimes they are based on what people have heard or read or imagine to be the case (subjective reality). When you think about it, don't you have strong positive or negative feelings about some places, including ones you have never visited?

Playing the mental game

Mental maps are tools that geographers use to display and analyze the impressions (subjective realities) that people have about different locations. The example provided in Figure 12-3 shows a mental map of the United States that reflects attitudes of college students who go to university in South Carolina (these rankings were collected over many years from my own students). Now, before somebody gets upset and writes a nasty letter to the publisher or me, remember that this map reflects attitudes that some people have toward places that perhaps they have never visited and about which they may have no direct personal knowledge. Thus, what they think they know about a given state or region may have no basis in reality. Moreover, just as some of these students have negative images about certain parts of the country, it's also true that people in other parts of the country have very negative attitudes toward their state.

Looking at the mental map in Figure 12-3 may cause you to ask, "What was going through the minds of those students?" In fact, like most mental maps, this one reveals a handful of factors that repeatedly determine place desirability or undesirability. They include the following:

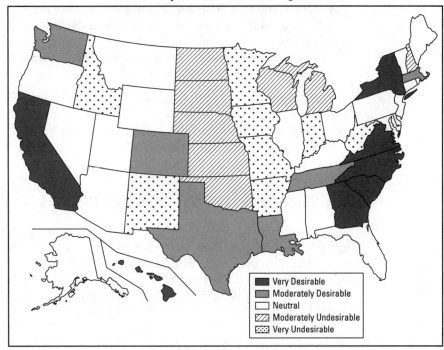

FIGURE 12-3: Mental map of state desirability as perceived by South Carolina college students.

(© John Wiley & Sons Inc.)

» **Home:** Dorothy said it best: "There's no place like home." No matter where you conduct a mental map exercise, you find that most people like where they are. Thus, New Yorkers like New York, Londoners like London, and so forth. Home, after all, is known and predictable and usually doesn't hold the nasty surprises that a move might. It makes sense that South Carolina and nearby states — known places like Georgia, North Carolina, and Virginia — stand out as more favorable.

» **Dislikable neighbor:** A dislikable neighbor can border one's desirable home. Thus, many people in Kansas have very negative perceptions of Missouri, and New Yorkers of New Jersey, and so on. What this factor says about the human condition is open to debate.

» **Physical environment:** Some people perceive some states to be more environmentally attractive than others. Most college students, for example, generally have positive views of states that they perceive as having warm beaches (Hawaii), hilly and/or mountainous topography (Colorado), and forests or related greenery (Washington). On the other hand, places that are cold, flat, and desert-covered are worthy of avoidance. Sorry Arkansas, West Virginia, and Missouri: These youngsters gave you no votes at all.

- » **Socio-cultural environment:** People come in different races, speak different languages, adhere to different religions, have different sexual orientations, and so forth. Thus, when people assess the desirability of a particular destination, they often ask themselves, "Will the people who live there accept me, or will I find it difficult to fit in?"

- » **Job prospects:** If you give a mental map exercise to people who are retired or about to be, then job prospects typically don't mean much. But because most college students are pondering employment after graduation, they are usually attracted to states that are viewed as offering good job prospects.

Getting an image adjustment

Mental maps can be a lot of fun to play around with, but they may strike you as rather trivial. In fact, they have a very serious side as predictors of potential migration. Most places seek to attract entrepreneurs and businesses that can create jobs and generate tax revenues. Mental maps can sometimes reveal that a particular place has what may be called an "image problem" in certain parts of the world or is simply not seen as a great place to do business.

This perception may encourage officials in that place to create ad campaigns or engage in other marketing measures designed to promote a positive image. If you routinely watch TV or listen to the radio, then sooner or later you are bound to see or hear a commercial that encourages people and businesses to move to a particular place. The reason is simple. Officials who live in that "particular place" know that shift happens. They also know that getting it to happen in their direction can add to employment, tax revenues, and political clout.

Putting your best image forward

I cannot end this chapter without a word or two about tourism — a short-term migration of sorts that has much to do with geographical impressions and subjectivity. Tourism is a multi-billion-dollar industry that continues to grow as more countries develop and more people have more disposable income.

TIP

How the world will bounce back from the sharp decline in travel stemming from the COVID-19 pandemic remains to be seen. Nonetheless, seemingly every country and locale wants a piece of the multi-billion dollar travel and tourism pie. They won't get a sliver, however, unless tourists perceive them as a desirable place to visit. And by desirable, I also mean authentic and different. Who wants to go to a place that is just the same as any other?

APPLIED GEOGRAPHY: MARKETING PUERTO RICO

Just a few years ago, Puerto Rico's tourism office embarked on an advertising campaign. Images invoked beaches, tropical jungles, and Spanish history, and some of the ads were stamped "passport" style. One of the embossed logos was encircled with the words "No Passport Required." This makes sense, right? People reading these ads need to know that political barriers for travelling to Puerto Rico don't exist, at least for them.

Here's my problem. The ads were directed to Americans living in the (mainland) United States. Of course, they don't need a passport! Puerto Rico is part of the United States. Puerto Ricans are U. S. citizens (anyone born there since 1941 has the right of birthright citizenship). Travelling there from Texas, or New Hampshire, or South Dakota is no different than flying in between those three states or any other. So, who was this advertising campaign directed to? The historically and geographically ignorant. In other words, those who have decided not to read this book.

Enter advertising, which is a major means by which tourists find out about and assess destinations that are anxious to attract them and assist in the disposal of their disposable income (see the "Applied Geography: Marketing Puerto Rico" sidebar).

In the world of tourism advertising, objective geography tends to give way to fantasy geography. It's not that tourism ads lie, but they definitely do put a certain spin on reality. Thus, I have yet to see a travel ad that shows a rainy day (even for Ireland, which isn't called the Emerald Isle for nothing). No matter the destination that is being touted, the landscape is always gorgeous, the locals are always smiling (and often dressed in traditional garb that nobody really wears anymore), and the tourists are healthy and fit. And why not? The goal is creation of a positive image that potential tourists will find attractive. Will tourists visit places they perceive as unattractive? Would you?

> **IN THIS CHAPTER**
> » Creating cultural diversity
> » Spreading culture, stopping culture
> » Considering religion and language
> » Heading toward a single global culture?

Chapter 13
Culture: The Way We Live

My wife had some work business in India not long ago, and lucky me, I was able to tag along. We travelled early, spending time in Delhi and then taking the train to Agra (home of the Taj Mahal) before a final stop in Jaipur for her meetings. While she worked, I played. Most days were spent jumping into a tuk-tuk, a small motorized three-wheeled rickshaw of sorts. I gave my driver the simplest instructions, which were to drive in one direction until I said stop, usually 5 to 7 miles from my pickup. And then I walked. I walked all over, here and there, hither and yon, until I arrived back at our hotel.

I was greeted in English, Hindi, and sometimes in Spanish. Some locals took my little beard to mean that I was from the Punjab. There were bright flower markets, noisy vegetable stalls, and an occasional cow mingling with the traffic. Smells equaled sights in their vibrancy and potency, and ranged from spices to the acrid smell of burning plastic to cook a meal. And then there were the snakes. Both of them. An Indian cobra visit was not on my itinerary that day, but there they were, bobbing along to flute music (see Figure 13-1 for a flavor of this show). Much of their bodies lay inside a basket, and the charmer welcomed people to get closer for a picture and to leave money. Part magician, part healer, the charmer entertained in the square and in the past may have been sought out to rid homes of snakes. From the dancing snake to the musical instrument, to the musical style, all the way to the coil pattern on the basket, I was witnessing the traits of a culture far different from my own.

FIGURE 13-1: Your friendly neighborhood snake charmer with his two basketed friends in Jaipur, India.

(© Jerry T. Mitchell)

Everybody possesses *culture,* a learned pattern of behavior that characterizes a group of people. In many respects, your culture defines your essence; and yet you tend not to think about it until perhaps, you find yourself in a geographic setting where your culture either doesn't work, or marks you as being different from others. Suffice it to say, that dancing snake made me suddenly and acutely aware of my culture.

Cultural geography is the field that seeks to describe and analyze the distribution of culture over Earth's surface, and is the subject of this chapter. But what can one say about the geographies of so many culture groups and their cultural traits in one chapter? The answer is "very little." Thus, the emphasis of this chapter is on key concepts of cultural geography and how they affect the two major culture traits of religion and language.

Being Different Thousands of Times Over

REMEMBER

Nobody knows exactly how many cultures exist on Earth today, and it's somewhat silly to place a number on them. Whatever the true total (Hundreds? Thousands? How do you count?), like birds, folk of a feather flock together. That is, people who are culturally similar tend to live in proximity to each other and in doing so, form *culture areas* — regions occupied by people who have something cultural in common. These may be quite large, like the Islamic culture region that extends from Dakar eastward across Northern Africa, through the Arabian Peninsula and Southwestern Asia. Culture areas may also be rather small, as in San Francisco's Chinatown, which occupies no more than a square mile or two.

In the act of practicing their culture, humans often transform the natural landscape into a *cultural landscape,* as when people convert a grassland to a farm. Because culture is diverse, so, too, are the world's cultural landscapes, which easily rival (and maybe surpass) purely natural landscapes in their richness and variety. Culture therefore distinguishes people as well as physical regions. In a sense, culture is the spice of life as well as place, differentially "flavoring" people and the land they inhabit.

How did we get so many cultures? Assuming human beings started out more or less the same way back when, then how did we end up so different? The answer is largely in three parts: the diversity of culture, effects of isolation, and adaptation to new surroundings.

Counting cultural diversity

Culture is extremely broad and complex, affording ample opportunity for people to be different from each other. Suppose you made a list of all the ways in which you are culturally different from people who live in Saudi Arabia, Thailand, and Tahiti. You could well end up with at least a couple of dozen items in each case.

REMEMBER

Comparison of those lists would reveal *cultural universals,* which are categories of traits that all cultures share, but whose specific manifestations vary from one culture to the next. Various cultural universals are given in Table 13-1. Language is an example. You speak at least one language. So do people in Saudi Arabia, Thailand, and Tahiti. Indeed, every culture has one, so it can be considered a universal. All told, more than 6,500 languages are spoken today. Likewise, religion is a universal, and hundreds, if not thousands, of them can be found throughout the world. You can add to them dress, architecture, sport, and all of the other universals, each of which come in many specific varieties.

TABLE 13-1 **Some Cultural Universals**

Language	Religion	Dress	Technology	Architecture
Government	Cuisine	Economy	Art	Gender roles
Medicine	Music	Sport	Agriculture	Education

Isolating people

Isolation is another reason why we have so many cultures. Communication generally breeds cultural homogeneity. In other words, the more people that share information and ideas, the more alike they tend to be. Geographic isolation, in contrast, breeds differences. Take a large number of people who have the same culture, divide them into, say, four groups that are isolated and completely out of touch with each other, and over time they are likely to go their separate ways, culturally speaking. Basically, as humans migrated eons ago from their common ancestral homeland, to ultimately occupy the world, that is what happened.

Table 13-2 shows the conjugation of the verb "to sing" in four languages. Visual comparison of the columns shows different spellings but also remarkable similarities.

TABLE 13-2 How People "Sing" in Four Languages

Verb Tense	English	German	Dutch	Swedish
Present	Sing	singen	zing	sjunga
Past	Sang	sangen	zong	sjong
Participle	Sung	gesungen	gezongen	sjungit

The explanation for the similarities in the spellings is that these modern languages have a common ancestral language that was spoken more than a thousand years ago in Northern Europe.

But there came a time when members of this culture began to migrate, ultimately to occupy lands that would later become England, The Netherlands, Sweden, and Germany. And in the process, groups of significant size became isolated from each other more or less permanently. Patterns of speech were no longer shared. Pronunciations began to drift apart, which accounts for the differences in spellings. Migrants encountered "strange" plants, animals, and physical environmental conditions for which there were no words, so they invented new ones. Over time, therefore, the once-common language developed different dialects, which became different languages.

Adapting to new surroundings

Adaptation refers to cultural adjustments that occur after a people migrate to a region whose physical environment is different than where they used to live. A reference was made in the previous section to how a single language became several languages in part because of human adaptation to new environments.

Here are examples of how adaptation has encouraged diversity in three other cultural universals:

» Traditional agricultural know-how may not work or is inefficient in the new setting. Immigrants borrow workable techniques from the local populace (if there is any), experiment with local plants for their food value, and otherwise adapt as best they can.

» Traditional modes of dress are either too light or too heavy for the new climatic venue. Adaptations may occur with respect to style, materials, and color.

» Traditional building materials may be absent in the new setting, resulting in adjustment in architecture. Also, climate may affect modification of housing with respect to thickness of walls, shapes of roofs, and openness to outside conditions.

In each case, adaptation that adds to the overall richness and variety of culture, increasing the ways in which people can be different.

Spreading the Word on Culture

REMEMBER

Culture creation and modification are not things of the past. Thanks to *cultural diffusion*, which refers to the spread of culture, the geographies of particular traits as well as cultural complexes that characterize groups of people continue to develop and change.

For example, 75 years ago few Americans had heard of yogurt, tortillas, tofu, tandoori chicken, sushi, and couscous, let alone actually eaten them. All those foodstuffs existed back then, but their geographies were pretty much limited to their respective native areas — Asia Minor, Mexico, China, India, Japan, and North Africa.

Today it's different, of course. Chances are you have heard of most or all of those foodstuffs and maybe even eaten them as those foodstuffs have spread here and are widely available in stores and restaurants. As a result, the geography of these foodstuffs, all of which are culture traits, has changed dramatically thanks to cultural diffusion.

The same is true of other traits. You, for example, speak one or more languages and may practice a certain religion. Chances are good, however, that none of these traits originated right where you live, but rather, like those foodstuffs, diffused from somewhere else. The manner of their diffusion may have varied, but probably has something in common with one or more of the three generally recognized modes of diffusion (see Figure 13-2): relocation, contagious expansion, and hierarchical.

FIGURE 13-2 A, B, AND C: The three standard modes of cultural diffusion.

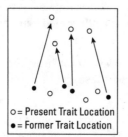
o = Present Trait Location
• = Former Trait Location

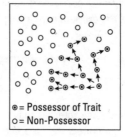
⊙ = Possessor of Trait
o = Non-Possessor

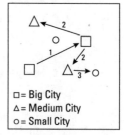
□ = Big City
△ = Medium City
o = Small City

(© John Wiley & Sons Inc.)

Relocating one's culture

Relocation diffusion is synonymous with migration (Figure 13-2a). When people move, they take their "cultural baggage" with them. As a result, the geography of culture may change because migrants impart their particular cultural characteristics to an area where perhaps it was not previously present.

Virtually every large American city, for example, has distinctive ethnic neighborhoods that exist because of the relocation diffusion of peoples from a foreign land. Small towns may also exhibit the same effect, and so, too, rather rural parts of America.

Chances are good that an example or two are near you wherever you live. One such example stared at me near daily when I lived in eastern Pennsylvania: gold-domed churches in nearly every small borough (town). Anthracite coal miners, migrated from Eastern Europe, brought their Orthodox Christian faith and its architecture with them. Other, (dis)similar examples dot America (see the sidebar "Applied Geography: Looking Swiss").

Coming down with culture

In *contagious expansion diffusion*, the geography of a trait expands because people who did not previously possess it, adopt it. Typically, this results from contact or direct exposure (hence, contagious) to the trait (Figure 13-2b). Thus, a farmer might "look over the fence" to see a neighbor growing some new kind of crop and adopt it as well.

APPLIED GEOGRAPHY: LOOKING SWISS

Drive into New Glarus, Wisconsin and you might get the impression that a little piece of Switzerland has been transplanted in the American Midwest. That's what the locals are hoping — and even more so that you will stop and spend some money. As in many other towns and venues in America, the people of New Glarus have taken to using their cultural heritage to promote tourism and the local economy.

In the 1840s immigrants from Canton Glarus, Switzerland settled the area, bringing their culture with them. The ensuing relocation diffusion of their German language, Catholic religion, dairying, and other attributes resulted in creation of a culture area in south-central Wisconsin that was distinct from neighboring lands peopled by immigrants from other European countries, as well as the local Native Americans. Other than dairying, however, little in the cultural landscape was particularly "Swiss-looking." That started to change a few decades ago when townspeople began erecting buildings in traditional Swiss styles and altering existing facades to render the same visual effect. The presumption was that if you turned the town into a "little Switzerland," then tourists would come and spend money. It worked like a charm.

You don't have to go to New Glarus to appreciate the concept, however. Lots of other towns and neighborhoods have done the same, albeit in all likelihood through a different cultural heritage. Perhaps you can think of an example or two in your own area.

That is not a far-fetched scenario. Efforts to increase food production, for example, often are an exercise in cultural diffusion, as when an *agronomist* or crop scientist seeks to encourage local farmers to discontinue a traditional way of producing foodstuffs and do something different. But whether they live in Kenya or Kansas, farmers tend to stick with things that work and try something new only when the likelihood of success is high.

Demonstration is a proven way to promote diffusion of agricultural innovation. Neighboring farmers "look over the fence," to see what's happening in the demonstration area, and adopt it. Their farms, in turn, become objects of observation by other farmers who look over the fence, adopt what they see, and so forth.

OK, so you're probably not a farmer. But at some point in your life, you probably adopted a certain cultural item because of direct exposure to advertising, or because you saw somebody doing something or wearing something that you found attractive, or because of something a friend said or did. You adopted a new hairstyle, began listening to different music, or tried out a new type of clothing. If so, then you have experienced contagious expansion diffusion firsthand.

Doing what the big boys do

In *hierarchical diffusion*, a culture trait is born in a large city, becomes adopted by a portion of the populace, spreads to other large cities, and then "trickles down" to medium-sized cities, small cities, towns, and villages in that order (Figure 13-2c).

Nowadays, cultural "fads" in particular tend to diffuse hierarchically. This is especially true of new clothing styles, body modifications (hair styling, tattoos, piercing, and the lot), and slang. Inhabitants of large cities generally are more dissimilar and accepting of personal differences than their counterparts in small settlements where, perhaps stereotypically, everybody knows everybody else's business and pressure to conform is comparatively high. "Being different" by adopting a new fad that may seem outlandish to some is easier in big cities. If the fad is successful and catches on, then its visibility and acceptability are likely to increase, which in turn increases the likelihood that it will trickle down the hierarchy.

History is replete with examples of these geographic processes. Take Blues music for example. Musical styles and instruments diffuse from Africa (relocation) to the United States. Adaptations ensue and these are shared among musicians (contagious expansion). Musicians travel (relocation) from the Mississippi Basin to larger cities such as Chicago and New York. The music matures further and diffuses from large places to smaller ones (hierarchical) and via radio (contagious expansion). The widespread nature of the Blues is similar to other culture traits. Sport is another example. NASCAR motorsport racing has its origin in the American South, but now there are races in California, Pennsylvania, and elsewhere. Helping to grow and spread its popularity are first television and later the internet, two mechanisms enabling contagious expansion.

Calling a Halt: Barrier Effects

Barrier effects are things that stop or inhibit cultural diffusion. When culture traits spread, they typically do not "keep going and going and going" like that battery-powered bunny of TV commercial fame. Instead, traits tend to diffuse outward from their areas of origin, achieve a certain geographic breadth, encounter one or more barrier effects, and then stop spreading. Were there no barrier effects, then culture traits would, in fact, "keep going and going," resulting in a rather uniform global culture. Thanks to barrier effects, therefore, Earth's cultural geography is a mosaic of culture areas instead of a monochrome.

Barriers may be *absorbing* or *permeable,* respectively stopping completely the spread of culture or selectively accommodating the spread of some culture traits, but not others. For millennia, the Atlantic Ocean was an absorbing barrier that stopped the westward expansion of European culture. More recently, societal decision making in Saudi Arabia has been a *permeable barrier,* allowing into that country Western technology related to oil drilling, but holding at arm's length other Western cultural commodities such as alcohol consumption. As these examples indicate, barrier effects may originate from the physical or social environment.

Getting physical

Physical barriers are natural elements that now or in the past inhibit cultural diffusion. These have historically served to isolate people, either preventing or seriously limiting access to agents of culture change. The following sections cover the classic examples.

Oceans

Oceans were formidable barriers to cultural diffusion for millennia. People didn't know what lay across them, or how far away places were. Similarly, they did not possess the technology to accurately plot a course to a particular destination or to return home whether or not they had discovered anything. On top of that, and for the longest time, ships were fragile and at the whimsy of wind and storm. Thus, until modern shipbuilding and navigation came along, oceans tended to inhibit the spread of culture instead of promoting it. To this day scattered Pacific Ocean islands are homes to people whose cultures have been altered less so by contacts with the broader world. In these cases, the surrounding ocean continues to serve as a formidable physical barrier that has insulated the islands from forces of culture change. Nearshore islands separated from the mainland can work the same way (see the nearby sidebar "The geography of Gullah").

Forests

Most of today's forests are mere remnants of their former selves. Five hundred years ago, nearly all of what is now the United States east of the Mississippi River was continuous forest — as was most of the Far West and Northwest. The same was true of virtually all of Western and Central Europe, as well as virtually all of humid Africa, Asia, Central and South America. And I don't mean the well-tended greenery you see today in many places. No sirree. I'm talking underbrush and thickets and dead limbs and all kinds of other stuff that limited visibility and mobility. It was ripe for disorientation and ambush. You could get lost in it.

And in a sense, that is what happened. Numerous peoples became separate, forest-dwelling societies whose woodsy surroundings provided isolation that contributed to development of distinctive cultures. Today numerous traditional societies inhabit regions of tropical rainforest, particularly in the Amazon Basin, but also in Central Africa and Southeastern Asia. Road building can be very difficult in these areas, and thus the forests, like the oceans in the case above, continue to isolate inhabitants from the outside world and promote cultural differences. In these environments, rivers — natural highways — have often served as avenues of diffusion.

Mountains

Rugged terrain, and particularly mountains, has historically tended to make communications difficult, and thereby encourage cultural diversity. For example, an estimated 800 or so languages are spoken on the island of New Guinea, which is about the size of Texas and Arkansas combined and has a population near 9 million. It makes no sense that so many languages coexist in such a relatively small space until you consider the topography. New Guinea has an extremely mountainous spine that has been eroded over the years into numerous steeply sided valleys that have no roads and few tracks between them. Add the dense tropical forest, and the results are hundreds of relatively isolated pockets of people that have, at least with respect to language, gone their own ways.

Also, the "thin air" that comes with the high altitudes of mountainous terrain has proven to be an impediment to diffusion of culture. For example, many facets of traditional Native American culture are alive and well in the Central Andes (primarily Bolivia, Ecuador, and Peru), where millions of people still speak Quechua, the language of the Incas. Although these peoples came under Spanish rule, the Spaniards themselves generally avoided settling in the high Andes because they found adaptation to the "thin air" to be extremely difficult. Accordingly, native culture in that area did not give way to imported culture.

Deserts

Outsiders generally were not adapted to desert conditions and therefore they found such regions inhospitable and avoided them. Accordingly, deserts have tended to isolate people and inhibit the spread of culture. For example, traditional culture groups continue to inhabit central desert areas of Africa, Australia, and Asia. The San people of Namibia and aborigines of Australia are historic examples, although members of these groups have experienced significant change in recent decades. Nevertheless, the long persistence of their unique cultures testifies that deserts create a formidable physical barrier.

Tundra

Tundra, which you can read about in Chapter 10, refers to very high latitude environments dominated by short grasses. The climate is sub-freezing for much of the year. Native peoples adapted to these harsh circumstances over the years and developed distinctive cultures. Like deserts, however, outsiders generally are not well-adapted to tundra, and therefore have found such regions inhospitable and avoided them. Thus, tundra has served as a physical barrier. Specifically, it has tended to isolate traditional peoples in northernmost North America and Europe from sources of culture change and in doing so encourage a world of cultural differences.

Socializing effects

Social barriers are human institutions that inhibit the spread of culture. These can be as formidable as physical barriers, and sometimes more so. The following sections discuss four of the more prominent examples of social barriers.

THE GEOGRAPHY OF GULLAH

Gullah is a dialect spoken by African Americans of the Sea Islands, which adjoin the coasts of Georgia and South Carolina. Closely related to an ancestral language (or languages) that arrived with enslaved Africans, *Gullah* is spoken not by people who learned it to get in touch with their African roots, but instead by folks who never lost those roots in the first place.

The geography of *Gullah* is a classic example of physical barrier effects. After emancipation, most of the black population of the Sea Islands stayed there, and for decades the islands largely remained unconnected by bridges to the mainland. The result was limited physical interaction with outsiders and mainstream culture. Under those circumstances Gullah continued as a viable language.

But times are changing. The Sea Islands, with their seaside settings, have become prized real estate for vacation homes, retirement communities, and resorts. Bridges now connect the mainland to many of these islands. As newcomers come and development physically transforms the islands, so, too, will they bring culture change that will threaten the continued existence of Gullah.

Why is this concerning? Just as we prize biological diversity, cultural diversity contains features that can be irreplaceable. Gullah has its own stories, figures of speech, and so on that will be lost to future generations if the language disappears.

Language

The simple inability to speak somebody else's language limits opportunity for cultural interaction and sharing. Though media and multilingualism are bringing people closer together, the continued existence of several thousand languages remains a powerful barrier to cultural diffusion.

Religion

Differences in religious beliefs may mark certain people as being "other" and nullify propensity to interact with them and adopt their cultural attributes. Religion may also manifest prohibitions (such as bikinis and beer) that deter exchange of materials and ideas.

Race and ethnicity

Many people have a deep "consciousness of kind." In some cases, that is code for racism and prejudice — a desire to be geographically separate from "them." In others, it may simply be a deep-seated preference for interaction with one's own kind. In any event, race and ethnicity tend to differentiate human groups — often very visually — and give rise to behaviors that limit interaction.

Historic events

Sometimes historic events, such as conflict and war, mark a people as "the enemy" and create wounds that refuse to heal. Intense dislike — if not raw hatred — of one group for another can be a powerful barrier to human contacts that would normally promote cultural transfer.

Getting Religion: How It Moves and Grows

Now let's see how the concepts mentioned up to this point come into play, beginning with religion. Looking at the geography of religion allows you to examine one of the more important culture traits and to make connections between cultural geography and contemporary matters. In Figure 13-3, a highly generalized map of the world's principal religions reveals culture areas that vary greatly in size. Christianity and Islam, for example, exhibit multi-continental expanses. Judaism, in contrast, is dominant in Israel, a comparatively small culture area, and in even smaller scattered urban enclaves mainly in Europe and North America.

Putting diffusion to work

With the exceptions of the traditional (or what some people might call "tribal") religions, virtually all of the distributions shown in Figure 13-3 are results of cultural diffusion. In that regard, Buddhism and Christianity are interesting in that their present distributions have little in common geographically with where they originated — Israel and the West Bank in the case of Christianity, and India in the case of Buddhism. The Jewish population of Israel largely consists of recently (post–World War II) relocated individuals and their descendants, while the other Jewish enclaves worldwide are to some extent latter day expressions of the ancient *Diaspora* (the dispersion of Jews after the Babylonian exile). In Africa, Europe and Southwest Asia, the Islamic realm is largely the result of relocation diffusion (migration and conquest) of people from Arabia coupled with conversion of peoples with whom they came into contact (contagious expansion diffusion). Another interesting geographic aspect of Islam is its dominance in Indonesia (which contains more Muslims than any other country on Earth), Malaysia, and the southern Philippines, which are products of ages-old trans–Indian Ocean trade between Middle Eastern and Southeast Asian lands (contagious expansion diffusion).

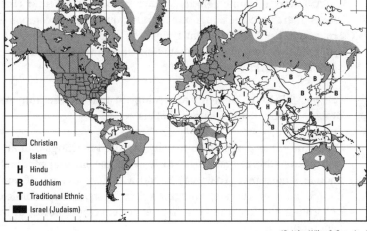

FIGURE 13-3: Predominant religious groups around the world.

(© John Wiley & Sons Inc.)

TIP

The widespread nature of Islam, Buddhism, and Christianity, as opposed to Judaism or Hinduism, is by design. All are universalizing religions that claim applicability to all people. These faiths are distributed widely because they have actively tried to do so through *proselytization* (the process of gaining converts).

Getting effects into action

To a certain degree, barrier effects are also evident on the map. The Himalayan Mountains, a formidable physical barrier, mark the boundary between parts of the Hindu and Buddhist realms. In West Africa, the interface between the Islamic and "traditional religion" realms coincide with the tropical forest fringe. A similar effect is seen in the Amazon, where tropical forests have isolated practitioners of traditional religions from agents of culture change.

Creating local character

In exercising their faiths, humans often impart religious character to the lands they occupy. The nature of these impacts vary from creating cultural landscapes to making psychological attachments to place, and from performing acts of solemnity to committing acts of bloodshed. Here are four ways adherents imbue locations with religious characteristics.

Places of worship

Practitioners of some religions build houses of worship that are magnificent works of architecture and, quite literally, outstanding components of the cultural landscape. Until fairly recent times, the spires of churches, towers of mosques, and domes of temples, dominated skylines. Indeed, in many smaller towns and some cities worldwide, they still do.

TIP

It is possible to see elements of theology in architecture from overly ornate cathedrals to minimalist Protestant chapels.

While the essence of place is usually captured in things that are seen, it may also be echoed in things that are heard. Thus, some cultural geographers speak of "the audible landscape." That may sound a bit cryptic, but if you have ever heard church bells reverberate through a valley, or heard the Islamic call to prayer ring out from a minaret, then you have experienced the power of sound as a cultural geographic characteristic.

Sacred sites

Most religions recognize sites that have special significance in the minds of believers. These may serve as destinations of pilgrimage (such as Mecca and Lourdes), places of great historical religious significance (the Wailing Wall), or places where a solemn ritual is to be performed (Varanasi or the Via Dolorosa). The bond between religion and sacred site may be so strong as to create an uncompromising sense of proprietorship and right to rule in the sacred area, perhaps coupled by exclusion of non-believers.

These places are known to modify human behavior, too. Consider how you might act just walking through a cemetery. Imbued with reverence, the place demands a hushed voice and quiet reflection.

Friction and flash points

Religions define behaviors that are pleasing and displeasing to God. Unfortunately, behavior (such as the proper way to call and worship God) that is practiced by members of some faiths may be displeasing or patently offensive to members of other faiths, leading to friction, if not outright bloodshed. It should come as no surprise, therefore, to observe contemporary conflicts that coincide geographically with the overlap or interface of different religious groups. Examples have included:

>> Northern Ireland (Protestant — Roman Catholic)

>> Kashmir (Hinduism — Islam)

>> Southern Philippines (Christianity — Islam)

>> Israel (Islam — Judaism)

>> Former Yugoslavia (Christianity — Islam)

Forbidden and favored foodstuffs

Religious dictates have changed the agricultural landscape of many areas by discouraging production of some foodstuffs and encouraged others. These have had a significant impact on the geographies of agriculture, animal husbandry, and cuisine. Among the better-known prohibitions are the Judaic and Islamic proscriptions against pork (pigs being viewed as unclean), the Hindu proscription against beef (the killing of cows being forbidden), and the Islamic proscription against consumption of alcoholic beverages. In some cases, in contrast, religious favor has encouraged diffusion of foodstuffs. Here are two examples:

>> **Turmeric:** This is a rather tasteless spice, grown throughout much of Southeastern Asia, whose principal purpose is to add a golden color to other foodstuffs. Its diffusion has been linked to the spread of Buddhism, in which gold is a symbol of enlightenment (hence the classic gold-domed Buddhist temples). The spice's purpose, therefore, is to bestow a desired aura — one of purity and prosperity — on food.

>> **Grapes:** The diffusion of Christianity, especially Roman Catholicism, in many parts of the world was complemented by the spread of *viniculture* (grape growing) to meet sacramental needs. In time, of course, grape production expanded to meet consumption needs of a more pedestrian sort. But

religious diffusion provided the impetus for agricultural diffusion. Interestingly, the religious connection is particularly visible in Germany in the guise of the famous "Blue Nun" label, and the popular white wine *liebfraumilch*, meaning literally "beloved lady's milk." The lady? Mary, mother of Christ.

Getting in a Word about Language

Language is arguably the most important of the cultural universals that were identified earlier. This is not to question the significance of religion or other traits, but language is essential to communicating and sharing many aspects of culture. The standard first step in analyzing the geography of languages is to produce a map of them. Unfortunately, on a page of this size, I can't possibly give you a map that shows the geography of the more than 6,500 languages that are spoken today, nor even a map showing the dominant language in different parts of the world. Simply too many are out there. But I can, at least, show the geography of English usage worldwide (Figure 13-4).

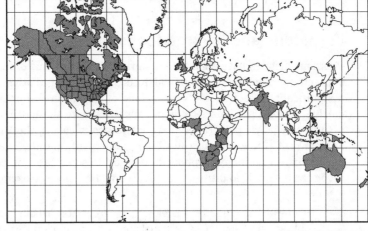

FIGURE 13-4: The geography of English is shown by the dark shade.

(© John Wiley & Sons Inc.)

What Figure 13-4 shows are large-scale English language culture areas around the world. In Britain, Australia, New Zealand, Canada, and the United States, English is spoken by the overwhelming majority of the population. In other countries, English is spoken only by a minority, even though it may be an "official language." The "big picture" map aside, consideration of language — like religion — affords opportunity to observe and apply diverse concepts of cultural geography.

Diffusing languages

The map of English is in large measure a product of cultural diffusion from Britain through its former colonies. The initial stage was largely limited to *relocation diffusion* (described earlier in this chapter). That is, large numbers of immigrants and officials moved from Britain to the colonies and, of course, took their language with them. Once there, they intermingled to different degrees with native peoples and non-English speaking immigrants, many of whom acquired English by *contagious diffusion* — contact with English speakers.

REMEMBER

English now enjoys the status of *official language* — the one in which government business is transacted and printed, as well as the language of publicly financed education — in virtually all of Britain's former colonies. In many cases it is also the *vernacular language* — the one that is spoken by the people of a particular locality. But official and vernacular languages are not always the same in a given area or region. English, for example, is spoken by most Americans, but literally millions of people living in ethnic neighborhoods, Indian reservations, and other enclaves across the land speak a different vernacular language. Look again at Figure 13-4 and you may get the impression that everybody in the United States, Australia, and New Zealand speaks English. Not so. People in parts of each country speak a different vernacular language.

Many countries that are former colonies have adopted the language of the colonizer as their official (or co-official) tongue even though, in many cases, only a minority of the populace speaks it. Examples include English in Ghana and Kenya; French in Senegal and Madagascar; and Portuguese in Angola and Mozambique. Typically, European languages are given official status in the former colonial realm for two reasons:

>> The country contains numerous ethnic groups, some of which have a history of friction. Elevating one local language to official status could lead to jealousy and unrest on the part of other groups.

>> Use of a European tongue stands to promote international trade and commerce more than would a local language, which may be spoken nowhere else on Earth.

Nevertheless, many (even most, in some cases) of the native peoples in these countries continue to speak their own tongue as the vernacular language. In most cases, use of the official language(s) is concentrated in the cities and larger towns, while the vernacular persists in the smaller towns, villages, and rural areas. To the extent that everyday use of the official language is gradually "trickling down" from urban to rural areas, its spread exemplifies the process of *hierarchical diffusion*.

Checking the physical effects

Language and physical geography may interact in various ways. The two most significant ways are through environmental terminology and linguistic refuges.

Environmental terminology

Languages tend to develop robust vocabularies that pertain to locally observed environmental conditions, and weak vocabularies that pertain to unfamiliar settings. English, for example, is weak in native terminology that pertains to deserts, the sub-arctic, very mountainous areas, and other characteristics that are not common to England. Thus, English has adopted environmental terminology from other languages to describe things that English cannot, or at least not very well. Accordingly, standard English dictionaries now include terms such as *arroyo* (from Spanish) to describe intermittent streams in desert environments, *taiga* (from Russian) to describe high-latitude coniferous forest, and *fiord* (from Norway) to describe steep-sided, glacially carved inlets of the sea.

Linguistic refuges

A *linguistic refuge* is an area where a language is insulated against outside change by virtue of remoteness, or the remains of a locale where a once widespread language continues to be spoken. Acting as physical barriers, aspects of the physical environment have served to isolate speakers of various languages and thus preserve their native tongues from outside agents of change. Heavily forested and extremely mountainous areas, as noted earlier in this chapter, have historically served that purpose.

The traditional Welsh and Irish languages, for example, at one time appeared to be on the brink of extinction, relegated to remote peninsulas, islands, and valleys in their homelands following the onslaught of English. However, nationalist aspirations and heritage awareness have led to campaigns to resuscitate these languages and promote their everyday use. Central to these efforts have been human resources — native language speakers — many of whom hail from villages and farms in linguistic refuge areas.

Playing the landscape naming game

Language may provide cultural character to the physical environment as well as to people. For example, what do New Jersey, Lake Okeechobee, Baton Rouge, and El Paso have in common? The answer is they are all *toponyms* or place names. People the world over have a habit of naming landscape features, be they mountains, hills, rivers, lakes, bays, seas, deserts, forests, cities, towns, streets . . . the list goes on and on. *Toponymy*, the study of place names, may provide diverse geographical insights. As per the four locales mentioned, toponyms may tell us something about where the settlers came from, who used to live here, and what language the settlers spoke. Toponyms may also tell us something about past religious distributions. Catholic settlers in North America, for example, had a propensity to bestow religious names on their settlements more so than Protestants, no doubt in part to solicit the protective favor of the Almighty in an often-difficult frontier setting. Thus, towns named for saints abound, especially in Quebec and California (San Diego, Santa Barbara, San Jose, San Francisco, and so on). In another example, there are two streets — Kolob and Nephi — side by side in my city. This tells me that some Latter-day Saints had something to do with the founding of that neighborhood.

Place names may also provide philosophical insights. For example, about two centuries ago American culture was affected by the *Classic Revival*, which involved a new reverence of ancient Greece and Rome. One manifestation is the existence in Upstate New York of literally dozens of cities and towns that were named or renamed in accordance with the classical theme. Examples include Syracuse, Rome, Utica, Ithaca, and Romulus. There's even a Florence in South Carolina, but that city eschews the classical for a more mundane reason. It's named after a railroad executive's daughter.

TIP

Toponyms can also speak to people's aspirations. My favorite local town name is Prosperity. That sure beats its original moniker, Frog Level. Another is Hope, Arkansas. Who wouldn't rather live in a prosperous place or one that was hopeful? That's all well and good, except Hope isn't about being hopeful. It's named after a railroad executive's daughter, too.

One of the most maddening things about toponyms is that they can be literally changed overnight, immediately rendering millions of maps and atlases out-of-date. The change of Burma to Myanmar and Zaire to Congo are examples. Prior to its dissolution, the USSR contained an estimated 20,000 places named for Stalin — mountains, cities, alleys, you name it (literally). When Stalin's legacy fell out of favor, so did toponyms in his honor. Few remain.

Creating a Single Global Culture

Is the number of cultures on Earth today increasing or decreasing? Will the cultural mosaic of your grandchildren's world have more pieces or fewer pieces? That is, will the world map contain more culture areas or fewer culture areas than exist today? The search for answers reveals powerful opposing sets of social forces.

Promoting cultural divergence

Cultural divergence is the concept and process of culture creation. I have shown how this works by virtue of isolation, adaptation, and barrier effects, as well as by the just-mentioned processes of nationalistic aspirations and heritage preservation. Because of modern means of communication, the first three of these are not nearly as powerful as they once were, though they are still players in culture creation. At the same time, cultural fads and technology changes are occurring with much greater frequency than ever before and help to create peoples who are culturally different from neighboring groups. Consider also the human spirit that rejects a world in which everyone looks and acts the same. Cultural divergence may have slowed, therefore, but no evidence suggests that a world of many culture areas will reduce to a single culture area any time soon.

Promoting cultural convergence

It seems pretty clear, however, that in the present day more cultures are dying out than getting born. As many as a thousand languages are likely to disappear in your lifetime because they are not being spoken by the next generation who are instead learning and speaking another language. This is an example of *cultural convergence,* the concept and process of culture destruction and attendant "convergence," or coming together of the world's people in the cultural sense.

This is largely happening because communication is overwhelming the traditional factors that isolated people and encouraged cultural differentiation. The key players in this trend are improved transportation (especially the advent of road-building in remote areas); expansion of social media, television, movies, and the internet; increased literacy and access to printed media; rising tourism due to increased disposable income and leisure time; rising foreign trade; the growing number of multi-national corporations; and desire for economic development.

In the future, culture will continue to be the spice of life and place, even if the range of flavors is significantly less than we have today.

> **IN THIS CHAPTER**
>
> » Dividing the world in different ways
>
> » Feeling tension over turf
>
> » Asking the question: Does size and shape (of countries, that is) matter?
>
> » Rigging an election and getting away with it

Chapter 14
Good Fences Make Good Neighbors

You may have head that line before from Robert Frost's poem, *Mending Wall* – "Good fences make good neighbors." The line basically means, "I know what's yours, and you know what's mine. Stay on your side, and we'll be fine." OK, so maybe I won't make my next career as a poet, but you get the point. We carve up the world with a bunch of imaginary lines for a variety of reasons.

You can divide up space and set up different types of boundaries in many ways. In addition to living in the United States, I live in the State of South Carolina and the City of Columbia, both of which have separate boundaries. I also live within a particular school district, police precinct, City Council District, U.S. Congressional District, South Carolina State Senate District, and so on. They too have boundaries, and none are the same. But that's just me. What about you? How many political lines on the map enclose your residence? And the government isn't the only one that has this "thing" about dividing space. The phone company has divided the country in area codes. The postal service has done the same, only their lines have different placements and define different areas — zip codes. Maybe you belong to a church that divides space into dioceses, presbyteries, or something else. Maybe you work for a company that divides space into sales territories. The bottom line is that Earth is divided by lines of all kinds.

REMEMBER

As all of these instances testify, *Homo sapiens* have proven to be a rather territorial species (as illustrated by the wall in Figure 14-1). Indeed, one of the most profound things we humans do to our Earthly home is divide it up among ourselves, thereby creating regions of sovereignty, control, or administration. We do this by drawing lines on a map, which begs a rather basic question: Where do you draw the line? This is a central issue of political geography, which is concerned with division and control of Earth's surface and is the subject of this chapter. In writing this chapter, I, too, had to draw the line — that is, make decisions regarding what to include and what not to. In the end, I decided to focus on two topics, political borders and voting district boundaries, which I believe are particularly timely.

FIGURE 14-1: The Great Wall of China is an enduring symbol of the human fondness for dividing and controlling Earth.

(© aphotostory / Adobe Stock)

Drawing and Re-Drawing the Boundaries of the World

A few years ago, I gave a short lecture to school teachers in an attempt to show them how using geography could add a new dimension to their history courses. I tossed a map of Europe on the screen and the murmurs followed. "That looks like Europe, I think, but all the lines are wrong." "Yeah, you can see Italy — the one that looks like a boot — but Germany looks pretty strange. Is that Germany? I am not sure." What confused the group is that the map showed country borders from 1945, not the present. I was showcasing data from the Holocaust, a rather grim subject to be sure, to make a point. Teachers tend to emphasize Germany in that discussion, but data on human losses shows both Poland (the location of Auschwitz) and Russia to be very important and tragic places, too.

What should be obvious to people but is often forgotten is that borders change.

Dozens of countries have come and gone over the centuries. Ones existing for hundreds of years rarely have the same borders now that they began with. The Soviet Union, East Germany, and Czechoslovakia are on a list of the disappeared, but you can also add the newly born like South Sudan or Timor-Leste. And the process continues today. Kurds want to be free of Turkey and Iraq. The Tamils want to be separate from Sri Lanka. Basques in Spain and France want a country of their own. And on it goes. If past is prologue, then I have no doubt that a comparison of today's world map with future ones will reveal a mix of new nations and lost ones.

Tensions over turf, like debt and taxes, never seem to go away. For starters, some ethnic group is always clamoring for independence or wanting to draw or re-draw boundaries in a certain part of the world. Without taking sides in such matters, let me simply propose a singular reason for much of this contentiousness. Namely, the people who drew most of today's international boundaries came from somewhere else, and they often gave little or no thought to the possible long-term consequences to the native population.

Typecasting Boundary Lines

Insight into the nature and consequences of boundary making is provided by Figure 14-2. Assume it is a map from more than a hundred years ago that shows the boundaries of five ethnic groups labeled A–F, as well as a river and a line of longitude. Also assume that you are an official who, in the language of the previous paragraph, comes from somewhere else, and must decide where to place an international boundary within the area shown in Figure 14-2. Three options are available, and they correspond to the features just mentioned. They are:

- » **The Ethnic Option:** The homelands of five ethnic groups, indicated by the letters A–F in Figure 14-2, occupy the area in question. One option is to divide the area on the basis of the ethnic boundaries highlighted on the map.

- » **The Landform Option:** A fairly large river flows through the area. A line running down its middle could be used as the border.

>> **The Graticule Option:** If you read Chapter 3, then you may recall that *graticule* refers to Earth's grid of latitude and longitude. A line of longitude runs through the region. This line potentially could serve as the international boundary.

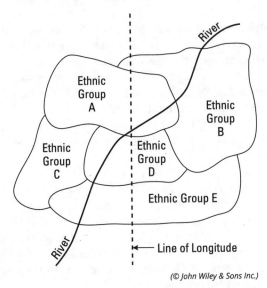

FIGURE 14-2: Three border options in a hypothetical region.

(© John Wiley & Sons Inc.)

I didn't simply dream up these possibilities. Instead, the three options represent the three kinds of borders that political geographers most commonly identify. Each type is discussed in greater detail in the following sections.

Ethnic boundaries

REMEMBER

This one's a no-brainer. Ethnic boundaries separate the homelands of neighboring ethnic groups. They are also probably the oldest kind of boundary there is. Viewed against the broad sweep of human history, the concept of a country with recognized borders and a centralized government is a fairly recent invention. Much older is the idea of an ethnic group possessing a distinct language, customs, and homeland.

For example, when the Europeans arrived in North America, they didn't find a country called the United States divided by lines into entities named New Jersey, Kansas, Arizona, and so forth. Those came much later. What they did encounter, however, was a continent inhabited by dozens upon dozens of Native American ethnic groups, each of which had its own distinct homeland rather like the areas labeled A through F in Figure 14-2. Similarly, when Africa was about to be divided into colonies, it did not consist of the countries we see today, but instead was composed of the homelands of hundreds of ethnic groups.

Typically, the boundaries of ethnic homelands in North America, Africa, and elsewhere were not fixed lines established by formal treaty. More often they were ill-defined zones that lay on the periphery of lands occupied by an ethnic group. Thus, they were fluid — prone to change. That is, if a particular ethnic group gained or lost territory as a result of some conflict, the ethnic boundary changed accordingly. For these two reasons — ill definition and impermanence — ethnic boundaries were rarely used as a basis for modern boundary making. Instead, the treaty makers who were responsible for most of today's boundaries opted for the types discussed in the next two sections.

Natural (physical) boundaries

REMEMBER

Natural boundaries are based on some aspect of *physiography* — physical landforms. While diverse aspects of landscape may be called upon, three are particularly common. The United States provides abundant examples of each type (see Figure 14-3):

- **Coastlines:** The Atlantic coast boundary of Florida and the maritime borders of Alaska are good examples. The shorelines of the Great Lakes states are relevant to some extent, though the border with Canada runs through the lakes rather than following the shoreline.

- **Rivers:** The New Jersey-Pennsylvania border (Delaware River); the Illinois-Iowa border (Mississippi River); and the Oregon-Washington border (Columbia River). Quite typically, the border is not the left bank or the right bank, but instead an imaginary line that runs down the middle of the waterway. This border can be problematic as we know that river courses change over time.

- **Ridgelines:** The crest of a ridgeline in the Bitterroot Mountains marks the border between Idaho and Montana. Similar crests in the Appalachian Mountains mark the boundaries between Virginia and West Virginia, and between North Carolina and Tennessee.

The principal advantage of natural borders is that they are based on things that can be readily seen, agreed upon, and mapped. In contrast to an occasionally questionable ethnic boundary, there should be no question as regards to the location of a river, coast, or ridge. They are not necessarily permanent, however, as noted for rivers. But the advantages of natural boundaries are powerful. As a result, many boundaries are based on this feature.

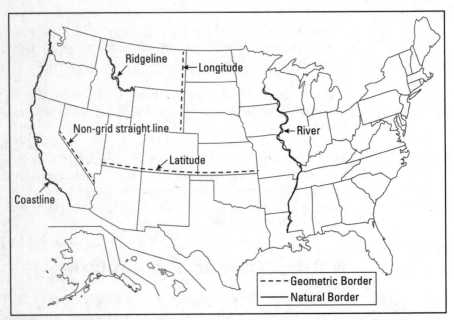

FIGURE 14-3: Examples of natural and geometric boundaries in the United States.

(© John Wiley & Sons Inc.)

Geometric boundaries

REMEMBER

Geometric boundaries are straight lines. In a majority of instances, they are lines of latitude and longitude. Several examples are evident on a map of the United States, including the latitude line that separates Utah, Colorado, and Kansas from Arizona, New Mexico, and Oklahoma (see Figure 14-3). Some straight lines are not part of Earth's grid, however. Perhaps the most noticeable U.S. example is the northwest-southeast line that constitutes most of the border between California and Nevada (see Figure 14-3).

Use of straight lines is not limited to the United States. Nearly all of the borders between the states of Australia are lines of latitude and longitude. In West Africa, similarly, the country of Mali has both a latitudinal and longitudinal border with Mauritania, and a non-grid straight-line boundary with Algeria.

The most attractive feature of geometric boundaries, as far as border-markers are concerned, is that they are based on abstract geometric space instead of the physical Earth or ethnic homelands. Therefore, they have a degree of permanency that the other types do not. Let it rain for 40 days and 40 nights, and the results may alter natural boundaries, but geometric boundaries will not be affected. Similarly, let every ethnic group on Earth have a bad day or whatever, and the results may alter previous ethnic boundaries; but again, geometric boundaries will be unaffected because they are abstract concepts. People can't pick up and move a line of latitude or longitude.

Living with the Consequences

After the boundary lines have been drawn, the citizens of those areas and the ones responsible for creating the lines have to live with the consequences. The world's international borders consist of thousands of linear segments that enclose some 200 countries. In some cases, the results have proven functional and harmonious. Other cases, however, are characterized by very different adjectives. Following are an overview and examples of some of the more challenging consequences of global boundary making.

Ethnic intrigues

As I explain earlier in this chapter, much of the world is divided on the basis of natural and geometric boundaries as opposed to ethnic ones. This is especially true of the many countries that once were colonial possessions. Over large parts of Earth, therefore, disparities exist between ethnic boundaries and country boundaries. Visual comparison of these border types in Africa, for instance, reveals two very separate sets of lines that seldom coincide. Though not necessarily bad, disagreement between these sets of lines has occasionally fostered significant ethnic and political tension in Africa and elsewhere.

REMEMBER

This unrest broadly falls into three categories. Before discussing them, however, I need to make a particular distinction between state and nation. For the purposes of this discussion, *state* is synonymous with country. It refers, therefore, to territory that is delineated by a border and has sovereign, independent status. *Nation*, in contrast, is synonymous with an ethnic group that has its own traditional homeland, history, and culture that causes its members to view themselves as distinct from other people. Sometimes, for example, particular Native American peoples are referred to as the Seminole *Nation*, the Hopi *Nation*, the Sioux *Nation*, and so forth. That is the meaning of "nation" used here.

If you think about terminology in this context, the United Nations should be called the United Countries or the *United States*. That might pose a problem, couldn't it?

Multi-nation states

A *multi-nation state* is a country that contains the homelands of more than one nation. Nigeria, whose population consists of the homelands of over 250 national (ethnic) groups, is a prime example. The main problem in this situation is trying to create a common identity and sense of unity among peoples who speak different languages, practice different religions, and, perhaps worse, have a history of mutual animosity, if not outright hostility. Consider the following scenarios, any one of which might result in a level of ethnic tension that results in unrest:

>> Appointed government officials, top military leaders, or chiefs of police are members of an ethnic group that is distrusted by members of other ethnic groups.

>> Political parties form along ethnic lines, heightening prospects that election results may lead to friction between nations.

>> Government declares official languages or religious holidays that promote the culture of one or more nations but not others.

>> The capital is located in (or relocated to) the homeland of a particular nation that is distrusted by others.

>> Economic development projects are disproportionately located in the homeland of a particular nation, leading other groups to feel they are being deprived of their fair share of economic opportunity.

Multi-state nations

A *multi-state nation* consists of an ethnic group whose people and traditional homeland lies within two or more countries. This situation may lead to *irredentism*, a foreign policy in which one country seeks to acquire foreign territory for the purpose of unifying a nation. The term dates from the 1870s when Garibaldi and his followers sought to incorporate "unredeemed" (*irredenta*) Italian-speaking territories into the Italy that you see on today's map. The potential problem, of course, is that a country that contains the coveted land is not likely to give it up without a fight.

The Somalis provide a case study. This ethnic group's traditional homeland includes much of the "horn" of East Africa (see Figure 14-4). In the latter part of the 19th century, Ethiopia, Italy, and Britain signed a treaty that divided this region between them. Following de-colonization, the present Somali Republic was established out of the territory controlled by Italy and Britain. But a large portion of the traditional Somali homeland, known as the Ogaden, remains under Ethiopian control.

Somali politicians have argued that dividing up their traditional homeland among foreigners without Somali consent was not fair in the first place; thus, the Federal Republic of Somali they say rightfully ought to annex Ogaden by any means necessary. Naturally, this argument echoes rather strongly among Somalis but alarms Ethiopia, which views the Ogaden to be part of its territory by virtue of a legitimate treaty. The result has been on-again, off-again debate and occasional armed conflict between the two countries over control of the Ogaden. Even the former colonial borders continue to threaten Somalia's stability as Somaliland, the former British area in the north, attempts to declare statehood.

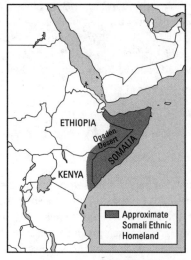

FIGURE 14-4: The horn region of East Africa.

(© John Wiley & Sons Inc.)

State-less nations

TECHNICAL STUFF

A *state-less nation* is an ethnic group whose traditional homeland is currently under the control of another country. Virtually hundreds of ethnic groups and their homelands are in this category. These peoples usually adopt one of three stances:

» **Do nothing:** A majority of stateless nations are comparatively docile about this particular circumstance. While some members may desire statehood, the overall level of support for autonomy has not — for whatever reason — reached a critical level.

» **Seek greater autonomy:** Some ethnic groups seek a greater degree of autonomy for their homelands as opposed to complete independence. The Welsh and Scottish peoples are examples. Although some hard-core nationalists long for complete separation from the United Kingdom, the majority — however small — favor continued union complemented by a high degree of self-rule.

CHAPTER 14 **Good Fences Make Good Neighbors** 247

THE KURDS

The Kurds, who number an estimated 30 to 40 million, are an example of both a multi-state nation and a stateless-nation. In the early years of the 20th century, their traditional homeland, Kurdistan, lay mainly within the Ottoman Empire, which ceased to exist after World War I. Kurdish nationalists hoped that an independent Kurdistan would be born of post-war boundaries drawn by the victorious parties. Sadly, for them, that did not come to pass. Instead, the modern map of the region gradually emerged, in which Kurdistan is now spread over six countries (hence, a multi-state nation), principally Turkey, Syria, Iraq, and Iran. The Kurds are not a majority in any of these countries, but they have not given up hope of independence for their homeland. Until that happens, however, they remain a state-less nation.

» **Clamor for independence:** Members of some ethnic groups seek complete independence of their homelands from the country to which they are presently bound. People in predominantly French-speaking Quebec, for example, have sought sovereignty for their province and taken the issue to the ballot box, where it failed to pass. In other cases, however, people are so aggrieved by the perceived injustice of their circumstance that they feel any means of achieving their goal — including extremely violent acts — are justified. Examples of such groups include the Basques, Tamils, and Kurds (take a look at the nearby sidebar on the Kurds for a more in-depth discussion).

Positional disputes

REMEMBER

Positional disputes concern disagreements over the location of a border. One of three factors typically underlies the resulting tensions.

Poor definition

In some cases, disagreement arises between neighboring countries concerning borders (often inherited from colonial times) that were poorly defined (see the sidebar "You're out of line!"). Here are two examples:

» **Kashmir:** At the end of the colonial era, a boundary commission sought to divide this Himalayan region between Pakistan and India, taking into account ethnic divisions. Nationalists on both sides disagreed over the recommended partition of the region, which is also the source of rivers that supply irrigation water to the countries' agricultural lowlands. Control of Kashmir continues to be a major bone of contention in Indian-Pakistani relations.

> ### YOU'RE OUT OF LINE!
>
> Don't be fooled into thinking that positional disputes can't happen in your own backyard, because it has in mine. Tennessee and Georgia, two of the oldest existing states in the United States have been fighting over their northern border regarding access to water, specifically the Tennessee River near Chattanooga. Georgia argues that the border was erroneously drawn too far south and if drawn correctly would include part of the river in Georgia. The upshot? A growing and sprawling Atlanta (in Georgia) has access to a new water source.

- **The Southern Andes:** During Spanish colonial administration, ridges and watersheds were used to define what would become the long border between Argentina and Chile. The area was poorly mapped, however, resulting in conflicting interpretations of boundary locations. Segment by segment, disputes over different parts of the boundary were resolved by direct negotiations, third-party mediation, the International Court of Justice, or — in one particular case — the intervention of the Pope.

  The north was no better with Peru, Chile, and Bolivia. Who needs to know the exact line in the middle of a desert, the Atacama? That's fine in theory until you find something you want, like copper or nitrate deposits. Good fences matter then, don't they?

Invisibility

In some cases, determining where a border is located is impossible. Such situations are potential powder kegs. Here are two examples:

- **Amazonia:** The periphery of the Amazon basin contains a couple of boundaries that pass through trackless expanses of jungle and rainforest. Thus, while the line is plainly visible on the map, no clue to its whereabouts can be found on the ground. The potential for trouble has been highlighted by discovery of oil and natural gas in the region. More advanced line-making using GPS (see Chapter 5) has been a help.

- **Arabian Peninsula:** Because of the vast emptiness of the Arabian Desert, the boundaries between Saudi Arabia and Yemen, Oman, and the United Arab Emirates have been approximated but not delineated, a case similar to that in the Atacama. Thus, on most detailed maps of Arabia, the borders are shown as dashed lines accompanied by "boundary undefined" or words to that effect. So far, the countries have felt no major ill effects, but situations like these have a way of turning nasty like when Yemen had commercial oil exploration take place on their behalf on "Saudi" territory.

"Acts of God"

An "act of God" is a cataclysmic event that results in disagreement over the location of a border. The most common (but hardly everyday) occurrence involves a boundary river that changes course following a major flood. The Mississippi River, for example, has changed course several times since it was established as the boundary between Arkansas and Mississippi. As a result, some land that used to be on the Arkansas side of the river is now on the Mississippi side, and vice versa. Similarly, a small portion of Texas ended up on the Mexican side of the Rio Grande years ago following a freak flood.

If you want to have some real fun, grab a map and look at Kentucky near the Missouri town of New Madrid (sound familiar? See Chapter 6.). That peninsula belongs to Kentucky, right? Or should it be Tennessee?

If a parcel of land "changes sides," then is it in the same jurisdiction it was before, or in a different one? Basically, it all depends on how the treaty was written. In any event, however, you can no doubt appreciate the potential here for political intrigue.

Functional disputes

Functional disputes concern policy disagreements regarding immigration, trade, customs fees, or some other matter(s) that apply to a particular boundary. Here are two examples.

United States–Mexico

Trade and daily exchanges across the U.S.–Mexican border have served to bring the two countries closer together than ever before. At the same time, however, immigration issues, terrorism concerns, and drug trafficking have caused many Americans to advocate a "hardening" of this border to make crossing more difficult. Thus, contrasting forces are respectively advocating a more open boundary and a less open boundary. On the American side, border patrols have been increased and physical barriers erected along some stretches. Critics view these physical barriers as virtual Berlin Walls, while proponents view them as necessary steps to maintain national security. Mexicans tend to view America's "hardening" policies as counterproductive to good relations and often contrast it to the U.S.-Canadian border, which is very open and largely non-policed.

United States–Canada

In the course of energy production at coal-fired power plants and goods at industrial plants, the United States produces air pollution. These pollutants can contain sulfur, an element that contributes to rain acidification (see Chapter 18). But this is more than an American problem as weather patterns see to it that Canada's forests are showered with lower pH rainfall causing substantive harm. This is an example of a *transboundary* issue.

No one is arguing over the location of the border but rather about a function across it, in this case poor air quality and acid precipitation. To deal with this issue, the two countries worked in the 1990s on an air quality agreement to reduce pollutants of concern to both the United States and Canada.

Resource disputes

REMEMBER

Occasionally an international boundary cuts across a natural resource whose use becomes a bone of contention between the neighboring countries. For Egypt and Ethiopia, the concern is water from the Nile River. Israel and Palestine? Water again, but now the source is the Sea of Galilee and the Jordan River. And what to do about nomadic peoples and grazing cattle in South Sudan that really don't care where you draw your imaginary lines?

Here are two additional examples.

The Rumaila Oil Field

This oil field straddles the border between Kuwait and Iraq, as shown in Figure 14-5a. Though it lies principally within Iraq, the nature of oil means that wells drilled in Kuwait can potentially suck up this resource as fast or faster than wells across the border. Sensitivities regarding "fair share" have run high. Indeed, increased production by Kuwait was one of the rationales offered by Iraq for invading its neighbor in 1990.

The Georges Bank

For many years this historically rich fishery, shown in Figure 14-5b, straddled a disputed marine boundary between the United States and Canada. Like the preceding example, each country had access to it and was sensitive to the notion of "fair share." The fish, who could care less about the boundary's location, habitually swam back and forth across the disputed area. That meant that over-fishing on one side of the border, while relatively enriching that country's fishermen, had the potential to deplete the entire fishery and, at the same time, deprive the neighbor of "fair share."

Remember the Exclusive Economic Zones discussed in Chapter 8 (if you've not read that chapter yet, flick back to get the general idea)? Those maritime boundaries come into play here with both countries claiming an overlapping 200 nautical mile zone. Adjudication in the International Court of Justice helped smooth over the dispute. But there is now a new political wrinkle. In 2016, then-President Barack Obama declared an ocean monument to protect sea life biodiversity, and it encompasses part of the Georges Bank and thus the ability to fish there under a new set of rules. Political lines governing behavior and who owns what are everywhere.

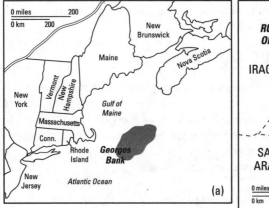
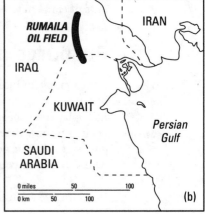

FIGURE 14-5 A AND B: The Rumaila Oil Field and Georges Bank.

(© John Wiley & Sons Inc.)

Sometimes resource disputes are resolved amicably by direct discussion between the parties involved, or with the assistance of an organization with which both are affiliated — like the UN or NATO. On the other hand, sometimes the problem simply festers or is resolved by armed conflict. Thus, resource disputes have diverse outcomes — some peaceful, some not.

Land-locked states

REMEMBER

A *land-locked state* is a country that is completely surrounded by land that belongs to one or more neighboring countries. Thus, it has no port of its own and therefore no immediate access to international waterways. Three dozen land-locked states are located throughout the world, including about a dozen each in Africa and Asia, two in South America, and the rest in Europe (see Figure 14-6). At least three potential problems emerge from this status.

Inhibited access to foreign markets

Lack of an ocean boundary carries with it a lack of unfettered access to foreign markets. You may be thinking, "Who needs a sea port if you have an airport? Why not import and export everything by airplane?" Aside from the fact that you can't just go flying over another country's territory without permission, transporting bulk trade items by ship is far and away the cheapest mode of importing and exporting goods. For all intents and purposes, therefore, access to foreign markets requires access to the high seas.

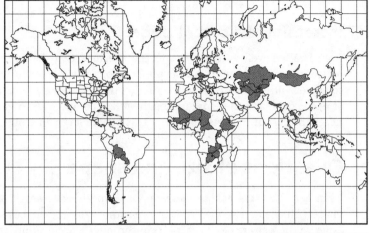

FIGURE 14-6: The world's land-locked states.

(© John Wiley & Sons Inc.)

In recognition of the peculiar position of land-locked states, international law has long granted them the *right of innocent passage* in matters of foreign trade. That is, land-locked countries have the right to transport imports and exports across the territory of neighboring countries provided such action does not pose a threat to the security of the host country or violate its laws. A land-locked state may even be allowed to lease dock facilities in a neighboring country, thereby giving it a port it would otherwise lack.

TIP

Being landlocked may be no better illustrated than being Bolivian. Losing their coastline to Chile after the War of the Pacific (1879–1884; ironically over a border tariff dispute), Bolivia finds itself shut off from the world's waters except for its own Lake Titicaca. But Bolivia has something it wants to get off its chest, or rather out of the ground, for the rest of the world: natural gas. This needs to be piped out somehow and the closest path would be through Chile. Historic animosities have made this impossible however, so the plan (still off in the future) would be to go through Peru — a greater distance and at greater expense. Borders and access matter.

Treading softly in foreign relations

Land-locked states find it beneficial to tread very softly in foreign affairs, especially in matters that pertain to neighboring countries that grant them the right of innocent passage. Tick off thy neighbors, and they may cut you off from the sea, bringing your ability to import and export to a halt. Obviously, the effect of this type of action on your economy may be devastating. So, make nice, even if you despise them. And you can scream all you want to about your rights. It turns out that the vaunted right of innocent passage is little more than some touchy-feely gentleman's agreement.

Increasing the cost of trade

Even if all is well with respect to the previous two points, being land-locked means that your imports and exports absorb extra transportation costs versus countries with a coast (see Bolivia, earlier in this section). That increases the cost of its exports (and therefore makes them less competitive in global markets) and raises the cost of imports, too. Neither is beneficial to an economy.

Questions of size and shape

I can hear the "tsk-tsking" already, but I'm going to say it anyway. Does size matter? How about shape? That is, are some sizes and shapes better than others as far as the welfare of a country is concerned? Clearly, countries come in all sizes and shapes.

REMEMBER

Russia, the largest country on Earth in area, contains nearly 6.6 million square miles. On the other end of the spectrum lies Monaco, which barely covers 1 square mile. Lots of other sizes can be found in between. Logically, we all probably assume that bigger is better. More territory usually means greater prospects for mineral wealth, more agricultural land, and so forth. But it may also mean remote regions, lots of boundary to defend, restless ethnic groups, large uninhabited areas, and umpteen challenges to cohesiveness. A small, compact country may face none of these challenges.

And that leads to a consideration of shape. Political geographers tend to agree that shape can make a difference in some cases. The following sections discuss the five types of country shapes, which are shown in Figure 14-7.

Compact states

A *compact state*, such as Uruguay, is roughly circular and is generally thought to be the ideal shape. Assuming absence of mountain ranges or other physical barriers, parts of the country may be readily interconnected with comparatively little investment in roads, railways, electronic cables, and so forth. Also, compactness minimizes the amount of international border that may need to be defended and may also minimize the extent of internal ethnic boundaries as well. All in all, compactness tends to promote internal cohesion. Imagine, too, if the capital city were in the center. Communication as well as the other benefits mentioned doesn't get much easier.

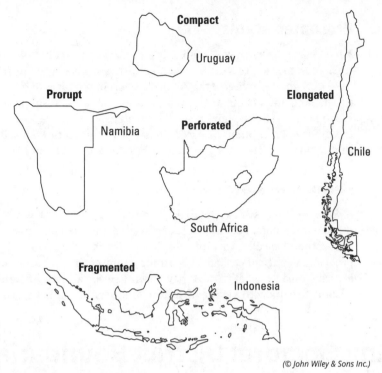

FIGURE 14-7: Different shapes of countries.

(© John Wiley & Sons Inc.)

Fragmented states

A *fragmented state* consists of scattered, disconnected pieces of territory, typically because all or part of the country is a chain of islands. Indonesia is an example, and the United States is another (Hawaii? Alaska? Puerto Rico? Guam?). Fragmentation makes it more difficult for government to impose central control and to promote cohesion and interaction between different parts of the country. Also, different fragments may be homelands of different ethnic groups that have little in common. Moving people into these spaces has been one strategy by some countries to lay claim to these lands.

Elongated states

An *elongated state,* such as Chile or Norway, is long and thin. In most cases this shape is detrimental to national cohesion. Some places may be remote from the capital, a lot of border needs to be defended, and elongation often results in a more diverse populace than a compact shape. On the other hand, elongation may result in climatic variety that is a plus to the country's agriculture and economy.

Prorupted states

A *prorupted state*, such as Namibia, is generally compact but has a noticeable protrusion. In the case of Namibia, a land protrusion gave Germany (once the colonial ruler in Namibia) access to the Zambezi River. In the cases of other countries, such as Thailand and Myanmar (Burma), proruption is the result of a peninsula. This shape is generally thought to be disadvantageous because it isolates a portion of the country. Proruptions also exist in national sub-units. Pennsylvania has a proruption in the northeast by the city of Erie for access to that Great Lake.

Perforated states

A *perforated state*, such as South Africa or Italy, is one that has a hole punched into it by another state. For Italy, it is San Marino and the Vatican. For South Africa, it is Lesotho. Basically, each of these countries is landlocked and comes with the problems identified earlier. This arrangement creates a host of diplomatic challenges. Consider what it must have been like for a black African government in Lesotho to work with a white African government in South Africa during apartheid.

Drawing Electoral District Boundaries

A very different, yet very important, component of political geography concerns the drawing of voter district boundaries. The governments of the United States, Great Britain, Canada, Australia and other democracies include legislative bodies elected on the principal of "one person, one vote" — equal representation for all citizens. This in turn necessitates creation of electoral districts that contain roughly the same number of people. Regarding the United States, drawing the boundaries of Congressional Districts immediately comes to mind, and so, too, the outlines of other political geographic entities — such as city council districts, state senate districts, and state house districts.

TECHNICAL STUFF

Making "one person, one vote" a geo-political reality involves four steps:

1. **Count:** Conduct a nationwide census that is as accurate as possible. The United States does one every ten years. Indeed, the Constitution requires it — not simply to count Americans, but for the specific purpose of acquiring data that permits creation of equal Congressional Districts.

2. **Map:** The next step is to map the collected information with as much geographic accuracy as the data permits. That means, for example, that when it comes to a large city, boundary-makers need population statistics on a block-by-block basis.

3. **Calculate:** The next step is to calculate the number of people who should be in each district. This is a fairly simple matter. Say, for example, that you are considering a given state's house of representatives that has 100 seats. In that case, you take the state's population total, as revealed by the census, and divide by 100. The numerical result is the number of people who *should* reside in each district.

4. **Draw:** The final step is to draw the map. In the case of the above, that means dividing the state in question into 100 polygons, each of which (as closely as possible) contains the same number of people, as calculated in the previous step.

Gerrymandering: Rigging the outcome

"Figures don't lie, but liars do figure." Maybe you've heard this cute little phrase, which alludes to the possibility of using or manipulating data to serve a particular interest. As regards political geography, it may manifest itself in *gerrymandering*, the drawing of voter district boundaries to benefit a particular group or political party.

Gerrymander pays mocking tribute to Elbridge Gerry, who, as governor of Massachusetts in 1812, sought to redraw the political district map of his state in such a way that assured the continued dominance of his political party in the statehouse. One particular political district was drawn in a narrow, sinuous shape that, according to one observer, "looked like a salamander." That prompted an observer to respond that it was no salamander, but instead a "gerrymander." The moniker stuck.

Then and now, gerrymandering typically occurs because a political party or voting bloc wants to maximize its representation in a legislative body and minimize that of another group. Figure 14-9 (later in this chapter) shows the two most common types of gerrymanders. By way of background, each diagram is characterized by the following:

>> The area in question consists of a city (inner circle) and its surrounding countryside.

>> An accurate census has placed the region's total population at 50,000.

>> Two categories of voters live in the district, the As and the Bs. You may think of them as members of different political parties or different racial or ethnic groups or whatever. The key thing is that they are on opposite sides of the political fence.

>> Every member of Group A lives in the city. Every member of Group B lives outside it. (This is a gross simplification of typical voting bloc geography, and is made strictly to facilitate understanding of gerrymandering techniques.)

>> Representatives of Group B are currently in power. They will resort to any legal means to stay there, including drawing voting district boundaries designed to promote election of their candidates.

>> The area in question is to be divided into five political districts, each of which contains 10,000 people.

"Diluting" a voting bloc

In Figure 14-8a, Group A consists of 20,000 people and Group B 30,000. Because Group A accounts for two-fifths (40 percent) of the voters, you may expect them to capture two-fifths of the seats being contested. That is, they may win two of the five seats. But they can quite possibly get nothing. In Figure 14-8a, members of Group A have been spread out over the five districts such that they are in the minority in every instance. In a manner of speaking, their voting power has been "diluted." Assuming voters vote solely along party lines, the results will be 5-0 in favor of Group B.

"Packing" a voting bloc

In Figure 14-8b, Group A consists of 30,000 people and Group B 20,000. Because Group A now accounts for three-fifths (60 percent) of the voters, you may expect a totally different electoral outcome. But there is a way to draw the district boundaries that ensures political dominance for Group B. And because these are the people who are drawing the lines, you can bet they will do it. In the diagram, two inner-city districts consist entirely of members of Group A. But by "packing" so many members of Group A into these two districts, the result is that they are in the minority in the other three. Assuming once more that voters vote solely along party lines, the results will be 3-2 in favor of Group B. Stated differently, the bloc that consists of two-fifths of the voters will control three-fifths of the seats.

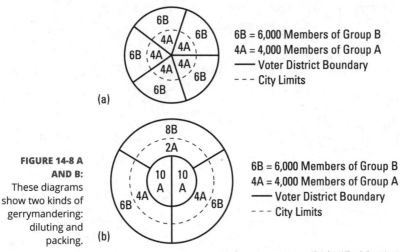

FIGURE 14-8 A AND B: These diagrams show two kinds of gerrymandering: diluting and packing.

(© John Wiley & Sons Inc.)

Meeting the letter and spirit of the law

You may be thinking, "That's not fair!" And you may be right. But more important is the question, "Is that legal?"

Some people say "Yes." Clearly, every political district in Figure 14-9 complies with the letter of the law. That is, the number of voters is the same in each district, so the constitutional concept of equal representation has been met.

Other people, however, believe that elected officialdom should be as diverse as the represented population. In their view, gerrymanders violate the spirit of the law, lead to a distrust of government, and may ultimately undermine the very fabric of American democracy.

A practical solution, in their view, is to create voting districts drawn to virtually guarantee the election of members of particular ethnic groups (Figure 14-9). Opponents view this solution as blatantly unconstitutional and say districts drawn to encourage election of particular peoples are as offensive as lines drawn to exclude them.

The courts, for their parts, have flip-flopped on the issue. Quite literally, the jury is still out. In the meantime, passions run high among folks who feel strongly about this issue, which is as much about geography as it is politics and law.

CHAPTER 14 **Good Fences Make Good Neighbors**   259

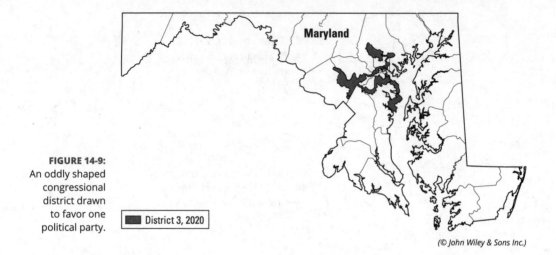

FIGURE 14-9: An oddly shaped congressional district drawn to favor one political party.

4 Putting the Planet to Use

IN THIS PART . . .

Earth is more than our home. It's also the source of our daily bread. Not to mention our daily aluminum can, job, gallon of gas, and everything else that sustains us. And at the end of each day, we take out the trash, which gets picked up by a big truck and goes . . . where?

Humans cannot help but interact with Earth's surface. On it we plant our crops, locate our places of business, and build our settlements and cities. From it we draw a host of resources that we either consume as they are, or transform into a host of products that define the various global economies and standards of living. And at the end of the day, we take out the trash. In doing all of these things, we sometimes modify our environment in ways that pose significant challenges for present and future generations.

In this part, you will learn some of the key concepts and phenomena that concern human use and misuse of planet Earth. Economic geography, resource geography, urban geography, and environmental geography will be the specific topics. All have a central bearing on this part's central theme — how humans interact with and impact the only planetary home we are likely to know.

IN THIS CHAPTER

» Finding different ways (and places) to make a living

» Choosing the best business location

» Navigating transportation systems and networks

» Moving to cyberspace or real space?

Chapter 15
Takin' Care of Business

When I was about 10 years old, I would sit on my front sidewalk under the landing approach of San Francisco International Airport. I would try to guess what kind of plane was overhead, say a Boeing 747, and whether I could identify the carrier. Oftentimes it was United Airlines. This made a lot of sense because this airport was one of the airline's major *hubs* (centralizing connection points for planes from all over). I'd wonder what was on each of those flights. Usually it was people, but cargo planes and all their goods routinely shot by as well. "What were they shipping?" "Where did it come from?" "Why did they send it by plane and not a boat?" All these questions about people, places, and their linkages populated my pre-teen brain, even working its way into a school project where I bothered some poor travel agent on the phone about the cost of plane tickets around the globe.

REMEMBER

The business of transportation is one aspect of *economic geography*. Geographies of trade, geographies of industry, geographies of agriculture, geographies of finance and real estate, and resources, and so on — each have spatial aspects. Think about the geography of services and where you would need to optimally locate an Amazon fulfillment center to get an item to a customer as quickly as possible. Where should we locate a Google data farm? Or a new Volvo factory that only produces vehicles with electric engines? (And yes, other companies and products do exist!)

Economic geography, the subject of this chapter, is a major subfield of geography and concerns the location of different means of livelihood. For a long time, economic geography was largely concerned with knowing where products came from and why. That's still important. This chapter is focused primarily on the economic geography of business activity. Remember that old saying about the three most important things for business? "Location, location, location?" If you don't believe that's true, why do so many apps on your phone ask to be geo-enabled? Businesses need to know where their inputs are, where their customers are, and which places charge more taxes, among other things. To be successful then, two modes of geographical thinking are crucial: an analysis of where things *are* located, and where they *should be* located.

This chapter addresses both perspectives with emphasis on where things *should be* located.

Categorizing Economic Activity

REMEMBER

Economic activities are all of the ways that people the world over make a living. Obviously, a complete listing plus descriptions would be encyclopedic. To simplify matters, and as a first step toward geographical analysis of livelihoods, geographers commonly characterize economic activities as a four-part scale whose components — primary, secondary, tertiary, and quaternary activities — become progressively more complex and detached from the natural environment.

Primary activities

Primary activities involve harvesting Earth or extracting raw materials from it. Examples include agriculture, forestry, fishing, and mining (Figure 15-1). If you think hard here, you realize that much of Earth's human landscape is really just Earth turned inside out. Rocks dug up become buildings. Seeds in the ground become timber and then become houses. Petroleum in the ground is pumped out to drive machines that build ever more things on Earth's surface.

Primary activities are resource-oriented and extraction intensive. In each case, the relationship with the environment is intimate and location-specific — meaning that you can't grow bananas, catch tuna, or mine coal just anywhere. Instead, you can only extract resources where they occur or, in the case of agriculture, grow particular crops where conditions are right. As a result, location options for many primary activities are limited.

FIGURE 15-1: This open pit gold mine in Australia is an example of primary economic activity.

(© John / Adobe Stock)

Secondary activities

Secondary activities change the form of raw materials in ways that add to their value. Examples include processing and manufacturing. Thus, a cheese maker may transform milk into cheddar, which also has a different form and higher price than its raw material. Similarly, an automobile manufacturer might bring together headlights, tires, upholstery, spark plugs, and so forth at an assembly plant and transform them into a vehicle that has a different form and higher cost than the sum of its parts.

As these examples illustrate, the raw material required by a secondary activity could be either a natural substance (milk) or products of previous processing or manufacturing (tires). Those secondary activities that require natural raw materials usually are located close to relevant primary activities in order to minimize transportation costs. In contrast, those secondary activities that require processed or manufactured raw materials are less intimately tied to the natural environment and have more location options.

Tertiary activities

Tertiary activities provide goods and services. They include retailers and wholesalers, who link producers and consumers; and personal and professional workers, including clerical personnel, financial services, entertainers, lawyers, and doctors.

CHAPTER 15 **Takin' Care of Business** 265

Tertiary activities typically have no direct relationship to environmental items (such as farming and fishing) and therefore have fewer constraints on possible locations.

The type of service is key, of course. Few architects have drop in traffic so they can be located just about anywhere that the population is large enough to support the service. But a gas station? There's a very specific and advantageous location for that service.

Quaternary activities

Quaternary activities involve research, information processing and dissemination, and organizational administration. "White-collar professional" probably summarizes this category as well as any single term can. Typically, these activities are urban-based and completely detached from the natural environment.

Activity distribution around the world

The relative presence of these four categories of economic activity varies geographically around the world. Generally, the economies of the least developed countries tend to have a very high percentage of their workforces engaged in primary activities, and progressively lesser percentages for the others (as shown in Figure 15-2). As national economies become more developed, the workforce tips toward the higher end of the range. And as a result, the most advanced economies tend to have a high percentage of their workforces in the tertiary and quaternary categories, and much smaller figures for the primary and secondary groups.

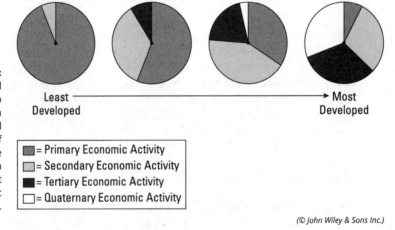

FIGURE 15-2: The generalized relationship between development and the proportion of the workforce engaged in different economic activities.

(© John Wiley & Sons Inc.)

Also, while these four categories of economic activity involve separate and distinct means of livelihood, they are nevertheless linked, especially in the more advanced societies. Thus, tuna caught by the crew of a trawler (primary activity) may be offloaded at a factory for processing and canning (secondary activity). Wholesalers then purchase large quantities of canned fish and distribute them to local grocery stores for sale (tertiary activity). Overseeing this buying and selling, as well as attendant marketing, are company executives (quaternary activity).

Putting Economic Systems into Place

In contrast to economic activities, which may be viewed as individual livelihoods, *economic system* refers to the organization and functional nature of a national economy as a whole. *Subsistence economies,* which are prevalent in the less developed countries, and *commercial economies,* which are prevalent in the more developed countries, are the two principal types of economic systems. Technically, a third category, *planned economies,* is characterized by government control of supply, demand, prices, employment, and distribution. Fewer of these planned economies exist today than did in the past, but even commercial economies have some planned aspects (such as farm subsidies or tax credits to stimulate production in a new energy sector).

Subsistence economies

Subsistence economies are characterized by consumption of goods by the people or groups that produce them. Buying and selling is rather limited, and tertiary and quaternary activities are rare-to-unknown. Traditional societies that rely on manual labor and possess low levels of technological sophistication are typical of this type of economy. Four kinds of subsistence economies are commonly recognized and are discussed in the following sections.

Hunters and gatherers

Hunters and gatherers subsist on wildlife, fish, and/or parts of plants (roots, nuts, or berries) that grow wild. Contrary to a one-liner about a very different kind of work, this is unquestionably the world's oldest profession. Indeed, eons ago, before the discovery of agriculture, everybody did this. Today hunters and gatherers are few in number and are found mainly in isolated areas of tropical rainforest.

Nomadic herders

Nomadic herders tend domesticated herbivores (including sheep, goats, cattle, and reindeer) that feed on natural foliage during the course of a controlled itinerary (wandering). *Steppe* (grasslands) and tundra are the traditional realms of these people, sometimes called *pastoral nomads*. The global population of these folk is rapidly decreasing as farming occupies more land, and as people turn to *ranching* — the raising of livestock in fixed locales for sale instead of consumption.

Also, governments in nomadic lands have engaged in campaigns to settle herders and bring them into the commercial (market) economy. Wandering people and cattle, as exist in South Sudan, can be unwelcome visitors to areas where national elites are hoping to extract oil.

Shifting cultivation

Shifting cultivation involves farming on cleared plots of forest until soil fertility plays out; after which, a new plot is cleared and the old one abandoned. This agricultural method is described in greater detail in Chapter 10. Millions of people in tropical parts of Latin America, Africa, and Asia engage in this endeavor for their livelihoods.

Intensive subsistence agriculture

Intensive subsistence agriculture is characterized by high amounts of labor brought to bear on relatively small land holdings. Cultivation of rice paddies, which require considerable inputs of labor for construction and maintenance, plus precise applications of water at precise times, is perhaps the quintessential example. Well over a billion inhabitants of South and Southeast Asia engage in this type of subsistence activity.

Commercial economies

REMEMBER

Commercial economies are characterized by production of goods and services for sale to others rather than for consumption by the producers themselves. Products do not enter into the distinction between subsistence and commercial economies and may, in fact, be the same. Rice, for example, is one of the principal foodstuffs that is grown and consumed by subsistence agriculturists. But commercial farmers, who eat little or none of their harvests, but instead sell it to companies whose boxes and bags of rice end up in your grocery store, also grow rice in quantity. As that example suggests, commercial economies may include primary economic activity, like agriculture. By and large, however, it's secondary, tertiary and quaternary activities (see "Categorizing Economic Activity" earlier in the chapter) that dominate.

BEING COMPLEMENTARY

Complementarity is the tendency for companies that produce goods and services for each other to cluster together, thereby maximizing mutual accessibility. New York City's Theater District, for example, consists of more than just venues for Broadway shows. Companies that act as agents for performers, produce sets, sell lighting and sound equipment, make costumes, coordinate ticket sales, and rent out sound stages and rehearsal studios can also be found on Broadway. You don't have to travel to New York to witness complementarity, however. It's a basic principle of economic geography. So if you live in a fair-sized city or town, chances are good you can find one or more examples of associated businesses that have clustered together for mutual benefit. For example, if your town contains a fair-sized hospital, then doctors' offices, insurance representatives, and medical laboratories are likely to be located close-by.

In commercial economies, laws of supply and demand determine price and quantity. Competition largely influences decision-making concerning production, distribution, and — most importantly as far as this book is concerned — facilities locations. Indeed, one of the distinguishing features of commercial economies is deciding where an economic activity should be located. "Being complementary" to other enterprises — see the nearby sidebar — is one important aspect of this decision-making process.

Understanding Location Factors

REMEMBER

A *location factor* is any aspect of an environment that influences a decision regarding the choosing of a site for business or industrial activity. Because competition is an essential characteristic of commercial economies, businesses of all sorts do things to give themselves a competitive advantage. Finding an optimal location for a business is a case in point, as well as an endeavor that has come to typify modern economic geography.

A good example of the importance of location concerns the decision to locate the plant that builds BMW motor vehicles in Greer, South Carolina, between the smallish cities of Greenville and Spartanburg. Theoretically, the factory could have been built just about anywhere. Indeed, after BMW corporate officials announced their intent to manufacture vehicles in a brand-new plant somewhere in America, several states lobbied to host the facility. And why not? The factory would create thousands of new jobs and provide a major boost to the economy of wherever it was built. You would also see other industry pop up. Someone has to make headrests and radio knobs to supply the car maker, right? Further, these types of developments can generate a *halo effect* whereby other industry sees success in a place and says, "Hey, I might be able to do well there, too." Just take a look at Boeing now making airplanes in nearby Charleston, South Carolina.

First you need to ask why BMW even needs to make cars in North America. Considering that the United States is a major market for their cars, and that cars are a *bulk-gaining product* (increasing in size and weight and therefore higher finished transportation costs when complete), it makes good cost sense to build a car closer to the customer. Since BMW wanted to build a high-quality product at the lowest possible price, several key characteristics dictated their factory location choice. Specifically, the chosen site would ideally have

- A large amount of inexpensive land on which to build
- Access to abundant and relatively inexpensive electricity
- A low tax environment compared to other prospective locations
- Connectivity to port, railway, and interstate highway systems
- An accessible location with respect to the various subcontractors who would supply parts
- An accessible location with respect to potential customers (domestic and overseas)
- An adequate supply of trainable employees

A number of prospective sites in different states were considered. Most were strong candidates with respect to some of the above criteria, but weak in regard to others. In the end, Greer got the nod because it satisfied most points or was able to play off one expensive factor against many less costly ones.

Economic activities vary, of course, and as a result, so do the factors that are most appropriate in deciding where a particular economic activity may best be located. Thus, what is important with respect to deciding where to build cars may be totally different than the location keys to growing mushrooms or merchandising children's clothing. Also, in the case of some economic activities, a single location factor may be all-important, while in another, a mix of several determinants may need to be considered. Not to be overlooked here is time. The locations that matter today are based on factors that are variable. A change in technology — for example, inventing a steam-powered cotton gin — can enable a shift of the textile industry from water-powered New England to the American South where the raw material, cotton, was located. So over time we can see shifts in the best locations for industry or any number of services (see the sidebar "Atlantic City as intervening opportunity?" for another example).

ATLANTIC CITY AS INTERVENING OPPORTUNITY?

TECHNICAL STUFF

An *intervening opportunity* is a closer source of supply to a point of demand. Clear as mud, right? Okay, suppose you occasionally need to purchase widgets, and the closest store that sells them is 10 miles away. Then one day another store that sells widgets opens up 2 miles away. From your perspective, the second store is an intervening opportunity because it's a closer source of supply (widgets) to a point of demand (you). Your travel costs go down and even better, you decide that these widgets are better quality. And because of that, you decide to discontinue traveling to the first store and become a patron of the second one. Naturally, the proprietors of prospective new stores have a keen interest in locations that offer competitive advantage. And for that reason, entrepreneurs often scope out the competition and then choose a location that represents an intervening opportunity for as many potential customers as possible.

A good real-world example of intervening opportunity concerns the advent of gambling in Atlantic City, New Jersey. Back in the early 1970s, gambling within the United States was legal only in the state of Nevada. Basically, it was a monopoly from which Las Vegas and Reno profited handsomely. But then a plan was hatched by New Jersey state legislators to permit casino gambling in Atlantic City, an old seaside resort that was a bit down in the dumps. The promoters clearly saw that Atlantic City would be an intervening opportunity for millions of Easterners for whom a few hours' drive to Atlantic City would be preferable to a trip of nearly transcontinental proportions. Indeed, when the casinos finally opened, business boomed. At least for a while.

Gambling was so successful in Atlantic City, that it gave people in other states the idea of doing the same thing. Gaming along the Mississippi Gulf took off and riverboat casinos began operating on the river of the same name. Other people on reservations were attracted to the idea because it promised income and because Native Americans did not need the approval of state legislatures and voters to open casinos (reservations are technically not state land). So casinos started popping up on reservations all over America, including the Northeast. And as they waxed, Atlantic City waned because it no longer enjoyed its status as the singular intervening opportunity for eastern gamblers. Taking another bite? Online wagering and state-based lotteries; the need to travel to lose your shirt has been greatly reduced.

Obviously, a good business location may be the key ingredient to profitability or overall success. Accordingly, nowadays two of the most important functions of economic geography are to help enterprises identify the location factors that are best for their particular business, and to sift through the list of possible locations and choose a favorable site. The following sections offer a discussion of some of the more important location factors.

Proximity to raw material(s)

Proximity to raw materials may be a critical location factor for economic activities that involve manufacturing or processing. Both functions typically require delivery of raw materials from source areas to the place of business. The cost of transporting raw materials to where they are needed may be a major business expense, however. Thus, locating as closely as possible to the source or sources of those materials is a good idea to minimize production costs, as well as producing a cost-competitive final product. Following are three examples:

- **Cheese:** Cheese production consumes large quantities of milk, which is expensive to transport long distances because it can be bulky and requires refrigeration. Therefore, if you want to make it in the cheese business, you'd be wise to locate as close as possible to a bunch of dairy farmers. It comes as no surprise, therefore, that Wisconsin, a major dairy state, also leads the country in cheese production.

- **Steel:** The traditional blast furnace way of making steel requires lots of iron ore and lots of coal, but not in just any amount. Generally, production of 1 ton of steel requires about 1.5 to 2.0 tons of iron ore and 2 to 3 tons of coal. Both ingredients are very bulky and expensive to transport from mine to factory. But because more coal than iron ore is required, an old adage is adhered to that goes, "Steel follows coal" — meaning that steel plants tend to locate closer to coal fields than to iron ore mines in order to hold down transportation costs. Figure 15-3, a map of the traditional American "Manufacturing Belt" in relation to coal and iron, very clearly shows the relationship between steel and coal.

- **Hogs:** If you want to seek your fortune in pork or bacon, then you should locate your hog operation as close to corn country as possible. That is because corn is the principal ingredient in pig feed, which is something you are going to need by the ton. Like many domesticated animals destined for your refrigerator, hogs nowadays are raised in specialized facilities where they are fattened for market. Central to this process is bringing feed (often by the ton) to the pigs rather than setting the animals loose in the countryside to forage. Being near the feed source (corn) holds down transportation costs and contributes to profitability.

Proximity to market(s)

Proximity to the market (the purchasers of products) for a business's goods or services may be a key location factor for some economic activities. The exact meaning of "market" varies from one enterprise to the next, however. If you're an ice cream vendor, then your market is people. If you're a zipper manufacturer, then your market might be companies that manufacture clothing. Here are three examples that illustrate the potential importance of proximity to market:

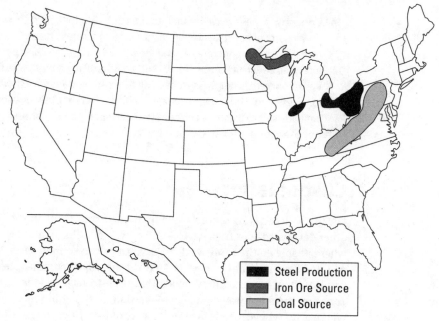

FIGURE 15-3: The traditional American "Manufacturing Belt" is located close to coal fields.

(© John Wiley & Sons Inc.)

REMEMBER

» **Pizza:** Pizza is a fairly low-cost item. Therefore, anybody who wants to make a profit in this business is going to have to sell a lot of it. An ideal location, therefore, is one that is close to a large number of people who are likely to want to have a fairly fast dining experience and not spend a whole lot of money. That could be a busy downtown area, a shopping mall, a freeway exit, a major highway intersection, or a site next to a college or hospital.

The number of people needed to support the business is called the *threshold* (lots of people like and buy pizza so that number can be much lower than say a BMW dealer).

» **Boutiques:** Boutiques typically specialize in high fashion designer wear intended for the upscale market. Unlike the pizza business, the main location strategy here is to get close to a particular variety of people (affluent folk) rather than lots of people in general. Locating in or near a high-income residential neighborhood or in a shopping district that caters to the well-to-do are good ideas. Here again, is the threshold concept. Important as well is the *range*. People are not willing to travel far (the range) to get a pizza, but they might be to get a good deal on that BMW coupe or a designer dress. This is one reason why we see many more pizza shops than high-end car dealerships.

» **Concrete:** Cities typically generate high demand for concrete, which is used for all sorts of construction projects. Once concrete is made at the plant and transferred to a concrete mixer, it needs to be delivered and poured at its intended construction site fairly quickly. Also, being a bulky and fairly low-cost item, concrete is not the sort of thing that can be transported profitably for a long distance. Thus, anyone going into the concrete business would be well-advised to locate his or her business within the city or on its edge as opposed to, say, out in the country.

Cost of labor

For many decades now, numerous manufacturing firms in the United States and other developed countries have closed down and relocated to developing countries where people do the same work as, say, American counterparts, but for much less pay and fewer (if any) benefits (see the nearby "Maquiladoras" sidebar). As a result, lots of manufactured goods that used to have "Made in the USA" emblazoned on them now say "Made in Malaysia," or the Philippines, or China or somewhere else. Believe me, if you aren't aware of this, then you have a major geography lesson waiting for you in your clothes closet. Check out the labels on your clothing and shoes. Chances are good that many, if not a majority, are imports. The two following sections give examples.

My smart phone

Consider the smart phone I used today (you probably used one, too — I bet your kids did!). Mine is made by Apple, and their product information frequently says something like "Designed by Apple in California." Clearly that is not the same thing as "Made in California," and it is giving you a hint that something more complicated is afoot. That phone may be constructed of glass made in the United States, a camera from Japan, a processor from South Korea, a battery from China (containing lithium from Chile or Australia), and so on, with all the design, research, and development taking place in Bay Area California.

But instead of having the manufacturing done in the USA, the company has all the components delivered for assembly in Taiwan or mainland China. There, workers perform for less money than the company would have to pay an American worker. Afterwards, the phones are sent to an Apple store in the U.S. for sale or marketed online. Even though it costs a bundle to send the raw and finished materials back and forth, the money saved by having the assembly done by comparatively inexpensive foreign labor more than makes up for the transportation costs.

TECHNICAL STUFF

MAQUILADORAS

Numerous American manufacturers engage in *outsourcing* — producing components or products abroad for domestic sale — to take advantage of lower-cost foreign labor. This tactic is exemplified by *maquiladoras,* American-owned manufacturing plants located on the Mexican side of the U.S.-Mexico border. Cross into Ciudad Juarez, Tijuana, Nuevo Laredo, and other Mexican border towns and you are likely to find plants bearing familiar U.S. brand names producing clothing, toys, leather goods, electronic items, auto parts, and other goods destined for sale in the U.S. Their presence is due in large measure to Mexican laws that allow U.S. companies to import components duty-free for assembly in factories located within (usually) a few miles of the border. Once completed, materials are sent back across the border for final assembly or distribution to retail outlets. Thus, the arrangement allows U.S. manufacturers to profit by lowering their production costs and Mexico to profit from job creation and (in some cases) technical training. One might characterize this as an unequivocal win-win situation, except that opening a *maquiladora* is often complemented by closure of a manufacturing facility in the U.S., resulting in people out of work. Frequently, critics also complain about lax environmental regulations, poor safety protections for workers, and long, strenuous work hours for factory laborers.

Strum roll, please

I own a Fender Stratocaster and a Martin dreadnought. These are electric and acoustic guitars, respectively. In small print on the head near the tuning keys, you can read "Made in Mexico" on the Fender. What gives with this iconic *American* music instrument maker? Simply put, they make different guitars at different price points and assembly abroad can reduce costs. Due to labor differences, a MIM (made in Mexico – in a maquiladora?) guitar will cost less than a made in USA guitar – and that is exactly what some buyers demand. In the sound hole of my Martin, it says "Made in USA." Manufactured in Nazareth, Pennsylvania from mahogany wood, this guitar sings in a way that no discount store bought model could, but there are still savings to be found. The Martin guitar case is made across the border just like that Fender guitar. Which border? Wait for it . . . Canada.

Accessibility

Accessibility concerns the ease with which a business can be reached, or with which goods can be transferred from one location to another. A downtown shopping area, for example, may be close to a lot of people, yet not be very accessible because of traffic congestion and lack of parking. Thus, ease of transportation is key to facilitating the accessibility of a particular location. The following sections give some examples of how businesses use accessibility to their benefit.

Hub-and-spokes networks

Airlines generally employ a hub-and-spokes network in which passengers or freight fly from different locations to a central point (*hub*), where transfer is made to another plane to reach the final destination (*spoke*). Point-to-point air travel exists, but many more routes are needed in this system and can therefore be cost prohibitive. While hub and spoke inevitably involves more travel time versus a non-stop direct flight between the points of origin and final destination, it also affords accessibility between all points in the route system and profitability to the carrier.

FedEx, for example, the company that pioneered the overnight package delivery business, operates a hub in Memphis, Tennessee (Figure 15-4). In the middle of the night, company aircraft from all over America bring their packages to Memphis, where they are swapped between planes that then return to their points of origin in time for morning deliveries. Thus, each night a company airplane departs from, say, New Orleans with packages destined for cities all over the country. But instead of delivering the goods "Santa Claus-style," going to each city one by one, the aircraft goes to just one place — Memphis, Tennessee.

FIGURE 15-4: Memphis, Tennessee, is the principal hub of the FedEx route system.

(© John Wiley & Sons Inc.)

Aircraft from dozens of other cities do the same thing, dropping off packages destined for New Orleans and elsewhere, and then picking up packages from New Orleans and elsewhere before making the flight back home. Thus, while it may seem a bit weird that a package sent from New Orleans to Phoenix makes the trip via Memphis, the network design promotes accessibility between multiple locations using a minimum of aircraft that fly a minimum number of segments. And that promotes profitability.

This model also works for FedEx internationally and they have other hubs in Paris and Singapore, among others. Future challenges to this model include how many people now shop online. If I want to order something from Target, an online and brick and mortar retailer, why would that company want to ship the item from a distribution center hundreds of miles away if they could just use a local deliverer servicing a nearby store? As you can see, the geography of business in constantly in flux.

This hub and spoke idea sounds great, but there's still one question that might be bugging you about FedEx as my example. That question is: Why did they choose Memphis? Remember where Memphis is located in the United States. Its centrality is important for shippers *and* receivers. Further, physical geography plays a role. The airport is closed far less often for weather like snowstorms, something that would kill a business dedicated to an over-night delivery promise.

Being in just the right place

The success of many businesses rests on its accessibility to a large number of potential shoppers. For that reason, nearly every Starbucks you'll ever visit is located beside a major roadway or by a busy intersection that accommodates a considerable volume and flow of traffic. Of course, all of those potential customers will need to leave their cars somewhere, so a decent-sized parking lot is at hand or there exists a drive-thru window — both key components of accessibility. In addition, the store will need to be located where disposable income levels are high enough to afford a cup of coffee that costs more per unit than a gallon of gasoline.

There is another strangeness to consider here. Have you ever noticed two Starbucks shops just a few hundred meters apart? There are two just like that not far from my house. One sits along a busy four-lane road and the other is tucked inside a Target store across the parking lot. They survive because they count on two different markets: one, a high-volume car-based clientele and a second group held captive inside a large big box retailer.

Containerization

As suggested in the discussion of labor costs, a company, such as an American athletic shoe manufacturer, may operate a factory in a foreign country in order to take advantage of low-cost labor. But if the finished shoes are not accessible to the

American market — that is, if they cannot then be shipped to the United States cheaply and efficiently — then the whole purpose of manufacturing the product overseas is rendered pointless.

Figure 15-5 illustrates *containerization,* the shipping of goods in large metallic containers that can be loaded and off-loaded from ships directly onto tractor-trailer trucks or trains for onward transportation to the final destination. This technology has revolutionized maritime trade by

- » Reducing loading and unloading time of ships from days to hours
- » Facilitating prompt delivery of goods
- » Reducing pilferage and breakage that occurred when goods were loaded and unloaded bit by bit from ship's holds

Special port facilities are required for the loading and unloading of containers. For that athletic shoe manufacturer, therefore, finding a country that offers cheap labor may not be enough. It may also have to have a containerization port facility to make foreign manufacturing sufficiently profitable.

FIGURE 15-5: A container ship unloads at port.

(© anekoho / Adobe Stock)

Cost of land/rent

Companies require land or rental space where they can conduct business. Within any sizeable geographical area, however, the cost of land or rent varies. Locating a business (or part of a business) with respect to this variation may have a significant impact on the company's bottom line.

Take, for example, a well-known bank that is headquartered in New York City. When you mail them your credit card bill, the letter might go to an address in South Dakota. Another well-known New York City department store may receive and process its credit card payments at an address in the state of Georgia. To some extent, geographic differences in the cost of labor would be at work here. But in these cases, another and more important location factor is at play: geographic differences in the cost of land and rent. Simply put, these operations require lots of office space, the price of which is much cheaper in towns in South Dakota and Georgia than in New York City (or even overseas!). You might quickly object, stating, "I pay all my bills online, so where these activities take place doesn't matter." That may be true to a degree, but even cyberspace is located somewhere. There's a computer and someone operating it in some physical space on this planet, and those spaces are not free.

I'm going to temper my discussion of this location factor because it appears again in the chapter on urban geography (see Chapter 17). Suffice it to say here that supply and demand largely determines the cost of land/rent; and that intense demand and competitive bidding for land/rent is greater in big cities than in small ones, leading to higher land costs in major urban areas (mostly due to accessibility!). Indeed, land/rent prices tend to "spike" downtown, and then trail off dramatically with increasing distance from urban cores.

Activities that locate downtown must be capable of covering operating costs that include high rents. If they cannot, then they are wise to locate at some distance from downtown, where land/rent costs are likely to be much less. Or perhaps even to a town in South Dakota or Georgia.

Taxes

They say that the only sure things in life are death and taxes. Maybe so. But taxes, at least, can be minimized. That is because tax laws and tax rates vary geographically. Thus, if the tax laws in a particular locale are so high as to seem unfavorable to personal or business finances, then one may consider moving to a more favorable location.

Local variation

Tax structures differ from one country to the next and, within the United States, from one state to the next. Some states have no income tax, while in those that do, the rate schedule varies. The same is true of corporate income tax and state property tax — some states have them and some don't.

On the other hand, every state has a sales tax, but rates vary substantially from one state to the next and even by county within states. Also, particular items may be exempt from sales tax in particular locales to encourage purchases there as opposed to somewhere else.

Enterprise zones

As the foregoing vignette suggests, "creative taxation" may be used to encourage business activity in particular locales. One fairly popular option is to designate corporate tax-free, tax-reduced, or tax-deferred areas that encourage job creation in particular locales. Called *enterprise zones*, *empowerment zones*, or *opportunity zones*, the typical goal is to counter unemployment and stimulate the economy of an area that has fallen on hard times.

To use a hypothetical example, suppose a certain city with an old industrial core has witnessed several plant closings, resulting in derelict real estate and an increase in unemployment. One turn-around possibility is to designate all or part of that area as an enterprise zone and mandate that any company that employs, say, at least 50 people and locates within the area will be exempt from corporate taxes for the next decade. Theoretically, at least, this is a win-win-win situation. The company gets a tax break that may increase its bottom line, unemployed people get jobs, and some of that income is returned to the government in the form of sales and income taxes.

Critics note rightly that the company with only 40 employees does not get similar treatment and others simply do not want that level of government involvement in the marketplace. But if these incentives were not applied, it would be hard to envision a BMW factory locating in South Carolina over another state that did offer them.

Climate

Climate is an obvious location factor for various agricultural and leisure pursuits. Thus, citrus fruits are produced in extreme southern (warm) parts of the U.S. because the plants are very sensitive to frost. Wheat, in contrast, is a hardy grass that does not require much moisture. Much of it is grown, therefore, in low-to-modest precipitation environments of the U.S., Canada, and Australia. Regarding leisure, ski resorts obviously call for a certain kind of climate as do most modern beach resorts. Other businesses, however, have climate connections that are a bit more subtle, but no less important. Following are two general examples.

Textile manufacturing

When raw cotton or wool is spun into thread, there is a propensity for the fiber to snap. This likelihood is particularly high in dry climates and comparatively low in humid climates. Thus, climates characterized by moderate-to-high atmospheric humidity are superb for textile manufacturing. Britain and the Southeastern U.S. are in this category. And both places are historically characterized by textile manufacturing. This matters less when climate controls such as air conditioning are placed on manufacturing facilities, but this is another business cost to be avoided where possible.

A matter of amenity

In the case of quaternary economic activities (and many tertiary ones, too), good personnel are critical to business success. If, therefore, a company can offer employees opportunity to live and work in a region that is generally perceived to have an attractive physical environment, then it may enjoy a competitive business advantage over competing firms. Thus, many high-tech companies have chosen their locations on the basis of a balmy climate (as much as any other location factor), in the hope of attracting talented employees.

The Austin, Texas metro area is one example. There you will find Silicon Hills — a play on California's Silicon Valley and the Texas Hill Country — where Texas Instruments, Dell, IBM, Amazon, and others are able to attract workers to a good climate, lower taxes, and a thriving culinary and music scene.

IN THIS CHAPTER

» Changing tastes and preferences

» Making a culture — resource connection

» Deciphering the geography of supply and demand

» Looking at the advantages and disadvantages of different resources

Chapter 16
Earth's Resources: Always Hungry for More

There's a small plot of land behind my home, over the fence beyond my property line. It is quite overgrown and is frankly a vegetative nuisance. Vines snake up and over the fence and into my trees and bushes, choking the landscape that I've neatly maintained. One day recently I heard a "jingle-jangle" on the other side of my fence. Peering over, I see four goats having a great time taking apart the jungle within a city. My neighbor rented the goats and managed to take this stuff and turn it into a resource – food. Human need, here the necessity to find grazing for a few animals, is a key aspect of defining what constitutes a resource. And we also eliminated the need for a lawnmower.

REMEMBER

People have an appetite for resources; and as their numbers grow, so does the cumulative thirst for Earth's bounty. The global cornucopia is varied, however. Resources differ in their definition, abundance, and longevity — meaning that some exist in finite supply while others are eternal. Geographical considerations further complicate the picture. People in different parts of the world have different appetites for different resources that typically are available in some regions but not others.

Supply and demand for Earth's resources is a geographically complex and fascinating story. The goal of this chapter is to touch upon and familiarize you with some of the key concepts of resource geography.

Energy and mineral resources serve as the focus of this chapter. This somewhat narrow perspective is certainly not intended to denigrate the importance of other kinds of resources, all of which lend character to different parts of the planet. Indeed, discussions of soil, fisheries, and forests appear in Chapters 7, 8, and 10 respectively. Likewise, much of Chapter 18 is devoted to the consequences of poor resource utilization and mismanagement.

For now, the key thing to remember is that *resource* entails a great many things that range from specific environmental items to the land itself. Energy and mineral resources, however, are of critical importance the world over, and embody the key concepts and challenges common to most resources.

Defining Resources and Assessing Their Importance

REMEMBER

A *natural resource* is any physical environmental item that people perceive to be useful for their well-being. That's a pretty broad definition, but an appropriate one. All kinds of environmental items are useful. Geographically, resources may be everywhere (air), pretty widespread (water), limited to relatively few areas (petroleum), or downright rare (diamonds). From an economic perspective, to be a resource it must also be accessible and usually profitably so. As the following sections will demonstrate, however, the geography of resources, as well as the wealth and power they provide, may have more to do with culture than the environmental items themselves.

The central role of culture

An energy crisis is currently underway. Fuel resources are becoming scarce, and demand for them is growing faster than new sources are becoming available. Prices are going up. People wish for self-sufficiency, but that is no longer possible. Increasingly, they must rely on fuel sources that are far away, under the control of somebody else, and require increasing amounts of time and money to procure. No immediate solution is in sight.

No, this is not about petroleum or the United States, but instead the firewood crisis in parts of sub-Saharan Africa. Much local domestic fuel use, mainly for cooking and boiling water, requires firewood. Populations are growing and with it the demand for firewood. But practically nowhere in the steppe and savanna realms, where scarcity of firewood is most acute, are there effective programs that plant or replant trees faster than they are cut. Woodlands that used to surround a village are gone, and therefore people must walk farther and farther to get what they need.

Consider, too, what this means for other economic activity. Wood is also used for furniture and building construction. In poor supply overall, wood will need to be imported from China, Brazil, and elsewhere. As a result, environmental problems such as deforestation may simply be exported to other places to satisfy wood demand. This issue clearly illustrates the globalized and integrated nature of our planet today.

REMEMBER

Resources are culturally determined. That is, whether or not a particular environmental item is useful depends on culture. Mesquite is one example. Early ranchers in the western United States considered this plant a pest that competed with grazing. Native Americans however, prized the plant for its beans, as a fuel, and for its shade. Same plant, two differing views. Today, many Americans could care less about access to firewood, but they certainly do care about access to gasoline. Lots of Africans feel the exact opposite. Cultural difference, particularly as it relates to technology, is the principal explanation. American culture is inseparable from a wide range of machinery that runs on gasoline or oil — cars for transportation, tractors to help produce food, furnaces that heat buildings and homes, and so forth. Over large parts of Africa, however, none of these matter very much, but firewood sure does. The bottom line is that American culture is different from varieties of African culture, which are different than Asian cultures, different than Eastern European cultures, and so on and so on. What this means is that people from different cultures use and evaluate the environment differently.

Culture change, resource change

All cultures change over time. As they do, certain items of everyday use (horse-drawn buggies) may gradually disappear, while previously unknown items rise in importance (automobiles). Because resources are culturally defined, it stands to reason that culture change may result in resource change. Uranium, for example, was not a resource barely a century ago. It was there in the ground all along, but it was just neutral stuff that humans did not have a need for or know how to use. Consider, for example, changes in energy resource consumption that have occurred in the United States.

In 1850, 90 percent of the United States' energy supply consisted of firewood. By 1910, the supply was almost 80 percent coal. And today? We are diversified among petroleum (35 percent), natural gas (34 percent), renewables (12 percent — hydro, wind, solar, etc.), coal (10 percent), and nuclear (9 percent). Firewood is now just a blip, used in only a few homes for heating purposes.

These changes had a great deal to do with culture change. The decreased use of wood over time is not explained by depletion of forests. In fact, more forest cover exists today in the United States than there has been for some time. Likewise, lack of reliance on oil 170 years ago was not due to lack of knowledge of its existence. Instead, there were limited means of putting these resources to productive use, and, therefore, they were in very limited demand.

But times change. Which is to say culture changed. And with it came a change in energy resource consumption as machinery was developed that made use of oil, gas, and uranium instead of wood.

Resources and power

REMEMBER

Fuel resources generate two kinds of power. On the one hand, they can be used to make heat that can be transformed into physical energy. In addition, resources may generate significant political and economic power for the countries that possess them.

Petroleum, for example, comes as close as any substance today to being the economic lifeblood of the global economy. Its geography is highly concentrated, however. The countries that border the Persian Gulf possess about half of all the world's known petroleum reserves. Saudi Arabia alone possesses just over 17 percent. Elsewhere, major reserves exist in Russia (6 percent of the world's total), Venezuela (17 percent), and Canada (10 percent). The United States possesses thousands of producing oil fields and about 4 percent of the share.

The essential fact is that few countries possess petroleum in great quantity. That virtually guarantees that those countries will enjoy substantial political and economic clout for as long as petroleum maintains its status as the most important energy source. This was vividly demonstrated in 1973 when, in the wake of an Arab-Israeli war, oil-producing nations in that region curtailed shipments to western countries. About 60 percent of the world's petroleum reserves then came from countries that border the Persian Gulf. Much of it was exported to developed nations whose economies depended on it to different degrees. When the tap suddenly was turned off, everybody got a quick and unmistakable lesson in political petrol-power.

But that kind of power is obtained only when a particular resource is sorely needed by countries that either have none or not enough. We have seen that resources are culturally determined, and that culture change brings resource change. Accordingly, as resources rise and fall in importance, so do the economic and political power of the places that possess them. One hundred years ago, the countries that border the Persian Gulf were of little economic importance to the outside world. In fact, many of them did not exist. They were colonial outposts of Europe (with borders drawn that create problems today — see Chapter 14 for more on this). Times have changed. Today, virtually every country consumes petroleum, and the vast majority of them cannot satisfy their thirst by domestic production. So they rely for their economic lifeblood on foreign sources — the Persian Gulf nations in particular — who accordingly assumed great power and developed landscapes that reflected their newfound wealth (Figure 16-1).

FIGURE 16-1: The explosive growth of Dubai – upward and economically – is a result of petroleum resources.

(© Rastislav Sedlak SK / Adobe Stock)

Resources and wealth

You may expect countries that possess lots of natural resources to be better off than those that don't. For better or worse, however, it doesn't always turn out that way by a long shot. In fact, some poor countries are rich in natural resources while some rich countries are poor in natural resources. This difference may occur for a number of reasons, perhaps the most important of which concerns the locations of people and facilities involved in the acquisition and processing of raw materials.

TECHNICAL STUFF

Turning to another resource for a moment, consider the geographical differences in employment and income that may occur when trees in "Country A" are cut down and sent to "Country B" for processing. Several people may be involved in harvesting the forest, from which timber is sent to a sawmill in "Country B." Many more people may be employed at a sawmill, where raw logs are cut into pieces of lumber, than were employed back in the forest to cut the trees. Also, because the factory employees are skilled workers, they are likely to command higher salaries as well. Later, skilled craftspersons might take pieces of that lumber and make high-priced furniture. Finally, company executives negotiate sales and contracts with transportation firms, wholesalers and retailers concerning the final disposition of finished products.

Thus, a single log can generate income for several people. But the amount of income and the location of the beneficiaries varies. Generally, the person who cuts down the tree probably earns less than the sawmill worker, and so forth. Therefore, in this example, the exported log generates much more wealth where it is processed and made into finished products than where it was harvested.

Though this example concerns trees, other natural resources — mineral ores and crude oil, for example — may exhibit similar outcomes. Formerly, Persian Gulf oil was drilled by foreign-owned companies that set the prices, employed relatively few locals, and sent crude oil home for processing and refining, resulting in a finished product worth substantially more than what came out of the ground in the first place. Now, however, the producing countries have a much greater stake (if not outright ownership) in production and processing, and are realizing close to the full economic potential of their resource base.

Differing Life Spans: Which Resources Are Here Today or Gone Tomorrow

Resources have different life spans. Some exist in finite quantity; so when they are used up, they're gone forever. Others can be replenished, as when seedlings are planted to replace a forest that was cut down. And some are virtually eternal (the Sun, for example), meaning they'll always be with us regardless of how humans use the environment. Consideration of energy resources that fall into the following categories provides insight into resource geography, the crux of the current energy situation, and the need for future planning.

Non-renewable resources

Non-renewable resources exist in *finite* (or fixed) quantity. You can think of them as coins in a global piggy bank from which money can only be extracted, not added. Once they have been used up, they are gone and cannot be renewed.

REMEMBER

Perhaps the most important energy-related fact of life is that the United States and other developed countries are overwhelmingly dependent on non-renewable fuel sources. That includes petroleum, coal, and natural gas. These are sometimes called *fossil fuels* inasmuch as scientists believe that they are the result of long-term decay and metamorphosis of organic matter. Thus, even as I write this, nature is at work doing whatever it does to turn, say, today's ocean bed into tomorrow's oil field.

But that process takes millions of years, in contrast to humanity's needs, which are exhausting the world's oil inventory in a figurative blink of the eye. The world's first commercial oil was drilled in Pennsylvania in 1859. Now, just shy of two centuries later, people are contemplating the end of the "petroleum era." This viewpoint has two dimensions. First, will the era end due to a dwindling supply? Or second, will the era of petroleum use end because society decides that its use is too costly to the planet (for example, climate change)?

Petroleum

Next to air and water, petroleum is perhaps the most essential resource of the moment, at least as far as developed countries are concerned. Arguably it can be considered the lifeblood of the American economy, if not the American way of life. Skeptics need only imagine waking up one morning to discover that every car, truck, motorcycle, and internal combustion engine no longer functions for lack of gasoline.

Presently, global petroleum reserves stand at about 1.7 trillion barrels, and are being consumed at a rate of 35 billion barrels per year. That means we currently have about a 48-year supply of oil assuming demand is constant. The COVID-19 disease dramatically curtailed economies, and petroleum use along with them, in 2020. This demonstrates that these figures can be subject to swings in the rates of supply and demand.

REMEMBER

In the years ahead, new petroleum deposits are likely to be discovered, but global demand for oil is also likely to rise. How this math will play out can't be accurately determined at present, but clearly the days of the "petroleum era" are numbered. Obviously, if a replacement for petroleum is not found, its price will rise even further as reserves decline. As that occurs, the economic and political power of countries that possess it will continue to increase. On the other hand, growing scarcity is also likely to increase development of alternative sources of energy. This last statement is key to a very bold one that I will make here: We will never run out of oil. I know what you are thinking. "Didn't Jerry just say that we only have 48 years of oil remaining?" As petroleum reserves become scarcer, it will become more costly to extract from ever-worsening reserves. At some point the cost will be too much to bear and we will move to another fuel source. We won't switch from oil because there's no more in the ground. It will always be there, but it will be simply too expensive to pump out.

Coal

The world's coal supply greatly surpasses petroleum in quantity and longevity. The estimated global coal reserve is in excess of 1 trillion metric tons with annual production levels keeping the supply close to 200 years at the current rate of consumption. The United States possesses about 23 percent of all of the world's known deposits, and is followed by Russia (15 percent), Australia (14 percent), China (13 percent), and India (10 percent).

If you're thinking, "Gee, that's terrific," then you need to temper your optimism for two reasons. First, most coal burns dirty, so its use comes at the expense of air quality (see the "Grades of coal" sidebar later in the chapter). As a result, substantial research dollars have been spent to find ways to burn coal without polluting the air — such as by removing impurities before being burned or by developing smokestacks that are akin to giant filter cigarettes.

TIP

This environmental question is a tricky one and shows why it pays to think geographically. It shouldn't surprise anyone that growing economies like China and India would want to use domestic coal as an energy source. After all, they have a lot of it. But if coal burning contributes to climate change, why should the United States limit its use of coal if India and China won't? One thing to realize, however, is that much of their pollution is being produced to make products for Americans. Their pollution is really our pollution, too. We merely exported it when American factories moved their production from North America to other parts of the globe to save on labor costs. Much of that Indian and Chinese pollution would have been produced in the United States if manufacturing relationships were different. One way or another, it at least makes you think (differently), doesn't it?

The second reason: coal is bulky and unwieldy. Although it can be crushed, mixed with water, and transported short distances by pipeline as slurry, you can't send coal down a pipeline a long distance or pour it into a gas tank — although it would be a much more convenient and widely applicable if you could. Thus, considerable research has focused on coal *liquefaction*. This process, however, uses a lot of water and still contributes substantive greenhouse gas emissions without some form of carbon capture.

GRADES OF COAL

TECHNICAL STUFF

Coal exists in several varieties, or *grades,* the difference being determined by the degree of carbon content. *Anthracite*, the highest grade, is almost pure carbon and therefore burns very hot and gives off little smoke or soot. *Bituminous,* the next highest grade, has somewhat less carbon content and a higher percentage of waste materials. As a result, it burns less hot (but still hot enough for most industrial purposes) but dirtier, generating more pollution. *Lignite* and *peat* are progressively lower grades, burning even less hotly, while generating more air pollution.

In nature, sadly, the most pure form of any substance tends to occur in the smallest quantity. Thus, all global coal reserves considered, anthracite accounts for the smallest portion. Being so pure and burning so hot, it is also the most sought after variety. The end result is that that particular kind of coal is mostly depleted. Thus, the vast majority of the world's remaining supply consists of "dirty varieties" whose consumption poses challenges to environmental quality.

In any event, coal, like petroleum, is very unevenly spread across the planet. While substantial reserves are found in North America, Europe, and Asia, South America and Africa are largely without any. Thus it has limited geographic prospects. In addition, and as discussed in the sidebar, some coal reserves are much more valuable and useful than others.

Natural gas

Natural gas exists in abundance and, as far as fossil fuels go, has attractive qualities. This was not always the case. In fact, natural gas was burned off as an unwanted by-product during initial petroleum production and this still happens today. Natural gas burns more cleanly (producing the most modest air pollution), requires little or no processing before use, and can be transported cheaply and efficiently overland by pipeline. Its versatility is also a point in its favor. In addition to heating a majority of the homes in the United States, natural gas is used to generate electricity, provide energy for some kinds of light manufacturing, and power an increasing number of motor vehicles.

On the downside, natural gas is not well-suited for transoceanic trade. Pipelines of that length are not feasible. Shipping is possible when *liquefied natural gas* is involved, but that requires great expense. Also, natural gas is, after all, a non-renewable resource, so someday supplies will be exhausted.

That is not going to happen anytime soon, however. Given current data on reserves and annual production, the world has a supply of natural gas somewhere between petroleum and coal. But people who know the pertinent geology all suggest that vast quantities are waiting to be tapped in various parts of the world.

APPLIED GEOGRAPHY: THE NUCLEAR DILEMMA

The "dilemma" concerns three questions. Should more nuclear power plants be built? If so, then where? And where should we store the nuclear waste that we need to deal with in any event?

Nuclear power is a technologically sophisticated means of producing electricity. Specifically, a nuclear reaction induces rods of enriched uranium to produce high heat but also toxic radiation. This occurs in a heavily leaded nuclear reactor chamber that captures and contains the radiation even as it becomes very hot. To keep the chamber at a tolerable temperature, water must continually swirl around it. Steam is a by-product; and under pressure it may be used to turn a turbine (propeller) that operates a generator that produces electricity.

The United States imports uranium — derived from non-renewable minerals — from Kazakhstan, Canada, Russia, and a few other countries to add to our own domestic supply. Proponents of nuclear energy see it as an important alternative to fossil fuels, but it does not necessarily diminish our reliance on foreign sources of energy — it merely redistributes it. The general public, however, is deeply concerned about the safety of these facilities (due in good measure to the Chernobyl (Ukraine) disaster in 1986 and the Fukushima (Japan) meltdown in 2011), so future construction is far from certain.

Two factors have historically guided the location of nuclear power plants: immediate access to a reliable water supply and proximity to users. Continuous water input is needed to guarantee continuous cooling of reactor chambers and with it continuous production of steam. Facilities therefore have tended to be located in coastal environments or along large rivers and lakes that have a history of not flooding. Proximity to users helps keep down costs of delivery and also conserve the electricity that has been produced. Not surprising, however, most people are reluctant to live close to one of these facilities. The dilemma, therefore, is trying to find locations that satisfy both the necessary water and the economics of proximity to large numbers of users.

The waste dilemma is perhaps even more daunting. After a certain period of use, the uranium rods lose their capacity to produce heat, yet continue to generate radiation at levels that may remain deadly for centuries. Finding a location or locations where this waste can be safely stored for a prolonged period of time is key. Storing this waste far away from people seems sensible, but even then, the waste must be safely shipped and you will, however briefly, encounter populous areas enroute to the storage site. So why not launch the waste into space? Or bury it in the ocean? Or drop it into a volcano? Rockets have the potential to explode and showering the waste on people is not a good idea. Irradiated ocean water *probably* won't give us Godzilla, but the health of those ecosystems is pretty important just from a fishing standpoint. So that leaves us with volcanoes. Drop the waste into the lava and all is well, right? Not exactly. We need to melt the fuel and remove its radioactivity, but volcanoes would need to double their temperature for that to be successful.

For now, Russia is home to about 20 percent of all global natural gas reserves. Iran possesses another 17 percent and Qatar has 13 percent. The United States possesses only about 7 percent of global reserves, but accounts for about 22 percent of annual worldwide consumption.

In whatever way the distribution may change, the benefits of natural gas — like fossil fuels in general — will be unevenly distributed. Some areas are certain to have lots of it and others none. But as far as non-renewable resources are concerned, the future of natural gas still burns brightest. Even some of the most-worried climate change activists see natural gas as a bridge fuel to something less damaging.

Mineral resources

I am going to interrupt our energy discussion here to toss in another non-renewable resource, minerals. Well, sort of interrupt. Some of these are still important for energy (see nearby nuclear energy sidebar and lithium below), while others are used in a whole host of non-energy applications.

Lithium is a chemical element with many resource applications. In recent years, its abilities in battery production are among the most important. From cell phones to electric cars, lithium has seen a marked increase in demand in the twenty-first century for an element not known to exist until 1817. There is a distinct geography of lithium reserves, with most being in Chile, Australia, Argentina, and China. Much like petroleum, the concentration of lithium reserves in a few scant places has elevated the economic power of countries and companies in a race to control supplies necessary for products with high global demand like mobile electronic devices.

Non-energy mineral resources include gypsum, phosphate, bauxite, and kaolin among many, but there are metallic ores of note, too. Let's just look at a familiar one — iron. Iron has been used for centuries in tool manufacture and when mixed with carbon, produces an alloy known as steel. Iron, then, has shaped many parts of the world by providing the tools to literally dig up — or shape — Earth and to erect steel-framed structures upon it. Further, landscapes and locations would not exist without this resource's identification and use. The Mesabi Iron Range in Minnesota would not be scarred from extraction and the city of Duluth, a major Great Lakes shipping port, would be an afterthought. Pittsburgh, conveniently located at the confluence of three rivers and in-between Minnesota's iron and Pennsylvania's eastern anthracite coal fields, might have been something different or nothing at all rather than a steel production powerhouse. Geography made this all happen as a result of conditions in places (e.g., iron reserves and people with need) and connections to other places (e.g., the Great Lakes bridging the reserves of iron and coal).

Let's take another detour into power, politics, and resource reserves — still using iron as our example — in the most unlikely of places: Antarctica. Iron can be found there and coal, too. You might surmise that there could be a few problems with access. First, no one owns Antarctica. Several countries lay claim to parts of the continent, but there is no "owner" to consult with. Further, international treaty bans mining under the need for environmental protection.

Let's lay those two issues aside and assume that you have permission to be there and to mine the land. It's time to get to work!

>> Step 1: Find and then dig up your iron under more than one mile of ice.

>> Step 2: Ship your iron across the treacherous Southern Ocean.

>> Step 3: Find a buyer willing to pay for your cargo even though it costs much more than iron found elsewhere.

This doesn't seem like a very good plan, does it? A resource that is far away from markets and practically inaccessible isn't really a resource. It will remain in Antarctica forever unless the iron is made available due to climate changes, all other reserves are used up, and international treaties are shredded. Don't hold your breath on this one.

Renewable resources

REMEMBER

Renewable resources are ones that can be replenished and, therefore, are theoretically inexhaustible. As mentioned at the beginning of this chapter, trees are an example because they can be replanted. Another is *biomass conversion,* which is the processing of organic matter into combustible liquids or gases. These forms of energy are "renewable" because the organic matter used to produce them can be grown (replenished) on a continuous basis.

In Brazil, a country that is poor in fossil fuels, sugar cane is used to make alcohol that is mixed with petroleum to form *gasohol.* Millions of vehicles in that country run on this substance, a mixture of gasoline and alcohol (up to about 25 percent of the fuel). This is made possible by growing sugar cane on extensive arable (farmable) land, and as result, may not work well in smaller countries. Elsewhere, organic wastes are locally collected and fermented in a low-tech way that produces a methane *biogas* (gas) used for heating, cooking, and lighting. China and India, for example, have extensive national biogas programs.

Of particular relevance to this book is the fact that biomass conversion may potentially break the bonds of location. Fossil fuels exist in limited, specific locales. In contrast, there are several regions that can and could grow, say, sugar cane or corn, in abundance for purposes of biomass conversion. The Corn Belt in the central United States is one possibility.

Perennial resources

Perennial resources are theoretically eternal. While they offer great promise, all have current limitations that relate directly to geography. Here are four examples.

Solar energy

Although astronomers tell us the Sun will eventually burn out, for all intents and purposes, it is an eternal, perennial source of power. Energy from the Sun can be absorbed directly by solar panels and put to use for heating. Even better, however, are *photovoltaic cells,* which convert sunshine to an electrical current. The Sun is fickle, however. Some days it shines and some days it does not, so a major engineering problem is how to store solar electricity from sunny days to carry us through cloudy days. Also, knowledge of climates informs us that solar intensity varies around the world (see Chapter 9).

Unfortunately, areas of maximum energy need often occur where the Sun is comparatively weak (or often obscured) and vice versa. Thus, solar and other energy sources that possess a measure of unreliability probably should not be looked to as complete solutions to future energy needs. Better perhaps that we view them as parts of a resource menu from which different resources can be called upon as conditions permit.

Wind energy

Wind is physical energy that can be harnessed to produce electricity. This is achieved by constructing rather specialized windmills that are tall poles fitted with a turbine at the top. When the wind blows, the turbine turns. This rotating motion is linked to and operates a generator that produces an electrical current.

Wind is a by-product of sunshine (see Chapter 9); and because solar energy is a perennial source of energy, wind power is categorized as the same. But like the Sun, wind is fickle. Some days it blows and some days it doesn't. Accordingly, for wind to become a major contributor to energy production, means must be found to store surplus wind-energy from windy days for consumption on calm days. Also, some places are inherently windier than others. Coastal areas, for example, tend to be particularly breezy because water and land absorb sunshine at different rates. Mountain and foothills areas also are especially favorable. The key is not to have an exceptionally windy place, but rather one where the wind blows consistently.

Windmill technology efficiency is improving rapidly and the cost of producing this energy is declining. As a result, we are seeing the development of wind farms worldwide. In addition to providing energy, landowners can receive rent from power companies that build the towers on their land. The small footprint means that other uses, such as agriculture (real farming), remain possible. These new "farms" clearly represent a new addition to the cultural landscape (Figure 16-2).

FIGURE 16-2: Wind farms are becoming more commonplace as an alternative energy source.

(© maksymowicz / Adobe Stock)

Geothermal energy

In the opinion of some, we are virtually standing on top of all the energy we will ever need. That is a reference to the very high temperatures that exist within Earth's crust (see Chapter 6). Because this heat literally radiates from Earth, scientists refer to it as *geothermal energy*.

Theoretically, all that is needed is to drill down to the level of very hot rock, inject water, and extract the resulting steam. This may be used directly to provide home heating and hot water. Under pressure, steam is also a physical force (witness the tiny whistling steam jet on your tea kettle, only imagine a huge one) that can be used to turn a turbine that operates an electrical generator.

Geothermal energy is free and non-polluting. But a major impediment stands in the way of its widespread use — cost. In most places suitably hot rock is encountered about 2 to 3 miles beneath Earth's surface. Drilling and operating steam-producing wells that deep would not be cost competitive with standard coal-fired plants. Some locations are exceptions, however. In lands near plate boundaries, the attendant fissures make it possible for interior heat to come close enough to the surface to allow for economical human exploitation. In Iceland, for example, nearly all home heating and hot water comes from geothermal energy. In parts of Northern California (and a few other locations worldwide), electricity is produced in this manner. In the future, geothermal energy may become generally available as a result of technological progress and the changing economics of energy production. For now however, its viability is governed largely by the geography of plate boundaries.

Hydroelectric power

Hydroelectric power (HEP) utilizes the movement of water on Earth's surface to produce electricity. It is most often associated with a dam that is built across a fairly narrow valley, causing a river that runs through it to back up on the upstream side. Eventually, a reservoir rises to nearly the height of the dam (see Figure 16-3). Intake vents near the lake's surface admit water that falls within large conduits, ultimately to strike turbine blades, which causes them to rotate at high speed and operate an electrical generator. HEP may be considered perennial because as long as the sun shines (practically eternity), water will be evaporated, fall to Earth as precipitation, collect in rivers, and be available for power production.

FIGURE 16-3: Dams such as Three Gorges in China may produce hydroelectric power while promoting flood control, irrigation, recreation, and drinking water supply.

(© marcaletourneux / Adobe Stock)

HEP supplies about 16 percent of the world's electricity. Not only is it a non-polluting energy source (nothing is burned), but the reservoirs may provide recreational opportunity, flood control, and water for people and agriculture. In addition, HEP projects may have important symbolic value. For example, China constructed the giant Three Gorges Dam on the Yangtze River and it became fully functional in 2012. While the finished project provides all of the benefits noted previously, the dam is perhaps even more important (to the Chinese at least) as a symbol of the nation's emergence as a major world power and of the government's ability to accomplish great deeds.

On the negative side, construction of HEP projects is expensive. Many millions (even billions) of dollars may be required before the first bit of electricity is produced. Indeed, this is doubly bad because the areas of greatest hydroelectric power potential includes countries in Africa and Southeast Asia that do not have the financial wherewithal to fund such projects.

TIP

Herein lies a new wrinkle: External funding to support these projects in some cases has come from China, creating new power dynamics with implications across the globe. Thirty-five African countries now have infrastructure ties to China. Projects include HEP energy production and transportation. The railway I rode in Kenya to get to the Galana River (see Chapter 1)? Chinese-built.

HEP development has additional geographical constraints. Obviously, you can't build a dam just anywhere. You need a river valley, and especially a narrow one for the simple reason that building across a narrow valley is cheaper than building across a wide one. Also, the eventual reservoir-related flooding upstream may cause serious displacement of people, agriculture, and transportation systems, all of which may occur at great cost. One last question: How well does this power source work during a drought?

Trading-off Resources: The Consequences of Resource Use

Further complicating resource geography is the fact that production and consumption of one natural resource may occur at the expense of another. The result may be deterioration of environmental conditions over specific geographic areas. Here are two examples from a wide range of possibilities:

> » **Fossil fuels and air quality:** Americans desire low-cost access to gasoline, which they consume in massive quantities. Americans also desire clean air. The atmosphere, unfortunately, is the gathering place of pollutants that result

from gasoline consumption. As a result, urban areas in particular, with their high concentrations of motor vehicles, often suffer high levels of atmospheric pollutants. Combined with environmental conditions such as tall mountains and temperature inversions, cities like Los Angeles, California and Santiago, Chile can see even worse effects.

» **Strip mining and soil loss:** In some areas large and valuable coal deposits are found just below Earth's surface. The safest and most economical way to extract these resources is to strip away the *overburden* (the soil and rock between the surface and resource) and then dig up the coal. This process is called *strip mining,* and its by-product is a deeply scarred surface.

Formerly, strip-mining companies were allowed to do their thing and move on without devoting any effort to repair the damage. Now, however, laws require major land reclamation. But even the best repair job cannot fully restore some lands to their former status. In a rather cruel twist of geography, some of the country's most valuable coal reserves lie underneath some of the best Midwestern farmland. Clearly, much care would be needed there and many people question whether the trade-off of coal for soil is worth it.

These two cases illustrate that when we pull resources out of the ground we also tend to push negative effects (such as air pollution, mine tailings, dirty water) somewhere else. We call these effects *externalities* and hope for wisdom in reducing or eliminating them where possible.

The global appetite for resources, and particularly energy resources, is very uneven. Both the United States (16 percent of the world) and China (26 percent) each out-consume Africa, South America, and Central America *combined* in energy use. The American economy and standard of living, as opposed to sheer number of people, explain the high consumption figure for the United States. For China, the number of people plays a larger role, but the importance of standard of living is becoming more influential annually.

Some people ponder the ethics of so few Americans or Chinese consuming so much of the world's energy. But perhaps the more critical consideration is all of the countries and peoples around the world who are striving to raise their economies and living standards, and in so doing emulate their resource consumption.

A simple formula developed in 1970 explains this issue: I = P x A x T. This is, unsurprisingly, called IPAT. I equals impact, P is population, A is affluence, and T is technology. Growing populations with increasing affluence tend to use a lot of resources, and technologies that increase efficiency can temper impacts. These variables express themselves differently from country to country showing how environmental impacts can differ. Assuming that other countries meet these two levels of consumption with at least modest success, and adding the virtual certainty of global population increase, the signs point to a growing appetite for resources in the years ahead and growing strain on the natural environment to provide them.

IN THIS CHAPTER

» Examining urban origins

» Learning how cities grow

» Acknowledging internal differences

» Dealing with dust domes and other problems

Chapter 17
Downtown to the 'Burbs: Urban Geography

If the year was 1790, then this could be a newspaper headline: 195,000 Americans Now Live in Cities!

Sounds preposterous, doesn't it? But in fact, in 1790, the first U.S. Census revealed that only 5 percent of the nation's 3.9 million inhabitants lived in cities, which actually were tiny islands in a sea of rural settlement and wilderness. In every census since, the urban percentage has marched steadily upward (as seen in Figure 17-1). Around 1920, it exceeded the rural percentage for the time, and has never looked back. Today around 80 percent of Americans may be categorized as urban. That amounts to about 265 million people out of a total population of some 332 million.

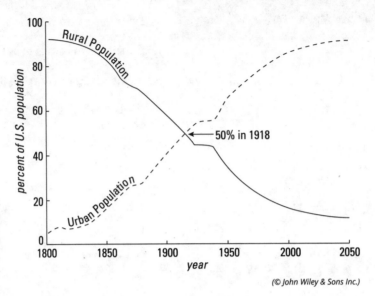

FIGURE 17-1: Urban-rural population trend in the United States, 1790-2050.

(© John Wiley & Sons Inc.)

Studying the Urban Scene

REMEMBER

As centers of population, political and religious power, trade, and commerce, cities have always been geographically significant. At no time, however, have they been more important than they are today — not only because so many people live in them, but also because cities have grown in number and size. Literally millions of square miles of natural landscapes have become "cityscapes." Indeed, cities have become so expansive that they have complex geographies of their own. No wonder an entire sub-field called *urban geography*, the subject of this chapter, has developed to analyze:

>> The reasons why cities develop at particular locations

>> The factors that underlie urban growth

>> The spatial arrangement of commercial, industrial, residential, and other kinds of land use within cities

>> The planning that is meant to improve the quality of urban life

>> The social and political tensions between people who live in different parts of cities, and between cities and fringing political jurisdictions

>> The environmental quality issues occasioned by urban development and growth

Each of these will be touched upon to some extent in the pages that follow. If you are wondering what "urban" means, take a look at the nearby sidebar.

WHAT DOES "URBAN" MEAN?

Obviously, urban means a lot of people bunched together in a recognizable town or city. But how many? The answer varies from one country to the next. As far as the folks who run the U.S. Census are concerned, the threshold figure is 2,500. Or is it 50,000? Both? Well, sort of. An *urban area* (UA) is a continuously built-up area with a population of 50,000 or more. But outside of those UAs, an urban place is one with 2,500 or more people. Thus, if you have a town with that many people, then its residents are considered *urban*. If a neighboring town numbers 2,499 souls or less, then they are classified *rural*. Depending on where you live, "urban = 2,500 people" may make a lot of sense or sound like a bad joke. Throughout the lightly populated Dakotas, for example, the definition seems entirely appropriate. In New York City, however, where you may have that many people and more in a single building, the number 2,500 may seem woefully lacking. Oh, there's something else, too. Did I mention that different federal agencies also have different definitions for urban and rural? Have fun getting those Census folks together with the Department of Agriculture.

Getting a Global Perspective

At the global scale, urbanization has broadly mimicked the American experience. No worldwide census data are available from back in 1790, but available anecdotal and statistical evidence suggests that the urban percentage for the world as a whole was about that of America's. Today, some 55 percent of the world's people live in urban areas, but that number masks significant variation. In the more developed world, about 80 percent of the populace (much like the American percentage) live in cities, although in some countries, the figure is much higher. Urbanization in Uruguay and Argentina, for example, is currently at about 95 percent and 92 percent respectively.

On the other hand, in the least-developed countries (by the United Nations classification in 2020) the rate of urbanization today is about 35 percent. It makes sense that this figure is so much lower because by definition a less-developed country is one whose economy is based on agriculture — so the populace tends to be down on the farm instead of downtown. But that statistic also masks variation. In Mongolia, for example, 69 percent of the people are urbanized, while in Bangladesh, the figure is 38 percent. Nevertheless, some of the world's largest cities have sprung up in developing countries, and with them some of the most challenging venues regarding urban planning and quality of life issues. Table 17-1 shows the top 10 largest areas. (Note: City is defined in Table 17-1 as a contiguous metropolitan area. That may include any number of municipalities that fringe a central urban area.)

TABLE 17-1 **The 10 Largest Metropolitan Areas in 2020**

City	Population
1. Tokyo, Japan	37.3 million
2. Mumbai, India	26.0 million
3. Delhi, India	25.8 million
4. Dhaka, Bangladesh	22.0 million
5. Mexico City, Mexico	21.8 million
6. Sao Paulo, Brazil	21.6 million
7. Lagos, Nigeria	21.5 million
8. Jakarta, Indonesia	20.8 million
9. New York City, United States	20.4 million
10. Karachi, Pakistan	18.9 million

Economic development brings with it employment prospects that are largely urban based (for more information on this, see Chapter 15). Now, in developed countries these opportunities tend to be widespread and manifested by country-wide urbanization. In developing countries, however, they tend to be much more concentrated in the geographic sense with the result that the populations of one or two cities (the principal development centers) skyrocket. That is the reason why, as Table 17-1 shows, most of these large cities are in developing countries.

Because of concentrated urban growth, an important aspect of developing country urbanization is the disproportionate occurrence of *primate cities.* No, this has nothing to do with *Planet of the Apes*. A city is classified as primate when it is far and away the largest city of a country. Classically, it tends to be at least twice as large as the second largest city. More often than not, it is also the capital. Thus, it stands as the overwhelming center of population, political power, employment, commerce and wealth.

Primate cities are powerful magnets for rural-to-urban migration, often resulting in sprawling areas of disease and squalor inhabited by poor folk who have come to seek work. Also, the concentration of so much power and prospective opportunity in a single setting may result in domestic political tension that pits inhabitants of a primate city against people in other parts of the country that feel deprived of their fair share of economic opportunity. On balance, therefore, the consequences of having a primate city are not entirely positive.

As suggested by the 1790 census data at the start of this chapter, the modern reality of so many people worldwide living in so many cities is at odds with the way things used to be. In fact, urbanization is at least a 5,500 year-old process whose early geography may be traced to a handful of locales.

Getting Started: Urban Hearths

REMEMBER

Cities originated independently in five regions (shown in Figure 17-2) often called *urban hearth areas* — places where the first cities developed. They include:

» **Mesopotamia:** The Tigris–Euphrates River valley that mainly lies in modern-day Iraq.

» **Nile Valley:** Especially in what is now Egypt.

» **Indus River Valley:** In what is now Pakistan.

» **The Lower Huang Ho (Yellow River) Region:** This is located in Northeastern China.

» **Mesoamerica:** The region that extends northward from Nicaragua to and including Central Mexico.

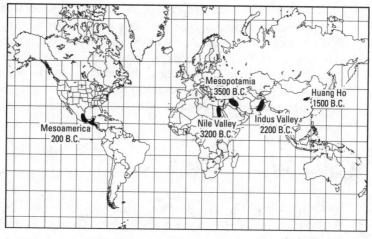

FIGURE 17-2: Locations of urban hearths.

(© John Wiley & Sons Inc.)

CHAPTER 17 **Downtown to the 'Burbs: Urban Geography**   305

Mesopotamia probably has the honor of being the first. Its oldest urban ruins date back about 5,500 years. Ancient Egypt appears to be a close second.

Trade and commerce played important roles in the growth and development of hearth-area cities. But so, too, did the need for centers for religious rituals and observances. As a result, imposing structures of ceremonial significance (such as pyramids and temples) dominated "skylines." These tended to be located alongside or at the intersection of broad boulevards aligned in directions that often had theological significance. Complementing these imposing structures were achievements of more mundane, yet fundamental, varieties. Yesterday's cities (yes, even without traffic lights and fast-food restaurants, they're still considered cities), with early populations that probably didn't exceed 25,000, showcased distinct residential, commercial, and institutional zones, as well as municipal services such as water supply and garbage removal. These prove that municipal planning and governments were at work even back then. Thus, the urban hearth areas were scenes of important experiments and innovations in human settlement, the results of which were passed down to succeeding generations, modified, and ultimately spread to other lands.

REMEMBER

Geographically, what is perhaps most important about these hearths is their locations. Each coincided with an area of high agricultural productivity. This positioning made a great deal of sense because the principal prerequisite for the appearance and growth of cities is an agricultural surplus. Why? Because that alone frees some people of the necessity to produce food and allows them to become artisans, traders, shopkeepers, soldiers, clerics, and so on, and cluster in settlements — cities — that promote the interaction of those services. Accordingly, each hearth (except Meso-America) is located amidst fruitful lands that roughly parallel one or more large rivers. The richness of the soils, whose fertility was regenerated by silts deposited by periodic floods, allowed people with even the most basic agricultural technology to produce good harvests (food surpluses) on a regular basis. Each hearth (including Meso-America) was also noted for impressive irrigation systems that maximized the agricultural potential of local waters and soils.

Finding Sites for Cities

REMEMBER

While the civilizations that produced those early cities are things of the past, the locations of urban hearth areas demonstrate a geographic principle that has persisted through time. Namely, cities don't just happen. Instead, they tend to spring up in particular locations because of some characteristic(s) of place that was attractive to settlers.

For example, and strange as it may seem today, the Spaniards settled the Los Angeles, California area for its agricultural potential (the landscape and climate also felt like home). Mountains and foothills come down to the sea along the West Coast, affording limited venues for agriculture and settlement. The Los Angeles area, however, offered a substantial expanse of relatively flat land complemented by a then-adequate local water supply that trickled down from the fringing San Gabriel Mountains. That was then. Today, few, if any, people live in LA because of its crops potential, and that makes an important point about urban location factors. Namely, the reason why a city began at a particular location may have nothing to do with the reasons why people live there today.

The list of factors that explain why cities began and developed at particular sites is lengthy. The four that are discussed in the following sections are the most popular.

Confluence

A *confluence* is a place where two rivers meet to form a combined flow. Pittsburgh, Pennsylvania, for example, began as a small settlement where the Allegheny and Monongahela Rivers join to form the Ohio River. Other examples are St. Louis, Missouri, near the point where the Missouri and Mississippi join waters; Khartoum, Sudan, where the White Nile and Blue Nile come together; and Manaus, Brazil, where the Amazon and Rio Negro combine their waters.

Rivers served as highways in olden times, so being at a confluence was like being at the hub of a transportation network. Confluences are prone to flooding, however, so availability of high ground is essential for safe settlement. Fortunately, plenty of high ground is available in Pittsburgh. In contrast, the land is rather low-lying where the Mississippi and Missouri meet. For that reason, St. Louis is not located right at that confluence, but a few miles downriver, where the banks are appreciably higher.

Protected harbor

Ever try to load or unload a sailing ship that is pitching to and fro in moderate-to-heavy waves? I didn't think so, and don't bother trying. It's very difficult and downright dangerous. What you need are calm waters, preferably in a *protected harbor* setting, where intervening land negates the big waves of the open sea. Because shipping has long been a principal means of moving people and goods, protected harbor settings have long been prime sites for settlement. Geographically, what do Boston, New York, Baltimore, Norfolk, San Diego, San Francisco, Seattle, Vancouver, Tokyo, Sydney, Mumbai, Karachi, Hamburg, Havana, and Rio de Janeiro have in common? You got it.

Head of navigation

A *head of navigation* is the farthest point upriver that a boat or ship can travel from a river's mouth. Washington, DC, for example, is located at the head of navigation of the Potomac River. That is, ships of good size can proceed up the Potomac from its mouth as far as the site of that city, where rapids make further navigation impossible. You may have heard of the *Fall Line* or *Fall Zone* before. This is a landscape feature in the eastern United States. This is an area where hard crystalline rock meets the softer, sedimentary rock of the coastal plain. The result is the rapids that make up the head of navigation. Augusta, GA, Columbia, SC, Raleigh, NC, and Richmond, VA, are each along this feature. In other instances, waterfalls, shallow water, or narrowness calls a halt to water-borne traffic.

In the settlement of North America, heads of navigation sometimes proved a logical spot for boats to stop, be unloaded, and later for settlements to develop. In some cases, these sites afforded other advantages for settlement. Where waterfalls or rapids impeded navigation, these features sometimes provided sources of waterpower for mills and other economic activities. Narrows or shallow water, in contrast, sometimes offered opportunity for bridge construction. Other examples of heads of navigation in the United States include Trenton, NJ (Delaware River); Louisville, KY (Ohio River); and St. Paul, MN (Mississippi River).

Defensive sites

Defensive considerations were paramount in the siting of several cities. The oldest part of Paris, for example, is the Ile de la Cite, a small island in the middle of the Seine River. The river served as a protective moat for the island's oldest known inhabitants, a tribe called the Parisii. Similarly, the oldest part of Montreal is an island in the St. Lawrence River that served a defensive purpose.

Some rivers exhibit dramatic U-shaped bends called *meander loops*. By siting a town on the inside of a loop, the river served as a protective perimeter. Examples of this include New Orleans, Louisiana, and Berne, Switzerland.

Acropolis sites, characterized by locally high ground, which is fairly easy to defend, have also been favored. Examples include Quebec City, Albany (New York), and Athens, Greece — whose famous Acropolis is the source of this category. *Peninsula sites* were also favored because a wall across a neck of land, complemented by the surrounding water (a natural moat), greatly facilitated defense. Boston and Mumbai, also noted for their natural harbors, are good examples.

Because of these and other site-related factors, settlements developed at various locations around the world. All were small at first. Many did not stay that way.

Getting Big: Urban Growth

At favored sites around the world, hundreds (if not thousands) of once-humble settlements have developed into cities of varying size. And the process continues. In fact, urbanization is not just a major human characteristic, but also a major trend. But the pace of urban growth and the reasons for it are not the same everywhere. Because space does not permit a truly global treatment of this truly global process, I'm going to focus on a prime example of the urbanization process — the United States.

Much of the growth of American cities has been due to immigration and natural population growth. But other factors have also encouraged cities to expand not just population-wise, but also with respect to the physical area taken up by them. The following sections discuss some of the principal factors.

Rural-to-urban migration

Over the world as a whole, the last century has witnessed substantial rural-to-urban migration. In the United States and several other countries, a couple of processes have been fundamental to this shift.

Mechanization of agriculture and farm consolidation

Agriculture was once a largely manual enterprise that required many hands to tend a farm of decent size. But with the advent of farm machinery (tractors in particular), the number of people needed to do farm work was greatly reduced. At the same time, mechanization also encouraged farm consolidation (in which one farmer buys out another, thereby increasing the acreage owned by the buyer) because tractors made it possible for a farmer to farm much more land than was the case before. Thus, the pre-mechanization rural landscape that consisted of a large number of small farms was transformed into one containing a comparatively small number of large farms. In 1920, (largely pre-tractor times) 6.5 million farms in the U.S. averaged 149 acres. Now about 2 million farms average about 444 acres. During the same period, farm population dropped from 31 million to about 2.5 million. Cities absorbed the brunt of the difference.

Changing economy

The mechanization of agriculture was symptomatic of a structural change in the overall economy that witnessed an increase in industrial and commercial enterprises that were largely urban-based. Manufacturing was particularly important because it afforded job opportunities for unskilled or semi-skilled individuals who were either recent immigrants or recently forced off the farm. Thus, in 1870 the overwhelming majority of African Americans were rural Southerners. Less than a century later, a majority were urban dwellers, many of them holding factory jobs in cities outside of the former Confederacy.

Changing means of transportation

REMEMBER

How big would a city be if every inhabitant had to walk to work? The answer is pretty simple: Not very big at all. And in fact, that's the way things were long ago. When overland transport was exclusively a matter of walking, horseback, carriage, and cart, it took a while to move people and products meaningful distances. As a result, cities tended to be rather compact.

Over the years, however, changes in transportation have allowed people to live progressively farther from their places of employment, thereby expanding the physical size of cities. Following are various "urban transportation ages," though not all apply to every city.

Pre-mechanization

Prior to about 1890, foot, carriage, or horse-drawn trolley accomplished most movement of people and goods. This resulted in compact cities, the largest of which could be completely traversed in a 30–45 minute walk. Attached or closely spaced housing was the norm.

Electric trolleys

Around 1890, introduction of electric-powered trolley cars provided a quantum leap forward in speed. While a 30-minute walk might allow a commuter to live perhaps as far as 2 miles from work, a commensurate trolley ride could extend the range of residential potential to 5 miles and more. As a result, the outer edge of urban areas expanded significantly.

Commuter railways

Starting around 1900, commuter railways developed using standard, instead of narrow, gauges. The result was another quantum leap in speed that expanded the range of commuting to 20 miles and more. Commuter trains can only travel where tracks are present, of course. Thus, cities expanded along rail corridors that were rather like tentacles, leaving countryside in between.

Pre-freeway

Advent of mass automobile ownership facilitated unprecedented freedom of choice with respect to residential location. Suburbs consequently developed and "filled in" substantial portions of rural land between the commuter rail tentacles described previously. Until roughly the mid-1950s, however, highways were largely of the 2- and 4-lane variety. Combined with red lights, stop signs, and traffic, commuting speed was somewhat limited and so too, therefore, the distance at which one may choose to live from the city.

Freeway

Multi-lane, high-speed, limited access (as in limited points of entrance and exit) divided highways first appeared in the early 1950s and, in the U.S., proliferated with the advent of the federally funded Interstate Highway System. Initially, at least, the effect of these arteries was to greatly increase the average speed at which a motorist could travel and therefore also increase the potential distance from residence to job. Complemented by construction of *beltways* — interstate-style ring roads designed to help people travel around the central city without going through it — this ongoing age has witnessed significant increase in the "filling in" of remaining rural land between commuter rail "tentacles," and pushed the outer edge of the city in multiple directions.

TIP

So massive have freeways become that some entrance/exit cloverleaf ramps are now larger than the entire "downtown" of some of the older cities in Europe.

Automobile ownership

By themselves, neither freeways nor roads of other sorts tell the full story of contemporary urban growth in the U.S. More telling is the number of people who own cars and use those roads and freeways to commute to work from a suburban residence. And that number is humongous. More than 285 million vehicles are registered in the United States with more than 90 percent of households having access to a vehicle, and about 24 percent of households owning three or more.

Low-cost fuel

The price of automobile fuel in the United States is relatively low. That prompts urban sprawl by encouraging automobile usage and increasing the distance people are willing to commute. Now, some people may recoil at the notion that gas is cheap in the United States. All they need to do, however, is vacation in Europe or almost any other part of the world to find out firsthand just how inexpensive American gas is. The principal reason for the price difference is low taxes on gas consumption in the U.S., which encourage automobile purchases, new home construction, and other consumer activity that contributes to the economy even as it contributes to urban sprawl.

Home mortgage deductibility

The U.S. federal government has long encouraged home ownership by allowing buyers to deduct from their federal income tax the interest paid on home mortgages. The effect is to encourage new home construction and urban sprawl. Canadians, in contrast, do not get the same tax benefit. Their cities also sprawl, but generally not to the extent of those on the U.S. side of the border. Several factors explain that, and mortgage deductibility is one of them.

Looking Inside the City

In addition to sites and sizes, cities around the world vary dramatically in their internal physical appearance. Numerous cities in Europe, North Africa, and Asia, for example, contain an "old town," characterized by traditional architectural styles, and a citadel, cathedral, or mosque that towers above historic low-rise buildings. In other cities, strict building codes produce an air of distinctness. Thus, while many modern cities are defined by center-city skyscrapers, Paris and Venice, to take just two of many examples, have uniformly low skylines that preserve historic character and visual dominance of leading landmarks. Taller buildings do exist, but they tend to be in the periphery outside the city center.

Internal physical appearances of cities may also vary as a function of wealth or income of residents. Urban neighborhoods of the rich and poor not only look different, but also tend to occupy different locations in the cities of different countries. Thus, in many developed countries, the urban poor tend to live in older housing in the inner city. The wealthy, in contrast, generally tend to live in the suburbs or urban fringe. In many developing countries, however, the pattern tends to be the opposite. That is, the poor tend to concentrate in slums that are located on the urban periphery while the wealthy are located much closer to the urban core. Accordingly, go up into the hills on the outskirts of many American cities and you are likely to see expensive homes that afford a great view. Do the same in most large Latin American cities, and you are likely to see some of the worst slums and shanties imaginable. If that distinction were not clear enough, find an aerial image of a city and look for green space and trees. That, too, will tell you where the wealthy are and are not.

Distinct areas of commercial, industrial, residential, and other kinds of land use came into being as cities grew. Their extent and arrangement vary substantially in different cities in different countries. In the United States, the factors promoting urban growth are largely the same throughout the country. To a fairly substantial extent, therefore, American cities exhibit a certain sameness with respect to their structure.

Specifically, at the center there tends to be a discernible "downtown" dominated by relatively tall buildings and mostly non-residential land use. On its fringe are high-density residential neighborhoods characterized by high-rise apartments and attached multi-family dwellings. As distance from downtown increases, population density generally decreases. Detached housing becomes the norm and lot sizes increase. Mixed here and there amidst suburbia are shopping centers of varying sizes, with occasional full-blown shopping malls on the periphery of larger cities (a phenomena on the decline in an era of increased online shopping).

If you live in a city, or are familiar with one, how well does that description conform to your experience? The sections that follow consider urban parts and processes in greater detail.

The central business district (CBD)

At or near the center of most cities is what students of urban geography call the *Central Business District* (*CBD*) and what normal, well-adjusted people call "downtown." This is the core of the city and in most instances is characterized by:

- The area where the city began
- The focal point of the main transportation arteries
- The tallest buildings
- City hall, court buildings, and government offices
- Major commercial, retail, and office buildings

In some respects, "downtown" is preferable to "CBD" because the former infers that much more is going on in the area than just business. Still, the concentration of business enterprises is such that commercial land use dominates the CBD and gives rise (quite literally) to its most outstanding characteristic: large and imposing structures, usually including the tallest buildings in town. Competitive bidding for real estate is what leads to those structures, and both the bidding and the buildings deserve at least modest description.

Competitive bidding and the rent gradient

TECHNICAL STUFF

Historically, downtown was *the* place to be as far as commercial enterprises were concerned. That is because it was the focal point of the transportation network and therefore readily accessible to a high percentage of the city's population, all of whom were potential customers or employees. As a result, demand for downtown business locations quickly exceeded the supply of real estate. When demand exceeds supply, whether you're talking real estate or rock concert tickets, the price rises. This *competitive bidding* for downtown real estate drove up land prices in that area relative to other parts of the city. As a general rule, competitive bidding for land generally decreases as distance from the urban center increases. As a result, land values tend to "peak" downtown and then decrease as distance from the CBD increases — sharply at first, and then more gently. Geographically, therefore, land values exhibit a gradient that declines from the CBD to the periphery of the city. The trend line is often called the *rent gradient*, even though land purchases rather than renting *per se* can be the predominant financial transaction.

Tall buildings

The quintessential result of the downtown peak in the rent gradient is tall buildings. And for good reason. Suppose you have just purchased land downtown and borrowed a hefty amount of money from a bank to do it. Now you've got a loan to pay off, along with its accruing interest. Also, given the high price you paid, you can bet your last dollar (if you have one) that the annual real estate tax on your property will be through the roof (if you can afford one).

How can you cope with this predicament and make a tidy profit to boot? Create as much floor space as possible on your property and rent it out to businesses that will pay a pretty penny to locate downtown. And the way to create that floor space is to go vertical — that is, build as many stories (thereby creating as much rental space) as you and your bankers deem feasible. In most cities, you find some tall buildings that are not downtown, but the rent gradient usually guarantees that the CBD will have the highest concentration of them. Where are these other tall buildings? Usually near the newest "accessible" space: a freeway offramp.

Residential areas

REMEMBER

Urban residents generally live in one of two areas: multiple housing buildings that fringe the CBD or detached, single-family residences of outer-city areas. The following sections discuss some of the areas that crop up when people choose to live in and around cities.

Rich folks-poor folks, inner city-suburbs

As a general rule (which has numerous exceptions) people of low-to-modest incomes tend to live in inner-city high-density neighborhoods while the more affluent reside in the suburbs. This geographic pattern is not as old as the cities themselves, but instead evolved over time. Specifically, an old high-density residential area usually adjoins a city's CBD. Most often the buildings date from pre-automobile times, when the need to be close to work encouraged dense settlement. As affluence rose and transportation improved, the middle- and upper-income people (largely white) tended to move to outlying areas, while low-income people (largely minority) moved into the newly vacant residential spaces. As the number of inner-city poor people rose, landlord incentive to spend money to maintain and improve their housing declined. Slums were the virtually inevitable result despite occupying land little more than a stone's throw from high-value CBD real estate.

Other factors have subsequently served to place suburban real estate beyond the reach of the inner-city less-well-off. These include local (suburban) decision-making not to build public or low-cost housing, and passage of ordinances that mandate large minimum lot sizes or house sizes. Also, lack of efficient (or any) public transportation in suburban areas decreases the likelihood that the poor will move in.

As a result, cities and suburbs generally end up being inhabited by people of different race and social class. These people tend to have different needs, different attitudes toward social services, different views on the role of government, different perspectives on taxation, different priorities about how tax money should be spent, and differences in environmental risk and amenities (see sidebar "Applied Geography: Lessons from Bhopal?"). Political tension between inner city and suburban residents is an almost inevitable side effect. This sometimes results in one political party being "the party of the city" and another being "the party of the suburb."

Ethnic neighborhoods

An *ethnic neighborhood* is a residential area in which people of common origin voluntarily live in close proximity to each other. (This is in contrast to a *ghetto,* in which the choice of residential location is involuntary.) These are common features of American cities (Canadian and European ones, too) inasmuch as they afford foreign immigrants opportunity to

>> Interact with others who literally and figuratively speak their language

>> Maintain contact with kinfolk

>> Have access to stores and eateries that specialize in familiar goods and foodstuffs

>> Worship in a preferred manner

The vast majority of these neighborhoods are found in high-density inner-city residential areas. That is a reflection of the (normally) limited disposable income of immigrants, proximity to real or potential places of work, and, again, strong desire to be among people with a similar background.

Ghettoes

As previously mentioned, a ghetto is an involuntary neighborhood where members of a group are forced to live. Typically, they are found in an inner-city, high-density residential area. *Ghetto* is Italian for "foundry," and reflects the fact that Jews in Venice of old were relegated to (and literally locked in at night) a small area in the factory district.

APPLIED GEOGRAPHY: LESSONS FROM BHOPAL?

Sadly, hearing about an explosion at a chemical plant is not necessarily big news. The explosion certainly is big, but we've grown accustomed to this risk as a tradeoff for all the benefits that come from products made from a whole host of chemicals. One accident stands out among those many. On December 2, 1984, an explosion at a chemical plant in Bhopal, India produced a cloud of toxic gas that settled over nearby residential areas. Some 5,100 people died, and many thousands more suffered permanent injures to lungs and eyes. Poor operational safety within the plant was the main culprit, but another issue was a lack of zoning and another concern referred to as *environmental racism*, the process by which minority populations are disproportionately burdened with health hazards like toxic chemicals, landfills, and other noxious industrial byproducts.

Zoning is the process in which urban planners (many of whom are trained in geography) allocate particular kinds of land use to particular parts of a city. With respect to Bhopal, why, some have asked, was a factory of that sort and a densely populated residential area ever allowed to exist side-by-side? An appropriate degree of geographic separation between plant and people, consistent with rational zoning, would not have prevented the accident, but no doubt could have resulted in far fewer casualties. One reason suggested for the lack of zoning is environmental racism. An objective look at these facilities anywhere in the United States and not just far off places like Bhopal clearly illustrates that it is not just horizontal geographic space that can increase your risk, but also vertical social space. In other words, *who you are* in a social hierarchy and *where you are* physically matter regarding the opportunities and threats that people face in their local environments.

Opinion varies (often heatedly so) about the appropriateness of the term in America, where it has been applied most often to the African American urban experience. While these people have never been literally locked in a particular area, in certain times and cities, they have been victims of *red lining*. This is a discriminatory and now illegal practice whereby real estate agents mark off on a map (classically with a red pen) parts of town in which housing may and may not be sold to members of this group. The result, for all intents and purposes, is a ghetto.

Leaving Downtown, Living Downtown

The end of the 20th century and early part of the 21st century have seen two rather opposing processes impacting the character of cities. One involves movement to the suburbs and outer fringe of large stores and other entities formerly located downtown. The other (and more recent change) entails the movement of middle- and upper-income people into downtown areas. Each process, described in the following sections has brought significant socioeconomic change — for better or for worse — to affected areas.

Moving out of downtown

Decentralization refers to the appearance in the outer city of functions and land use formerly associated almost exclusively with the CBD. The quintessential example is a large downtown department store that closes, perhaps after years of being part of the CBD, and relocates to a newly opened suburban shopping mall. The most profound effect is to move commerce and retailing from the center city to the outer city (hence decentralization), often resulting in vacant real estate downtown and loss of jobs for inner city residents. Here are some principal manifestations and results.

Suburban shopping malls

Enclosed shopping malls consisting of two or more major department stores and dozens of lesser-sized shops of considerable variety are perhaps the highest expression of decentralization and were a hallmark of American commerce at the close of the 20th century. They come about for one or more of the following reasons:

» The cost of land or rent is cheaper in the suburbs than downtown.

» Proximity to people with high disposable income stimulates business, and more people fit that description in the suburbs than in the central city.

> » Malls are more convenient to get to by auto than is downtown. Specifically, getting to a suburban mall is likely to involve less traffic, fewer lights, and lots of free parking.
>
> » Shoppers (especially suburbanites) tend to prefer mall shopping because of the juxtaposition of different kinds of shops (it's the equivalent of a modern-day bazaar), their comparative cleanliness and perceived safety, and their enclosed climatically controlled environment.

Nowadays, many of these malls and other large, big-box retail locations stand empty. People are finding local shopping more fulfilling or are turning to online sources that can ship goods directly to their home.

Office parks

An *office park* is a cluster of multi-story office buildings (glass facades are particularly popular) that typically occupy more than enough acreage at or near a major suburban intersection. Trees and other landscaping (hence, park) complement the architecture. The suburban location often constitutes a preferable commute for employees who might otherwise have to drive downtown. Similarly — and just like shopping malls — the locations facilitate accessibility by existing near potential middle- and upper-income clients, who are more likely to reside in the outer city than the inner city, and who prefer the former setting to the latter as a place to take their business.

Edge cities

Comparatively small edge cities sometimes appear on the outer fringe (or edge) of large cities. Sometimes this is simply the result of merging together due to urban growth. More often, however, edge cities are newly incorporated areas that arise in response to the needs of suburbanites for a wide range of goods and services, and their preference to obtain them in car-friendly places that are closer to home than the old CBD. This is the location of the secondary set of high-rise buildings outside the CBD. Edge cities can grow to be quite large and include Tysons Corner outside Washington DC or La Défense in greater Paris.

Jurisdictional spillover

Urban sprawl often results in a contiguous urban area that exceeds the original city limits and spills over into the territory of other cities, counties, or even states. As a result, hundreds or even thousands of people may earn a paycheck in one jurisdiction, but pay taxes and spend their disposable income (thereby encouraging job-creation) in another jurisdiction. This is cool as long as it all "evens out," but it seldom does. The jurisdiction that "loses" is the one with the highest number of workers who live somewhere else, and typically that is the central city.

Some cities attempt to enact a commuter tax in these cases. The argument goes that workers use city services (police, fire, and so on) and infrastructure without paying for them. As with most taxes, this is rarely popular.

Moving back downtown

In recent years, a rather provocative geographic trend has been occurring that rather flies in the face of urban sprawl and decentralization. Called *gentrification*, it's the process by which middle- and upper-income people (who you would normally expect to see settling down in the suburbs) buy and refurbish central-city housing.

Causes

The causes of this are diverse, but typically include one or more of the following:

- Preference for social diversity and cultural amenities, which typically are in greater supply in the CBD as opposed to suburban settings.
- Desire by people whose jobs are downtown to forgo the time and cost of commuting from the urban fringe.
- The often-favorable cost differential between buying and refurbishing inner city property versus the purchase of property with a home on it in the suburbs.
- Availability of large rental spaces downtown pursuant to commercial and manufacturing concerns that have decentralized or gone out of business.

Characteristics

Having been set in motion, gentrification may be manifested in ways that may include the following:

- Social change takes place in the inner-city gentrifying area as more affluent people move in.
- Housing stock becomes refurbished.
- The continued influx of buying power leads to competitive bidding, which drives up housing and residential rent costs.
- Spending by newcomers has a spillover effect, increasing the value of commercial real estate as landlords find they can command higher rents.

Consequences

Once it has firmly taken hold, gentrification may have significant consequences that include the following:

- Rising rents cause displacement of long-term commercial and residential tenants, giving rise to homelessness.
- Low-rise buildings are bought and demolished to make way for high-rises.
- "Mom and pop" stores become boutiques. The former unique commercial character of the area becomes increasingly "placeless" as chain stores and franchises become more a part of the local landscape.
- The urban tax base expands.

The bottom line is that gentrification is a double-edged sword. It can encourage social and physical displacement of those least able to afford it even as it encourages revitalization of the inner city and reinvention of the old CBD.

Facing up to Environmental Issues

Just as city growth has had diverse social and economic consequences, so, too, has it given rise to a host of environmental challenges. The following sections cover a few of the biggies.

The urban heat island

The atmosphere over a city tends to be significantly warmer than over rural areas, a phenomenon called the *urban heat island*. While the effect may be a modest blessing during the cold of winter, it can also amplify summer heat waves, resulting in higher rates of health emergencies and heat-related fatalities than in non-urban areas. Here are three reasons for this phenomenon:

- Asphalt, concrete and other hard materials that dominate city surfaces have rather high capacities to absorb solar energy, causing urban areas to warm up much faster than countryside.
- Heat is a by-product of motors and machines, which abound in cities. Ironically, air conditioners are a major contributor to urban heat islands during summer.
- Exhaust from automobiles and engines is warm in and of itself, but also has high concentrations of gases that absorb and retain solar energy with great efficiency.

Dust domes and smog

Exhaust from internal combustion engines and other sources is high in particulate matter and smog-producing chemicals. As a result, air pollution problems, attendant to resulting *dust domes* and *smog,* tend to be particularly acute in urban areas, as manifested in part by disproportionately high rates of asthma and other respiratory problems.

The extent of dust domes and smog vary from city to city, and not simply as a function of sizes. Terrain may be a factor. Mexico City, Los Angeles, and Santiago de Chile for example, occupy lands that are bounded by high hills and mountains that concentrate air pollution. Existence and enforcement of environmental law, particularly as it relates to engine emissions, is also a factor.

As a high schooler in Los Angeles, I vividly remember sirens that pierced the air to warn us to head indoors to our classrooms on high smog days. You could look down the hillside of the Palos Verdes peninsula and watch orangish-brown air wafting its way upward toward us. In that regard, the Golden State sometimes wasn't.

Finding a place for refuse

Large cities produce large quantities of refuse and sewage and have few solutions regarding their safe disposal. Pursuant to passage of environmental protection laws, ocean dumping is no longer an option, and many cities are running out of sites for landfills within their boundaries. One of the more provocative consequences has been substantial interstate commerce in refuse, in which typically a poor rural county agrees (for a price) to serve as the final resting place for big-city garbage and sewage. When that fails, sometimes the scale changes and waste finds its way to a less developed country.

IN THIS CHAPTER

» Linking the parts of ecology

» Messing up a planet in so many ways

» Turning small problems into big ones

» Making decisions that matter

Chapter 18
Only One Home: Impacts on the Environment

During the swim portion of a recent Ironman triathlon in Maryland's Choptank River, I was met by thousands (yes, thousands) of jellyfish. I and more than 2,000 other athletes were stung repeatedly in the arms and face over the 2.4-mile swim. Imagine how fun it was to then bike and run another 138.2 miles.

But there was something else concerning in this river that feeds into the Chesapeake Bay. The Chesapeake is the largest *estuary* (an inlet where the salty tide meets freshwater current) in the United States and perhaps the most economically important one, too. The Bay is a world leader in blue crab production. Striped bass, perch, shad, flounder, oysters, and a host of other species help comprise this historically rich and robust fishery. So, naturally, anything that could upset this ecological system is a worry. And in the Chesapeake Bay, that worry is phosphate.

The resulting problem with phosphate is algae, tiny aquatic plants that takes in oxygen from the water. Occasionally, algae reproduces by the gazillions, creating algal "blooms" that deplete oxygen over large portions of the Bay and suffocate large numbers of fish. These blooms are caused by periodic build-up of phosphate in the water. Algae, it seems, just love this substance, and reproduce like crazy in its presence.

That, of course, leads one to inquire into the origins of the bloom-producing chemical. The Bay is not an isolated body of water. Instead, rain and runoff connect it with a broad land expanse (see Figure 18-1). The phosphates originate mainly in agricultural fertilizers and manure that run off from chicken production and other farms into several Chesapeake tributaries. Indeed, virtually hundreds of watercourses from six states drain into the Chesapeake, collecting chemicals from many, many sources along the way.

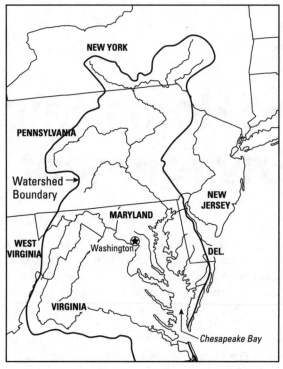

FIGURE 18-1: Hundreds of watercourses from six states drain into the Chesapeake Bay.

(© John Wiley & Sons Inc.)

Maybe you're saying, "OK, but what does this have to do with geography? This sounds to me more like chemistry or biology." Geography is concerned with human and natural phenomena that give character to different parts of Earth's surface. Often, of course, important characteristics that we observe in a location or region are products of interaction between people and nature. While these relationships are sometimes harmless, they may also have negative consequences, as in the Chesapeake Bay example.

Environmental geography is the sub-field that focuses on human impact on the natural environment, and is the topic of this chapter. Previous chapters have referred to negative environmental impacts without mentioning environmental geography by name. Devoting an entire chapter to it is meant to emphasize both the magnitude of society's ability to alter the face of Earth, and the existence of a discreet area of geography that is devoted to the topic. Although talking about pollution may seem like a downer, it affords the opportunity to think critically about the role of geography in assessing the present and planning for the future.

Grasping the Basics — Environmentally Speaking

Environment refers to the myriad of natural characteristics and conditions that affect and are affected by humans. *Environmental geography*, therefore, is the study of the characteristics of locations and regions that are the result of human-nature impacts. This is closely allied to the field of *ecology*, which studies the complex connections that link the following elements:

- **Atmosphere:** the climate surrounding Earth
- **Hydrosphere:** water on Earth in all its forms
- **Lithosphere:** the solid earth including soil and other loose surface particles
- **Biosphere:** all life on Earth

The emphasis on location and place is what most distinguishes environmental geography from ecology. Thus, while an ecologist might study agriculture-nitrate-algae-fish linkages in the abstract or at the scale of microbiology, environmental geography studies them as large-scale events that characterize locations or regions. In that regard, geography is very much at home with the study of *ecosystems,* the living things that occupy particular areas together with the inorganic elements on which they depend. In Chapter 2, for example, there was a discussion of the African lion that basically was about an ecosystem. Specifically, lions (biosphere) inhabit areas where the climate (atmosphere) results in rainfall (hydrosphere) characteristics that produce grassland (biosphere). These grasses attract grazing animals (biosphere) like wildebeest, zebra, and impala that in turn attract lions. But the grasslands also result in soils (lithosphere) that are amenable to agriculture. We cannot forget people in this discussion, however. They are a part of this ecosystem, too, and people convert grassland to farms and grazing, depriving lions of the habitat they need to thrive.

The impact of people on the environment — of which we are very much a part and not separate – has become so great that some scientists have been describing our age as the *Anthropocene*. Earth is currently in a geologic epoch called the Holocene. Though unofficial as a measure of geologic time, the Anthropocene is used to describe this time as one where humans have permanently altered the planet. Changes in methane and carbon, adding radioactivity to the air and soil, reduction in biodiversity, soil degradation, and so on: These are markers for some that we are now in a totally new environmental time and space. For more on how people differ in their views on the environment, see the sidebar "The role of attitudes toward nature" later in this chapter.

Contributing Factors: Pollution on the Move

Impacts on the environment take several forms. In the sections that follow, several kinds are described together with natural phenomena that often (and perhaps surprisingly) help to turn small environmental problems into major ones.

Making an impact

Perhaps the most obvious human impact involves pollution, the introduction of substances (pollutants) that are harmful to the environment. These may result in *degradation*, the reduction in quality of natural environmental elements, or in *depletion*, the reduction in quantity. Human impacts can also be manifested by acts of removal or addition. Thus, the wholesale removal of forests (*clear-cutting*) may not only deplete forest resources, but perhaps also contribute to soil erosion. On the other hand, introduction into the environment of some element for which no natural counterbalance exists may have an effect no less profound than, say, accumulation of phosphates in the Chesapeake Bay. A good example is provided by kudzu, a fast-growing vine that was let loose in the American South and has proceeded to dominate the natural vegetation over large areas.

Geographically, pollution may originate as a location-specific *point source* or as a larger scale *non-point source* (sometimes called an *area source*). In the Chesapeake Bay example, a pipe at a sewage treatment plant that discharges directly into the Bay would exemplify a point source. Crop fields in, say, Maryland, on which fertilizers have been applied and then run-off during rains would typify a non-point or area source.

Spreading the mess

By themselves, pollution sources *per se* generate limited environmental impact. The major problem, rather, is that pollutants tend not to stay put because they are released within an environment characterized by several kinds of motion. In other words, nature has mechanisms that "spread the mess." As a result, pollution that starts out as a confined geographic event may end up affecting broad areas and having extensive consequences. Previous chapters have described these mechanisms in some detail. Here is a brief recap with examples.

The water cycle

The water cycle (see Chapter 8) is a major "culprit" for spreading the mess. Rainfall originates, falls to earth, and runs off the land to create streams that flow to the sea. In so doing, the water cycle acts as a pervasive and efficient mechanism that "spreads the mess."

This process picks up the phosphate discharge, carries it away, and turns a local event (discharge) into a problem of Chesapeake-sized proportions. This process also produces run-off on an agricultural field that picks up fertilizers and carries them away. And of course, the Chesapeake Bay does not have *a* tributary, but instead more than a hundred of them, which together collect pollutants from thousands of point and non-point sources.

Wind

In innumerable instances, smoke and pollutants from a manufacturing or electrical generating plant go up a chimney (point source) and end up in the atmosphere. Here, too, the resulting environmental impact would be nominal if nature just stayed put. But, of course, it does not. Wind is another component of the dynamic environment. In Chapter 9 I discuss how solar energy leads to creation of high- and low-pressure systems that cause wind. And as a result, pollution that starts out as a very localized phenomenon becomes geographically general.

For example, as sanitary landfills fill up, incineration has been used as a means of solving garbage disposal problems, especially in big cities. But even the most efficient incinerators generate pollution that is released to the atmosphere, and then spread over wide areas, thanks to wind. Thus, a source of point pollution (the incinerator chimney) may affect the health of humans, vegetation, and property values over a wide area.

Currents

In many times and places, oceans have been convenient receptacles for human refuse. Also, as the global economy has grown and become more inter-connected, maritime traffic has increased, and with it the possibility of incidents that result in the release of cargo that is harmful to the environment. At this point the familiar story line repeats: If nature stayed put, then the impact would be minimal. But we live in a dynamic environment. The oceans are restless. Surface currents (see Chapter 9) can carry things far and wide while tides and waves can affect every coastal nook and cranny. Once again, therefore, nature can spread the mess, so what starts out as a local event is turned into an issue of greater geographical proportions.

The possibilities were amply demonstrated on April 20, 2010, when the Deepwater Horizon oil rig exploded in the Gulf of Mexico, and ultimately released over 4 million barrels of oil into the sea (Figure 18-2). Currents in the Gulf slowly began spreading the mess — in this case, the oil slick. Coastlines in Texas, Louisiana, Mississippi, Alabama, and Florida were threatened, impacting wildlife and the coastal tourism industry. Turtles, marine mammals, birds, plants, and basically anything else living within about a 70,000 square mile expanse of Gulf suffered or were killed by the thousands. Ten years later, samples of some fish still showed traces of oil.

U.S. Coastal Waters Affected by the Gulf Oil Spill

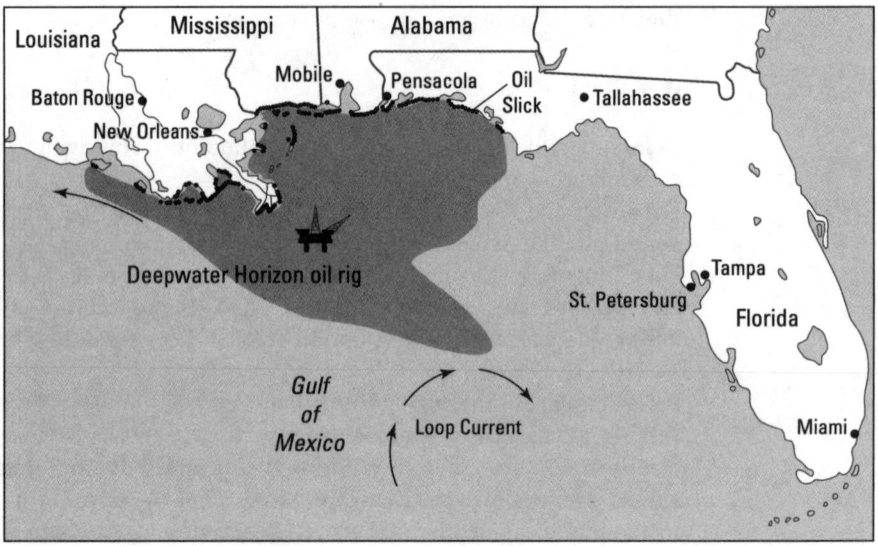

FIGURE 18-2: Currents in the Gulf of Mexico spread the Deepwater Horizon oil spill over a great distance.

(© John Wiley & Sons Inc.)

Migration and trade

Human beings are a species on the move. We migrate (see Chapter 12). We travel for business and leisure. We engage in global trade and commerce (see Chapter 15). And in the process, we sometimes purposefully or inadvertently transfer a particular species from its current home to a part of the world where it was previously unknown, possibly with negative environmental consequences. Following are two examples.

Rabbits were unknown in Australia until 1859, when a dozen or so were purposefully introduced to serve as a source of food. Lacking natural predators in their new homeland, their numbers went ballistic, reaching about a billion within a century. The problem? Sheep are important to the Australian economy, and five rabbits eat about as much grass as one sheep. Poisoning campaigns have subsequently killed off literally hundreds of millions of rabbits, but people now fret about the impact of those substances on local food chains (a concept discussed in the next section).

In 1988, an East European freighter in the Great Lakes dumped some ballast water (whose sole purpose is to add weight to a ship that would otherwise bob like a cork), and with it probably introduced the zebra mussel to North America. This diminutive critter with cute little stripes is native to the Black and Caspian Seas, where certain duck species and crayfish keep its numbers in check. Lacking a large predator population in the Great Lakes — plus the tendency of adult female zebras mussels to produce more than 30,000 eggs per year — the population soared at a rate that would make an Australian rabbit swoon. The mussels attach to hard surfaces, and in the process have caked and stopped up water intake vents of waste and industrial treatment plants.

Focusing on food chains

While one may decry negative environmental impacts on ethical or scenic grounds, the repercussions for human well-being is perhaps of paramount importance. Food chains provide an excellent example of how negative impacts may come back to haunt humans big time. No, not *fast* food chains that provide hamburgers, tacos, fried chicken, pizza, or other staple foodstuffs; but instead the food chains that involve consuming and passing along injurious substances in ways that prove ultimately harmful or fatal to humans and animals.

The "chain" of events

A *food chain* is a sequence of living things through which energy and other matter move in an ecosystem. That mumbo-jumbo clearly calls for a diagram — two in fact, inasmuch as there are *terrestrial* (land) food chains and *aquatic* (water) ones (see Figure 18-3). Though their specific components differ, the overall structures are quite similar.

> » At the base of any food chain are a number of *primary producers* that convert solar energy to organic matter — green plants in the case of terrestrial ecosystems, and aquatic plants or algae in the case of aquatic systems.

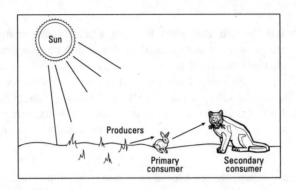

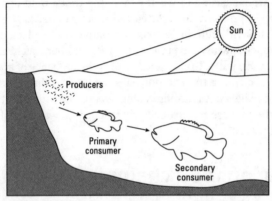

FIGURE 18-3: Terrestrial and aquatic food chains link diverse life forms.

(© John Wiley & Sons Inc.)

» Any animal that feeds upon a primary producer is a *primary consumer*. This may be a rabbit or squirrel (any herbivore will do) in the case of a terrestrial food chain, or a small fish in the case of an aquatic food chain.

» Any carnivore that feeds on a primary consumer is a *secondary consumer*.

» Any carnivore that feeds on a secondary consumer is called a *tertiary consumer*, and so forth. Humans can be primary, secondary, tertiary, or whatever consumers depending on what they are eating and whether that something had itself previously eaten another consumer.

» All living things ultimately die (a major bummer) and decay. *Decomposers*, which consist of various microorganisms and bacteria, feed on organic waste and matter at all levels in the food chain and, as their name suggests, aid in decomposition. Ultimately, they disintegrate organic matter to an elemental chemical level that makes it suitable for reuse by plants and animals — completing the food chain.

TIP

Thus, every food chain may be thought of as a set of linked organic units in which matter is constantly used and recycled.

The potential for danger

While food chains sustain life in all its forms, they may also pose grave (perhaps fatal) harm when toxic substances are let loose in the environment. The scenario may easily follow these steps:

1. Toxic waste from a chemical manufacturer enters a lake and taints microscopic forms of life.
2. Small fish eat the tainted matter and store the harmful stuff in their fatty tissue.
3. Bigger fish eat the smaller fish, and with each meal increase the concentration of toxic matter in their own fatty tissue, a process call *biological amplification*.
4. Fishermen catch the affected fish, eat them, and so acquire the toxic substance, perhaps with fatal result if enough contaminated fish are consumed.

TECHNICAL STUFF

Fallout is another way in which hazardous materials can enter the environment. Fallout is a quaint term that most folks don't fully appreciate. When a big explosion occurs, a lot of solid matter is instantly pulverized and becomes very fine particles that are thrust into the air. Being teeny, they can remain aloft for a long period until, because they have weight, they "fall out." Before that happens, however, wind may "spread the mess" over a wide area. If the explosion involves radioactive substances, then the fallout will be radioactive in nature. That alone may taint grass that is eaten by cattle and other livestock. Rainfall, however, may percolate radioactive matter into the soil, where it is taken up by roots and thus rather thoroughly contaminate the plants.

The Fukushima disaster provides an instructive real-world case study. On March 11, 2011, a nuclear power plant in that Japanese prefecture (the equivalent of a state in the United States) suffered an explosion that resulted in discharge of a large quantity of radioactive matter into the atmosphere and Pacific Ocean. The explosion had an environmental trigger, an earthquake and subsequent tsunami (See Chapter 6, "Shape-shifting Earth") that flooded the facility and knocked out a number of measures designed to protect against a nuclear meltdown.

Measurable amounts of radioactive cesium isotopes were still found in the Sea of Japan several years after the accident. Fears that a contaminated food chain would lead to adverse human health impacts have focused on fisheries. Public health officials will monitor for problems in the years ahead.

THE ROLE OF ATTITUDES TOWARD NATURE

People generally have kindly attitudes toward some natural environments and less kindly attitudes toward others. These differences may result in concerted efforts to preserve some components of the natural world even as others are degraded or destroyed by societal actions.

Consider, for example, *forest* and *swamp*. If you are like most people, then these two terms probably evoke different feelings. More specifically, most Americans have a rather benevolent and protective attitude towards forests and a much less benevolent and protective attitude toward swamps, even if the latter is technically defined as a forested wetland. Forests, of course, are places we humans can readily walk through and even make our home. Swamps, in contrast, are places we cannot similarly experience and are much less likely to call home. Wildlife may also play a role in explaining why these habitats are perceived so differently. Cartoons and other media have generated a generally positive mindset towards chipmunks, squirrels, deer and other forest animals (a phenomenon that has been called *the Bambi Complex*), but a less cuddly set of attitudes towards the reptiles and amphibians that tend to populate swamps. The same can be said for large mammals like giraffes and elephants, what some refer to as charismatic mega-fauna. Which do you prefer? Lion cub or baby cobra? You get the point.

This, of course, is bad news for swamps, which provide unique and necessary habitat for a great number of species. But how can one protect natural habitat and successfully lobby legislators to that effect if people simply have negative attitudes toward the thing that you seek to protect? In this case, the solution has been to drop *swamp* from the environmental vocabulary and replace it with a term used a few sentences ago: *wetland*. Accordingly, a substantial body of environmental law aimed at wetland protection has been enacted during the past few decades and enjoys broad public support. People, it seems, are perfectly willing to protect a wetland rather than a swamp, even if the two places in question can be one and the same.

Going Global: Environmental Issues Affecting Us All

The Fukushima and Deepwater Horizon accidents discussed earlier in the chapter illustrate how a one-time point-source release of pollutants in a dynamic environment can have wide-ranging effects. However, both of these events are still more localized in impact than other environmental issues that have the potential for global reach. Though treated as discrete concerns here, many of these problems are interrelated to one another. The first three, deforestation, biodiversity loss, and soil degradation prove that point. Let's take a look at them in that order.

Deforestation

Deforestation is the result of clearing land for agricultural, commercial, industrial, or residential purposes. This loss of green cover has a wide range of impacts that can outweigh the gains made by a small group of people benefiting from the newly cleared areas. Forests purify our air and water and serve as habitat for a many different animal and plant species that not only have utility for humans but some would say deserve to live simply in their own right.

Forest loss contributes to erosion, flooding, the loss of a carbon sink (more on this later in the chapter), the loss of habitat, and in some places, a diminution of opportunities for indigenous people whose livelihood is the forest. Deforestation solutions include government regulations on clear cutting and persuasive campaigns aimed at consumers to reduce forest product use such as paper.

The geography of deforestation is uneven presently, mainly occurring in less developed tropical regions such as in the Amazon or Southeast Asia. Before patting yourself on the back, realize that the lower activity in North America and Europe are because those areas already cut down much of their forests over the last few centuries.

Biodiversity loss

Sometimes I see eye rolls when discussing biodiversity loss. After all, there are lots of birds out there so who cares if one kind of pigeon disappears? And would it be so terribly bad if mosquitoes could just find an exit? That one hits home each time I go backpacking.

Protecting biodiversity protects us. *Biodiversity* is more than just the number *of a* species or the number *of* species. It's also genetic variability, and over time losses in each of these can diminish the health of ecosystems. Primary causes of biodiversity loss include habitat destruction (ahem, deforestation, described previously in this chapter), pollution, and over-exploitation. Often unseen are the ecosystem services that these plants and animals provide. Some help to form soil, some to break down pollutants, and some serve as energy (such as food). Ensuring ecosystem health relies on policies that reduce overhunting, protect habitat (for example, parks and refuges), and restrict pollution, among others.

Soil degradation

Farms and the food they produce demand soil to grow crops for you and crops for the animals that you eat (assuming you are a carnivore). This resource — high-quality topsoil — can take centuries to create and only years to destroy. Soil filters water, serves as habitat for earthworms and ants, is a carbon storage site, and is an all-around good thing to have.

Due to poor management and over tilling, some soil simply blows or washes away. This would be considered soil loss, but there are other ways that humans are diminishing this resource. Another is chemical degradation that may result in unproductive soils either by building up harmful pesticides, making soil more saline, or by growing crops that remove nutrients. A rotation of crops that add nutrients like nitrogen back to the soil or tempering chemical usage can be helpful. Finally, soil compaction can occur in the presence of heavy machinery, and reduces soil drainage and breathability. In many locations worldwide, soil degradation is a byproduct of deforestation, often for agriculture. Sustainable agricultural practices would go far in alleviating the stresses of soil degradation.

TIP

Starting this section with deforestation shows how cleared forests can lead to reduced biodiversity and poorer soils. Each example demonstrates that we are nested within systems that are nested within systems. All are interconnected, as are much of the remaining issues that follow.

Ocean acidification

This may be a new term for you, especially considering that when you are at the beach you probably do not consider that blue expanse in front of you to be an acid. It's not really, but the process of *ocean acidification* does refer to the ocean as becoming more acidic over time. Remember learning about pH in chemistry class? A reading of 7 is neutral and ocean water is usually higher than that, mostly over pH 8. So, ocean water is slightly basic but it is moving in the other direction, toward the acid end of the scale.

Why is this? Because humans have pumped a whole lot of carbon dioxide into the atmosphere from industry and transportation, and since the oceans absorb a lot of that CO_2 as part of the carbon cycle, well . . . you're going to end up with higher concentrations of carbon dioxide in the ocean. That means lower pH levels and more carbonic acid. This impacts calcium carbonate, a building block for coral and hard-shelled organisms such clams. Many of these sea creatures are *keystone species*: key to maintaining food chains and coastal ecosystems.

As the acidification process is closely tied to climate change, addressing those concerns should be a help here, too. This is clearly as threat to ocean health, but it's not the only one . . . let's take a look at overfishing next.

Overfishing

More people on Earth means more mouths to feed. As explained in Chapter 11, this concern was identified by Malthus and others many years ago. Agricultural output is key, but so are livestock raising and fishing. One serious problem with the latter is *overfishing*. In that case, more fish are caught at once than the breeding population is capable of reproducing. Compounding the problem is the *bycatch*, or the unintentional capture of other sea life. This is commonly the case when trawling with large nets, especially near the sea floor. Coral and sponges among other sea life suffer.

Why is overfishing a problem? One estimate has indicated that only 65 percent of fish stocks are biologically sustainable. Over time this impacts not only ocean ecological health, but also the livelihoods of fishers across the globe.

Reducing this problem generally relies on fisheries management, but this requires cooperation among nations as fish don't recognize political borders. Designating fish-catching seasons and how many fish can be caught are two management options. Enforcing fines for violators and making marine sanctuaries can also reverse losses over time. The tough part? That wide open ocean is very difficult to monitor.

Acid precipitation

REMEMBER

Acid precipitation has an unusually high acid content. It has been identified as a major cause of forest demise in various parts of the world, and also of contamination of lakes and estuaries. It may originate far from "the scene of the crime," but thanks to the environment's ability to "spread the mess" any one and all of the myriad point sources contribute to global impact.

The principal cause of this phenomenon is the burning of fossil fuels to operate motor vehicles, manufacturing facilities, and electrical utility plants (see Figure 18-4). A lesser but important cause is the smelting of ores that contain sulfur. Two unfortunate by-products are sulfur dioxide and nitrogen dioxide. When sulfur dioxide interacts with atmospheric moisture, it transforms into a dilute solution of sulfuric acid, a very corrosive substance. Nitrogen dioxide is transformed into nitric acid, which is not particularly corrosive, but is harmful to organic matter.

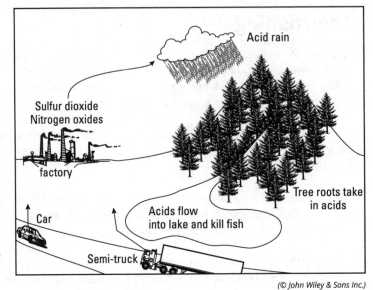

FIGURE 18-4: The acid precipitation that destroys forests and contaminates lakes largely originates in personal, commercial, and industrial consumption of fossil fuels.

(© John Wiley & Sons Inc.)

The range of impacts

After sulfur dioxide and nitrogen dioxide have formed in the atmosphere, they may remain there for long periods and be carried by the wind far from their source areas. Ultimately, they fall to earth with snow, sleet, hail, or rain. Whatever the source, negative impacts may be manifested in the following ways:

» **Terrestrial impacts:** Terrestrial impacts concern vegetation and soil. After acid precipitation has seeped into the ground, it may be taken up by the root systems of plants. If a sufficient amount of acid is taken in, even as small dosages over a long period, the stress may be sufficient to kill individual trees and perhaps destroy entire forests. Obviously, the effects of such outcomes may be devastating for wildlife habitat and recreation. Loss of vegetation results in decreases in the maze of root systems that hold soil in place, perhaps leading to erosion and mudslides. In addition, the acid may also have a fatal effect on bacteria and microbes that break down organic nutrients in the topsoil and contribute to soil fertility. Thus, acid precipitation may have a harmful effect on both the quality and quantity of soil.

» **Aquatic impacts:** Acid precipitation runoff may adversely impact streams, rivers, lakes, and oceans. Aquatic plants may be affected in the same manner as terrestrial ones. If they die, so does habitat for all kinds of animal species, as well as potentially significant elements in food chains. In addition, fish and other animal species may perish directly from prolonged exposure to the acids, or indirectly from silt laden runoff attendant to loss of roots that held soil in place.

Small lakes that collect runoff, but have no outlet, are particularly susceptible to harm. Literally and figuratively, they are dead ends, the final collecting place for acid precipitation. Acid concentrations may be such that lakes become, for all intents and purposes, sterile. How geographers work to protect water sources is described in the sidebar "Applied Geography: Watershed management."

» **Material impacts:** Material impacts principally concern deterioration of stonework and *statuary* (sculptures and statues), particularly ones made of marble. The corrosive effect of acid rain may literally eat away at these to the point where — to use an admittedly extreme example — statues become undistinguishable slabs of rock. Building facades are of more immediate importance, and thousands of them around the world are slowly dissolving.

APPLIED GEOGRAPHY: WATERSHED MANAGEMENT

A *watershed* is an area that drains into a river, lake, bay, or reservoir. As we saw with the case of the Chesapeake Bay, the quality of a body of water may be intimately related to the quality of the watershed that feeds it. Watershed management concerns a wide range of efforts that monitor and ensure the quality of watersheds and the receiving bodies. This is a significant environmental undertaking in any case, but especially so when municipal or regional drinking water is at stake.

Various geographical techniques are routinely applied to this endeavor. Global Positioning Systems (see Chapter 5), for example, are used in field surveys to pinpoint and map pollution sources. Aerial photography and satellite imagery may be used to assess the health of vegetation that anchors watershed soils.

But perhaps the most intriguing technology is *digital terrain modeling,* in which watershed managers use geospatial technology to simulate all kinds of eventualities across geographic space. Want to see what will happen if a small pollution source becomes a big one? Or if acid precipitation should reduce forest cover by 55 percent (or any other percentage)? Or if 2 inches of rain should fall anywhere or everywhere in 25 minutes?

This geospatial technology facilitates simulation of all kinds of scenarios, with one goal being watershed management that promotes and preserves the quality of drinking water.

The geographical dimension

To some extent, acid precipitation happens everywhere, but its intensity is very uneven (Figure 18-5). In general, the world's principal industrial regions tend to the major source areas. That includes the southeastern Great Lakes region, Western Europe in general, plus the Ukraine and southwestern Russia, Eastern China, and Japan. Acid precipitation tends to be greatest in these immediate regions plus those directly downwind. Given the general west-to-east movement of the atmosphere at mid latitudes, the typical result is an elongated impact area that tails off easterly from the source region.

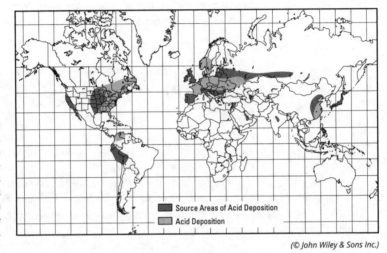

FIGURE 18-5: While the effects of acid rain are global, the greatest impacts are felt in regions downwind from major source areas.

(© John Wiley & Sons Inc.)

Thus, in North America, relatively pristine areas in northern New York, New England, and Canada's Maritime Provinces have been impacted. Forests have been harmed over large areas and numerous lakes, especially small ones, have become lifeless. The situation is similar in Europe, where the Black Forest and similar tracts have been affected.

Sometimes a state or country finds itself on the receiving end of pollution that originates in another state or country. This is called *trans-boundary pollution*. Thus, the nitrates that cause algae blooms that kill fish and hurt the economy in Maryland's portion of the Chesapeake Bay may have their origin in fertilizers that were spread on agricultural fields in Pennsylvania. Similarly, acid precipitation that kills forests in eastern Canada may have their origins in power plants in the United States.

People who live in areas that suffer the effects of pollution (such as acid precipitation) tend to be among the most committed to finding a remedy. But how can you affect a solution to a problem that originates outside of your jurisdiction — in another state or country? Assuming no treaties or protocols have been broken, about all one can do is encourage your neighbors to clean up their act. In the case of acid precipitation, that is going to take some doing because so much of the problem is associated with fossil fuel consumption that is likely to continue for decades to come. Reducing nitrogen oxide emissions from cars through catalytic converters and adding "scrubbers" to smokestacks to remove sulfur from power plant emissions are two technological solutions.

Climate change

Quick-fire question time! When was the first academic paper published quantifying the impact of increased CO_2 to planetary warming?

1896. "What?", you say, "I thought this was only a recent concern." Not so! Some even earlier experimentation took place in the 1850s courtesy of American scientist Eunice Foote.

Our Earth climate has and will continue to change, and a variety of those natural change mechanisms were described in Chapter 9. Several ice ages came and went long before modern-day humans emerged as a major instrument of global change. That means that natural mechanisms capable of effecting significant global temperature change over time must exist. Climate scientists know this. But they also know something else: There is a discernable human signature in the warming that we are seeing now.

This warming trend appears to be the result of atmospheric changes for which humans are responsible. Since the beginning of the Industrial Revolution in the mid- to late-1700s, humans have consumed (burned) large quantities of coal, petroleum, and natural gas to operate vehicles and machinery, provide power for manufacturing facilities and electrical utilities, and to heat up and cool down homes and buildings. In the process, huge quantities of certain gases have been added to the atmosphere. Most reputable scientists believe that these substances are the cause of the current warming trend.

The global greenhouse

REMEMBER

The commonly held scenario for why the atmosphere is heating up likens the atmosphere to a greenhouse, whose glass panes allow sunshine to penetrate the structure, but then contain the solar energy's heat and prevent it from escaping. Of course, no giant glass roof hovers over planet Earth. But gases in the atmosphere do act in much the same way (Figure 18-6). When these gases are scarce, a considerable amount of solar energy is able to radiate off Earth's surface back into the coldness of space. When they are abundant, however, they produce an atmosphere that absorbs this same heat, resulting in a warming atmosphere. The gases that accomplish this are called *greenhouse gases* (such as CO_2, methane, and water vapor) and the over-all process is referred to as the *greenhouse effect*.

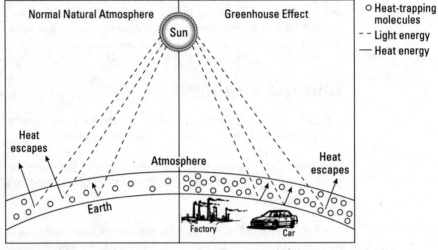

FIGURE 18-6: Large quantities of carbon dioxide in the atmosphere greatly contribute to the global greenhouse effect.

(© John Wiley & Sons Inc.)

Following are two of the primary contributors to global warming:

>> **Effects of the Industrial Revolution:** Carbon dioxide (CO_2) is the most prolific greenhouse gas. Although it's a naturally occurring substance, substantial quantities of it are released to the atmosphere when fossil fuels, wood, and other organic matter are burned. Increased fuel consumption, especially as regards to fossil fuels, has been a hallmark of the Industrial Revolution. Before it began, the carbon dioxide content of the atmosphere measured about 274 parts per million (ppm). By the beginning of the present century, that number had risen to more than 360 ppm. In 2020 the level reached nearly 420 ppm! Because global consumption of fossil fuels continues to rise, one can only expect increasing levels of greenhouse gases in the atmosphere, and with them the likelihood of increasing temperatures.

Emissions vary geographically. Because economic development and fuel consumption go hand-in-hand, understandably developed nations generate a much greater portion of annual emissions than do developing ones. And because countries such as China and the United States are among the world's leaders in fossil fuel consumption (see Chapter 16), it makes sense that they also lead the world in greenhouse gas emissions. Many would say that they should also lead in solutions to mitigate the problem.

» **Effects of deforestation:** This era of increased fuel consumption has been complemented by a general process of deforestation — first in Eurasia, then in North America, and now in the tropical rainforest areas. Trees take in carbon dioxide from the atmosphere and give off oxygen. Thus, when trees are cut and consumed, they are no longer available to affect this exchange. And if they are burned, they emit the CO_2 stored in them, increasing the atmospheric store of that gas. Experts disagree about the impact of the current decline of tropical rainforest on global atmospheric balance, but the majority opinion is that its contribution to the greenhouse equation is measurable, if not significant. Unknown, too, is just how much more CO_2 the oceans can store (see the section on ocean acidification earlier in this chapter) as opposed to the gases concentrating themselves solely in the atmosphere.

Winners and losers

If you live in an area that experiences a cold winter, then you may be excused for thinking that global warming can't be all bad. If continued increase in global warming comes to pass, then clearly some lands will benefit. For example, large parts of Canada and Siberia that are now too cold for agriculture may become productive land (assuming, of course, that water availability and soil conditions permit — agricultural success is more than just temperature) and some ports that freeze over could become viable year-round.

But climate models and current conditions are predicting that large, productive areas will dry up and become desert. Diseases that rely on warm locales may see their geographic range increase. Foods that grow easily in one place may shift to others if they are viable at all. Environmental hazards such as drought, wildfire, and hurricane intensity seem to be on the rise. Clearly, sea level will rise, resulting in coastal inundation and infrastructure damage. And with that will come salt-water intrusion of neighboring aquifers, resulting in an overall decline in the global supply of potable water. For these sorts of reasons, most models forecast more losers than winners should the warming trend continue. And that is likely given continued and growing consumption of fossil fuels.

REMEMBER

Most climate change solutions focus on reducing fossil fuel use, improving energy efficiency, conservation, promoting alternatives, reducing resource consumption, protecting vital Earth spaces such as our forests, and educating people about the consequences of their actions on people in other places. Let's say for the sake of argument that you think global climate change is overblown. Is reducing the number of resources we consume really a bad thing? Even if it did nothing to impact our warming climate, how are we worse off by having less trash, cleaner air, and rich, biodiverse landscapes?

Taking on the Challenges of Tomorrow

This chapter began by describing environmental events in Chesapeake Bay in terms of its connections to a much broader geographical expanse.

In a manner of speaking, the same applies to you. Wherever you live, you are part of an environment that, by virtue of dynamic mechanisms, is connected to other places that may be miles away or next-door. While the results may affect you, you also have the capacity to affect results. Specifically, as citizens of a democracy, you have the opportunity to participate in personal and community decision making that can affect at varying scales the quality of the world in which you live.

Having a geographic perspective, where you understand the interconnected relationships across space, is one step closer towards enacting the changes necessary to protect the only place humans have ever lived — Earth.

5
The Part of Tens

IN THIS PART . . .

Lists sure are popular these days. The 20 best so and so. The 15 worst such and such. Ten, however, seems to be the most popular number. As in "The Ten Best Dressed Celebrities," or "The Ten Most Beautiful Beaches."

In this part, you will encounter 10 geographical occupations, 10 geographical organizations, 10 great online resources, and 10 geographical things you can forget (this one is purely for fun of course!).

IN THIS CHAPTER

» Tackling acronyms aplenty!

» Learning with the best at AAG, AGS, NGS, and RGS

» Teaching the next generation with NCGE

» Surveying the landscape with the USGS

Chapter 19
Ten Organizations for Geography in Action

A number of public and private organizations provide information, services and products of interest. You've learned by now that geography appears *everywhere*. Nonetheless, here are ten of the most noteworthy.

American Association of Geographers (AAG)

The AAG is the pre-eminent organization of professional geographers in the United States. The organization's goals are to promote professional studies in geography and to encourage the application of geography across all areas, including business, education, and government. The AAG publishes two scholarly journals (*The Annals of the American Association of Geographers* and *The Professional Geographer*), a newsletter, and annually hosts a national conference and several regional meetings. Membership is not recommended for the average novice, but if you are interested in furthering your geographic education at the university level, then the AAG should be kept in mind.

TIP

If you want to see how varied geography and its practitioners are, take a look at "Specialty Groups" under the "About AAG" tab on their website (shown at the end of this entry). You will find academics investigating Africa, Cultural Geography, Mountain Geography, Queer and Trans Geographies, Tourism, and Water Resources. Those are just six of the 76 (!) different groups looking at different aspects of our world.

You can learn more about AAG at www.aag.org.

American Geographical Society (AGS)

Founded initially with a strong interest in polar regions — the Arctic and Antarctic — and later Latin America, the AGS is now truly global in its concerns. A journal, *Geographical Review*, is the society's primary publication, but the heavily read *Focus* appeals to a more lay reader audience. A timely and much needed initiative called EthicalGEO has promoted a better understanding of privacy in an era of increased location tracking and how governments use geospatial data to police and surveil its citizens.

TIP

If you really want to understand what is going on in the world today, visit their website and sign up for DailyGeo, a digital news blast that highlights geography news in public health, climate change, and resource management among just about any other topic that you can imagine.

You can learn more about AGS at www.americangeo.org.

National Aeronautics and Space Administration (NASA)

"Now, just wait a minute," I hear several of you saying. "NASA is about space and *planets*." Ahem, what is Earth, exactly? In addition to getting humans back to the moon or even Mars, NASA is interested in a variety of "Earth topics" (just look at their website for that same section), and these include air, climate, hazards, water, ocean, and ice.

The agency had been helping Central America prepare for hurricanes, investigating air quality from wildfires, and more, often from a global perch in space via satellite. Who else has the best radar to monitor earthquakes and volcanoes from space? Antarctic ice break-up, contrails from airplanes, the changing nature of glaciers — this is where you'll find some of Earth's best scientists looking after some of Earth's most interesting and often pressing issues.

You can learn more about NASA at www.nasa.gov.

National Council for Geographic Education (NCGE)

Teachers, take note! NCGE seeks to enhance the status and quality of geography education and learning. Familiarization with this organization is recommended for anyone interested in geographic education at any level, kindergarten through graduate school. The *Journal of Geography*, the NCGE's principal periodical, is published six times a year and is devoted to research on geographic education. The council's other journal, *The Geography Teacher*, is focused on classroom-ready lesson plans and activities. The organization holds an annual meeting, the venue of which changes each year, and has an extensive online library of lesson plans.

You can learn more about NCGE at www.ncge.org.

National Geographic Society (NGS)

National Geographic is more than just a magazine. Founded in 1888 "for the increase and diffusion of geographic knowledge," the Society is by far the most visible symbol and promoter of popular geography in the United States. In recent years, the Society's reach and influence have been augmented by its television channel and social media channels such as Instagram, where its renowned photography is viewed by millions. The old stalwart, *National Geographic Magazine*, continues to rank among the top periodicals in terms of circulation, and is translated into many foreign languages.

The Society has become a force in geographic education, as evidenced by its pivotal role in promoting the National Geography Standards (with NCGE), and as a major marketer of maps, globes, atlases, and textbooks.

You can learn more about NGS at www.nationalgeographic.com.

National Oceanic and Atmospheric Administration (NOAA)

NOAA is the principal agency of the U.S. government concerned with collection, analysis, and dissemination of information about the ocean and atmosphere. As such, it offers a wealth of data regarding non-terrestrial aspects of physical geography. Its oceanic concerns involve coral reefs, tides and currents, marine sanctuaries, coastal development issues, and the effects of chemical and oil spills. Given its mandates regarding safe navigation and transportation, NOAA also engages in ocean mapping and is the country's major publisher and supplier of nautical and aeronautical maps and charts.

Atmospherically, NOAA is probably best known as the parent agency of the U.S. Weather Service, whose official forecasts, advisories, and satellite images are standard fare for weather reports. Climatic phenomena, such as El Niño and La Niña, global warming, climate prediction, and *paleoclimatology* (ancient climate patterns) fall within its scope, as well as topics about atmospheric quality and human-atmosphere interaction.

You can learn more about NOAA at www.noaa.gov.

Population Reference Bureau (PRB)

The Population Reference Bureau provides timely information on population trends and their implications. While it markets some very fine resources pertaining to American demographics, PRB's signature product is its *World Population Data Sheet*, an annual update of demographic variables for most of the world's countries. Other projects include Kidsdata, about children and youth well-being, of course, and an education program that produces high-quality lesson plans and classroom resources for the elementary and secondary school levels.

TIP

PRB is completely apolitical. It does not endorse or advocate positions related to population policy or controversial population issues. In short, the PRB has no agenda other than to provide the most reliable information possible, and to let the data speak for themselves.

You can learn more about PRB at www.prb.org.

Royal Geographical Society (RGS)

The RGS is the UK's "learned society" for geography, but their outreach is global just like their field of study. As a counterpart organization to the AAG, they provide a number of similar resources for schools and professionals such as workshops and webinars. Their primary publications include journals such as *Area*, *The Geographical Journal*, and *Transactions of The Institute of British Geographers*. Among many of their useful services is a section of their website titled "I am a geographer" where you can read about people performing important work in law, coastal systems, business logistics, and Earth monitoring via satellite (see the end of this entry for the web address).

TIP

Although the AAG and RGS are noted in this chapter, don't stop there. Many other countries have national geographic organizations that do much to promote a better understanding of Earth. Egypt, Slovenia, Italy, Saudi Arabia: The list goes on from there!

You can learn more about RGS at www.rgs.org.

United States Census Bureau

This is the place to go for information on U.S. population geography. The census happens only once a decade, but the U.S. Census Bureau is a perennial organization with a lot to offer. Naturally, most of that is data derived from the census and organized by various geographic levels — national, state, county, city, census-tract (rather like a postal zip code), and occasionally city blocks. But the bureau is not just demographics. It provides data on housing, businesses of all sorts, foreign trade, and other matters like community resilience (how neighborhoods respond to the impacts of disaster). Educational resources for teachers also are available. Part of the Bureau's mission is to make its data readily available to the general public. Accordingly, reams of information can be accessed and downloaded from the agency's website.

You can learn more about the Census Bureau at www.census.gov.

United States Geological Survey (USGS)

Created by an Act of Congress in 1879, the USGS is the major science agency of the federal government concerned with our physical planet and its Earthly processes. While the name rather matter-of-factly recounts the agency's original mission, its scope has expanded to include ecosystems, endangered species, environmental quality, and natural hazards of all sorts. The USGS is also renowned for its handsome and exquisitely detailed topographic maps of the United States, which are available to the public.

Like many other government agencies, the USGS has a public outreach mandate. They monitor volcanoes. They explore water levels and streamflow. They update the public on landslide potential after wildfires and heavy precipitation. And they promulgate earthquake activity from across the globe.

You can learn more about USGS at www.usgs.gov.

IN THIS CHAPTER

» Mapping a career in geographic information systems

» Researching a job in the health, planning, or transportation services

» Analyzing employment prospects in other fields

Chapter **20**

Ten Interesting Career Paths for Geographers

Hundreds of geography departments in colleges and universities around the world offer courses and degree programs that prepare people for exciting and rewarding careers. The American Association of Geographers (www.aag.org) has a great overview of the possibilities. Often geographers are working without having the actual title of "geographer."

Here are ten examples of how a geographic education gets put to work.

Area Specialist

An *area specialist* is a person who possesses a high degree of expertise in a particular part of the world. Titles like Latin Americanist, Africanist, and Sinologist (China expert) give you an idea of what they do. Government agencies and international business concerns are the principal employers.

Virtually hundreds of such people are hired by the U.S. State Department and Central Intelligence Agency to "keep current" on their region so as to be able to provide the best possible information and advice to policy makers.

In the business world, area specialists act in much the same capacity, focusing particularly on economic geography, human/cultural geography, and climatology. As the global economy becomes more interconnected, job opportunities for area specialists in the business world are likely to grow.

Educator

The prospects for landing a job as a geography teacher in this country are the best they have been for some time and keep getting stronger. Sure, a long time ago geography largely disappeared from the curriculum as a stand-alone subject and became submerged in that academic goulash called social studies. Times have changed. The National Standards movement and a course in Advanced Placement Human Geography have not only resuscitated the teaching of geography, but have also mandated in many states that those who teach it have solid academic preparation in the subject.

Sadly, the number of people who are so qualified is still too low. The positive flip side of that statement is that opportunity awaits tomorrow's teachers who specialize or major in geography.

Environmental Manager

Environmental managers monitor and protect natural resources. Environmental protection agencies hire these professionals, as do public and private enterprises involved in waste disposal, water supply, forest conservation, wetland and coastal zone management, and other resource-related pursuits.

Monitoring compliance with environmental laws is a major concern of environmental managers, as are preparation and review of environmental impact statements. Because environmental phenomena characterize different parts of Earth's surface, they are an important field of study for geographers. Geography programs in colleges and universities offer a range of environmentally related coursework.

Also, expertise in geographic information systems, remote sensing, and/or spatial data analysis greatly enhances career prospects in this field.

GIS Technician

Geographic Information Systems (GIS) technicians are concerned with production and analysis of maps and geospatial data management. These maps are indispensable tools for displaying information about places and regions and government agencies of all kinds use them for analytical and display purposes, as do construction and utility companies, architectural and engineering firms, and a host of other employers.

The National Geospatial-Intelligence Agency alone has thousands in its hire, and that is just the tip of an employment iceberg that keeps growing. Chapters 3, 4, and 5 in this book, all about mapping, make it clear how crucial it is to keep this job sector filled.

Health Services Planner

These planners seek to optimize delivery of and access to health services. Some of these professionals analyze and recommend locations for clinics, hospitals, and emergency response units. Others help to decide how to allocate specialized services (such as coronary care and burn units) among existing hospitals and health facilities. Still others analyze patterns of disease and health risks with a view to geographically targeting outreach and education programs aimed at prevention.

While most of these planners have degrees in public health, many also turn to geography programs to acquire critical skills in geographic information systems and spatial statistics.

Location Analyst

An old saying states that the three most important determinants of the success of a business are location, location, and location.

Although this is not literally true for every kind of business, the right location is often the difference between success and failure. Location analysts identify the factors that most affect (for better or worse) the success of a particular business, evaluate the viability of prospective sites with respect to those factors, and recommend a site to decision-makers.

Many large (and not-so-large) retail chains maintain in-house think tanks of location analysts to help make all-important decisions regarding the locations of future stores. Numerous generic consulting firms do the same.

Location analysis is an important component of economic geography courses and most professionals in this area increasingly have a solid background in geographic information systems. Chapter 15, "Takin' Care of Business," provides more detail on this type of work.

Market Analyst

No, not the stock market, but instead, the prospective buyers of particular goods and services. People have different tendencies to purchase different things. Such variation may be a function of age, culture, race, ethnicity, income, or some other factor that may vary geographically. Haven't you noticed how many apps on your smart phone want to enable location services to better serve you (ahem, to sell something to you)? Determining the geography of the market for particular products and allocating merchandise accordingly is the job of the market analyst. You think this isn't important?

Recently, a nationwide retailer (that shall remain nameless) sold decent merchandise, but nevertheless went belly up, in part because of poor market analysis. Mind you, their location analysis was great. They had nice stores in superb locations. They were just the wrong-sized stores (in many instances) with the wrong product mix in the wrong locations. Good market analysis, a staple subject of economic geography, could have made a difference.

Remote Sensing Analyst

Satellite images are key sources of information about Earth's surface and, therefore, important tools for geographic analysis.

Numerous government agencies employ thousands of interpreters and analysts to monitor goings on in foreign countries as well as to keep tabs on domestic agriculture, forestry, and other items of environmental interest.

City and regional planning authorities, as well as civil engineering and consulting firms, also hire specialists who use these tools to acquire information about their employers' respective areas of interest and provide input into maps and databases that aid planning.

All kinds of maps produced under government auspices these days contain information provided by air and satellite images.

Transportation Planner

Transportation concerns movement of goods and people from one location to another. Thus, transportation planning is inherently geographical and attracts people with training in geography.

Several specializations are involved. Surveyors who plan for future road construction typically are well-grounded in modern cartographic skills. Route planners concern themselves with devising optimal itineraries for public buses, planning for evacuation of areas in the face of diverse emergencies, and how to best get a product ordered online from a fulfillment center to the buyer.

Meanwhile, in the realm of manufacturing, "just in time" planning, which seeks to bring together product components at a factory just in time for assembly, can reduce significant warehousing costs. All of this involves managing spatial relationships. As a result, the job outlook for transportation planning specialists is excellent.

Urban Planner

Cities consist of countless land parcels devoted to residential, commercial, industrial, cultural, recreational, transportation, governmental, sanitation, and other uses. Urban planners geographically allocate the various kinds of parcels in ways that promote orderly development of cities and enhance their attractiveness as places to live and work.

Virtually thousands of counties and municipalities in the United States employ urban planners. The fact that cities keep growing and that the overwhelming majority of Americans now live in them bodes well for the future of urban planning as a career field.

Cartography, geographic information systems, image interpretation, spatial database management, and the ability to make educated population projections have emerged as critical tools of the planning trade.

IN THIS CHAPTER

» Booting the Bermuda Triangle

» Leaving The Land of the Midnight Sun behind

» Shuffling aside the Seven Seas

» Canceling membership to the Flat Earth Society

Chapter 21
Ten Things You Can Forget

If reading through this book to this point (good for you!) has overfilled your brain with new geographic content and concepts, then you may wish to free up some memory by tossing out these ten geography-related items.

The Bermuda Triangle

The *Bermuda Triangle* is an area off the southeastern coast of the United States that has allegedly witnessed a disproportionately high number of strange and unexplained disappearances of boats, ships, and airplanes.

Disagreement abounds as to its exact location. Most commonly, the Triangle is depicted as a part of Atlantic Ocean bounded by a line that extends from Bermuda to Miami to San Juan, Puerto Rico and back to Bermuda. The U.S. Board of Geographic Names does not recognize "Bermuda Triangle" and has never indicated where it may be. This gives you complete license to place the Bermuda Triangle anywhere your little heart desires, and that is precisely what some people have done. Indeed, on some maps the Triangle isn't even triangular.

The term first appeared in a February 1964 article ("The Deadly Bermuda Triangle") in *Argosy* magazine. Several other articles on the subject subsequently appeared (along with books and TV documentaries), many suggesting the region is mysterious and dangerous. If you've seen some of this stuff and found the whole thing rather spooky, rest assured you are not alone. Given widespread public interest in paranormal activity, the Bermuda Triangle has become something of a geographical poster child for enthusiasts of the occult.

TIP

In reality, the Bermuda Triangle is much ado about nothing. The vast majority of the supposedly abnormal disappearances have very normal explanations, and some of the most famous incidents occurred well outside the Triangle. About two decades ago, a researcher compiled an extensive list of sinkings and crashes, based on Lloyds of London accident reports and similar reliable sources, and plotted them on a map. If anything, the data show that the Triangle is remarkable for its lack of mishaps and is practically the safest part of the western Atlantic Ocean.

Cold Canadian Air

It happens several times each winter. I'm watching a weather report on TV and the map indicates frigid temperatures up in parts of Canada, such as Alberta, Saskatchewan, and Manitoba. With squinting eyes and stern gaze, the usually smiling weather-person glumly warns that "cold Canadian air" is on the way. Cold Canadian air — have you ever heard that term? If you live in the U.S., pay extra special attention to weather forecasts between November and March. Sooner or later you're going to be told that cold Canadian air is on the way. I guarantee it.

Given the imminent arrival of cold "Canadian air," it seems to me that you have every right to say "Darn those Canadians! Why don't they keep their cold air to themselves?" Of course, "Canadian air" does not exist. Neither does German air, Pakistani air, Ethiopian air, etc. Air is air. It doesn't belong to anybody. "Air space," the atmospheric area that is vertically overhead a country, is another matter. That belongs to Canada, according to international law. Thus, aircraft of foreign countries are not supposed to fly over Canada without the Canadians' permission. But Canada owns the physical space, not the air itself.

If the air over Canada is very, very cold, rest assured the Canadian people had nothing to do with it. And if that same cold air is coming your way, again rest assured that there is no nefarious neighbor to your north who is responsible. So as to cold "Canadian Air," forget about it.

"Coming Out of Nowhere"

Watch a football game on TV and sooner or later somebody will "come out of nowhere." Typically, a quarterback drops back to pass, spots a wide receiver in the open far downfield, and heaves the ball. It looks like a sure completion. But at the last millisecond, the cornerback or safety lunges to deflect the pass and save the day. The crowd screams and so does the play-by-play announcer: "Wow! He came out of nowhere to make that play!" It seems athletes have this knack for "coming out of nowhere" to do something spectacular. And it's not just football. Hockey, baseball, and basketball players do it, too. Usually, it involves a great play on defense, but not always.

In the years before instant replay, I wondered how somebody could be nowhere, yet come out of that non-location to perform a feat of great athleticism. It was one of the great mysteries of geography. I figured, maybe the person coming out of nowhere was in one of those parallel universes they keep talking about on *Star Trek* reruns. But then along came instant replay and guess what? That defensive back was there all along. So there was no great mystery of geography after all.

Your nowhere is always somewhere to someone else.

"The Continent"

Europe is sometimes called "The Continent." It's one thing that Europe is even referred to as a continent, but *The* Continent? Puh-leaze! Europe is nothing more than a peninsula of Asia. It was the ancient Greeks who coined the name and first placed it on their maps. And indeed in those days travel between, say, Southern Europe and Central Asia took so long and involved such perils that, for all intents and purposes, they might as well have been separate land masses. Culturally, Europe and Asia were separate worlds, too. But physically separate? Not on my map, nor even on that of the ancient Greeks. So is Europe a continent? As a matter of standard usage, yes. As a matter of geographic reality, no. And *The* Continent? Sounds to me like a matter of self-esteem — or lack thereof.

Want to have some real fun? Are the Americas one continent or two? Hmmm. Africa is connected to Asia just like Europe, so is that just one big mega-continent? If this sounds kind of silly, it is. Forget about it.

The Democratic Republic of . . .

Every country has an official name, and it usually differs from the name by which it is commonly called. Thus, Iceland is really The Republic of Iceland, and Thailand is really the Kingdom of Thailand, and so forth. Some of the "real" names are wonderfully evocative. Examples include The Hashemite Kingdom of Jordan, the Plurinational State of Bolivia, and The Grand Duchy of Luxembourg. Great names!

But then there are other official names that I look at and think, "You gotta be kidding me." Examples include The Democratic People's Republic of Korea (commonly called North Korea) and The Lao People's Democratic Republic (Laos). Each of these countries is currently under a virtual dictatorship.

In fact, what "democratic republics" seem to have in common is a lack of characteristics that most people associate with democracy, like periodic honest elections, secret ballots, and two or more honest-to-goodness political parties. Are we talking bona fide countries? Yep. But should they be called "Democratic Republics"? Let's take a vote.

The Flat Earth Society

Some people just don't get it. Which is to say yes, there really is a Flat Earth Society. See for yourself. Get on the internet and do a key word search. All sorts of things will pop up under "Flat Earth Society" including an honest-to-goodness organization by that name (I wonder how they feel about having members from across the *globe*?).

Many (most?) members are motivated by literal interpretation of Scripture. In the Gospels (Matthew 4:8; Luke 4:5), for example, Satan tempts Jesus by taking Him to the top of a very high mountain to see "all the kingdoms of the world." That would be theoretically possible on a flat Earth, but not on a sphere-like one.

There's a certain logic to the flat Earth idea. Basic personal experience argues for it. And despite everything you know and trust about gravity, there's still something odd about being upside down at the South Pole, isn't there? Bona fide flat-Earthers feel the same way. They also believe all that photography from space showing a sphere-like Earth is a hoax — ditto for the spacecraft, at least with respect to orbiting Earth.

TIP
In any academic discipline, dissent and contrary thinking have a funny way of proving healthy. Goodness knows how many incredible discoveries have been made because somebody had the wisdom and the courage to go against the flow. But The Flat Earth Society? With all due respect to the faithful, it's time to find another conspiracy theory.

Land of the Midnight Sun

Alaska is called the "Land of the Midnight Sun." The future 49th state became a U.S. territory in 1867 when, in the greatest real estate deal since Jefferson purchased the Louisiana Territory from Napoleon, Secretary of State William H. Seward negotiated its purchase from the Russian government for $7.2 million. People thought Alaska was worthless and Seward was nuts, so they called Alaska "Seward's Folly" and "Seward's Ice Box." Either might have become the moniker, so you got to admit "Land of the Midnight Sun" sounds pretty good. But as far as truth-in-geography is considered, that's another story.

For part of the year the sun does indeed shine at midnight north of the Arctic Circle. But only about 25 percent of Alaska is within that realm, and likewise only about 1 percent of Alaskans. Thus, if you want to get picky about it (and of course I do), the State's nickname does not apply to 75 percent of its land and 99 percent of its people. Certainly, nights are very short over most of Alaska during the summer. Indeed, nighttime may be more like a prolonged twilight. But midnight sun? Sorry. That only works north of the Arctic Circle.

And there's one more detail that has been conveniently overlooked. Any locale that experiences midnight sun at one time of the year will also experience noon-time darkness at another time of year. "Land of the Noon-time Darkness?" Put that on your license plate! Actually, yes, I do get the point of it all. "Land of the Midnight Sun" is about image and tourism and putting your best foot forward. But as far as truth-in-geography is concerned, forget about it.

"The Rain in Spain Stays Mainly on the Plain"

I hate to be the one to tell you this, but "Wrong! Wrong! Wrong!" The plain (or *meseta*) that occupies central Spain is among the driest parts of the country. Madrid gets about 18 inches of rainfall on average per year. Valladolid, a city in another part of the plain, averages a little less than 19 inches.

But it's another story in Galicia, the mountainous area in extreme northwestern Spain. Santiago de Compostela, the world-famous Galician center of pilgrimage, averages about 49 inches of rain per year. As moisture-laden air enters Galicia from the Atlantic, it rises to cross the mountains, and in doing so cools, condenses, and forms rain. By the time this air reaches the central plain, it's low on moisture, so rain is relatively scarce.

I know it makes for poor poetry, but "The rain in Spain stays mainly on the mountains in Galicia." You don't have to memorize that if you do not want to. But as regards to "The Rain in Spain Stays Mainly on the Plain," unless you're preparing for a role in your town's production of *My Fair Lady*, forget about it.

The Seven Seas

"The Seven Seas" is a colloquialism that roughly means "all of the world's oceans." You would figure that whoever coined the phrase must have counted something, but what? There are literally dozens of "Seas" around the world. Maybe it's simply a matter of "The Eighty-four Seas" (or whatever) just not having the same poetic impact of "Seven Seas." Some suggest the term refers to the constituent seas of the Mediterranean Sea (blame the Greeks!) — like the Aegean Sea, the Ionian Sea, the Adriatic Sea, and so forth. Trouble is, my atlas shows the Mediterranean contains at least eight lesser seas, so forget that. And wouldn't those ancients also know about the Arabian Sea? And maybe the Baltic, too?

Then there are the oceans. My atlas shows the Arctic Ocean, the North and South Pacific Oceans, the North and South Atlantic Oceans, the Indian Ocean, and . . . that's it. Six. Some people speak of the "Antarctic Ocean" or the "Southern Ocean" (same water body), which would make seven. Indeed, there is a lot of water surrounding Antarctica. Its degree of coldness makes it ecologically distinct from adjacent oceans and is the prime rationale for separate status. But neither "Antarctic Ocean" nor "Southern Ocean" appear on most world maps. So, what about "The Seven Seas?" My advice is that you simply say, "the oceans." And even then, isn't there really only one?

Tropical Paradise

Tropical + paradise. They go together like rocket + science, or banana + split. The genesis of this verbal union is, of course, The Book of Genesis from the *Bible*. Adam and Eve wore no clothes, so the Garden of Eden must have been warm. And since the Garden was a garden, it must have had a fair amount of rain. Warm + rain = tropical. Therefore, Paradise (capital P) was tropical.

Over the ages, lots of people believed the biblical Paradise was real. That meant not only that the Bible was literally true concerning the Garden of Eden, but that it actually existed and was awaiting rediscovery. Some medieval world maps even showed an island off East Asia named Paradise. So when the early Spanish voyagers, thinking they had reached East Asia, encountered lush Caribbean islands, they truly believed they had found the tropical Paradise.

Today, of course, "tropical paradise" has much more to do with tourism than theology. Typically, it appears in ads that encourage you to dispose of your disposable income by vacationing in some tropical destination. But tropical locales have many of the same social and economic problems present in the vacationer's home country. So, a "paradise"? You might want to drop that connection and just enjoy the warmer weather.

IN THIS CHAPTER

» Exploring the world through your computer screen

» Discovering facts and figures online

» Playing hide and seek with GPS

Chapter 22
Ten Great Places for Online Geography

Geography-related websites abound in cyberspace — why do you think they call it the *World* Wide Web? The problem isn't finding ten good sites, but instead having so many from which to pick! In most cases, the sites presented here were chosen for their collective breadth rather than their individual greatness.

TIP

If you have not already done so, I encourage you to check out Chapter 19. Each organization listed there maintains a website, some of which, had they not already been mentioned, would surely be included in this chapter.

Any County/Local GIS Department

It's been said that all politics are local, so I guess the same can be said for geography, specifically the importance of geospatial data to local citizenry. I can't give you a direct website here, but this is a top recommendation, nonetheless: visit your town, county, or regional geographic information system (GIS) department's website. I say regional as there are some rural counties that sometimes band together to provide these services when they can't fund them on their own. My own county's website has maps of voting districts, school districts, and topography. They support 911 dispatching, make planning maps, report on tax delinquent properties, and record real property sales and transfers. This is your one-stop shop for local geographic decision making and it should be a bookmarked tab in your web browser.

Geocaching

Global positioning systems (GPS) can be used for a lot of fun if you've got a bit of a treasure-seeker in you. *Geocaching* is an outdoor recreational activity where players use a GPS receiver to both hide and seek out small containers (called caches). These caches can be as large as an ammo box and contain a sign-in log and toys as a prize. Others are "virtual" caches that simply take a user to an important site where they learn about a historical event.

Imagine, for example, a puzzle whose answer is coordinates that take you to a historic marker depicting events at a Revolutionary War battle. The toy is not the prize, but rather new knowledge about that place.

TIP

Geocaching.com is one website where you can download an app to your phone and let the fun begin. I use this app with college-age students and it's very user friendly.

Geoguessr

"Shall we play a game?", asks the supercomputer (aka Joshua) in the 1983 movie, *WarGames*. Yes, we should play a world-based game, but with far lesser consequences than shown in that film.

Geoguessr uses the street view feature from Google Maps, dropping users randomly across the globe. Your task is to look for clues in the environment to identify your location. You receive a higher number of points if you can accurately place your location on a world map.

This mildly addictive geography geek-out has a limited free version and a subscription that provides more features. Give it a go and see how well you can do to use trees, road signs, soil types, architecture, and other environmental clues to place yourself on this planet.

Geoguessr is available at https://www.geoguessr.com.

Geoinquiries

Esri is the world's leader in GIS software. They've been at it since 1969. But one thing that I've admired about this company for a long time is their commitment to education, especially in K-12.

One way they help create the next generation of geographers is to showcase geographic thinking across all disciplinary areas while using GIS. Geoinquiries is a great example. These are short, inquiry-based activities that any teacher can display from a single classroom computer or modify for student use. And when I say interdisciplinary, I mean it. They've got American Literature, Government, U.S. History, and Mathematics among six other areas, with 15 to 20 distinct lessons in each area. Teachers, this one is for you!

Geoinquiries is available at https://www.esri.com/en-us/industries/education/schools/geoinquiries-collections/.

Google Earth

Well, this here is the game-changer, I suppose. But like many online resources, this didn't start with Google. Back in 2001, a company called Keyhole started it all and were acquired later by our Google friends.

Google Earth is primarily a geovisualization tool where satellite imagery is draped over a digital globe. It sure looks 3D, but realize it's not, of course. You are using a flat, 2D screen to view it, after all. There are a number of layers that one can view over the imagery, including roads and renderings of some important buildings and landmarks.

So why has Google Earth become some immensely popular? I think in some ways it taps in to the inner geographer in all of us, a desire to place ourselves within a larger world (and how cool is it that we can do so from our phones!).

Google Earth is available at `https://earth.google.com/web`.

Google Lit Trips

TIP

You can do a whole lot more with Google Earth than zoom around the planet looking into other people's backyards. So, here's a shout-out to Jerome Burg, a retired English teacher from California who said (this is a re-enactment, by the way), "Every story has a setting, and a setting is a place. Let's map them!" Anyway, that's how I picture his mind processing this idea.

Google Lit Trips were a result, a series on online ventures into your favorite literature, but plotted on Google Earth. Follow *The Grapes of Wrath* with the Joad family from Oklahoma to California! How about *The Catcher in the Rye*? *Anne Frank: The Diary of a Young Girl*? *Make Way for Ducklings*? *A Long Walk to Water*? More than seventy titles are here to explore the world of these books beyond the printed page. Do it!

Google Lit Trips is available at `https://www.googlelittrips.org`.

Perry-Castañeda Library Map Collection

Thousands of digitized maps are available online in The Perry-Castañeda Library Map Collection at the University of Texas at Austin. On the home page, you can click on various world regions to access map menus that pertain to different countries. In many instances, this leads to another menu that offers a mix of contemporary and historical maps, some centuries old.

Need a map of Burkina Faso? Fifteen of them are present here. The library has over 250,000 maps, but only about 20 percent have been scanned for your use. That should still be plenty for most of you!

The Perry-Castañeda Library Map Collection is available at `https://maps.lib.utexas.edu/maps/index.html`.

World Bank Open Data

The World Bank is not without critics. As an international financial institution, it has provided loans and grants to developing nations, but not everyone likes the strings that can be attached. Nonetheless, even its detractors can admit that they freely provide a wealth of world data useful to the average and above average person (such as the readers of this book). This data was used for the world population and demographic maps in Chapter 11.

Twenty different data categories are provided (e.g., gender, trade, climate change), and within those categories you can find a fascinating array of information. In just the Education category there are 41 different data sets. What's more, you can look at this data in a table, as a chart, or as a map — with most data going back to 1990.

World Bank Open Data is available at https://data.worldbank.org.

World Factbook

Clandestine intelligence gathering is a key aspect of spy craft, but *The World Factbook*, produced by the U. S. Central Intelligence Agency, has all its information right out in the open.

It would be hard to find more basic information about individual countries elsewhere. The factbook has world and regional maps, plus detailed descriptions of country landscapes, their government and economy, and even audio files of languages spoken.

Curious about the Chilean national anthem? Listen here! You can even spend time exploring territories and countries you've never heard of (Bouvet Island? Akrotiri? Eswatini?). A host of demographic information is available, with population pyramid graphics being among the most useful.

The World Factbook is available at https://www.cia.gov/the-world-factbook.

Your State's Geographic Alliance

If you are interested in geography education in your state — particularly at the K-12 level — then you should become familiar with your state's geographic alliance. The alliance network is a product of The National Geographic Society's

Geography Education Foundation and was created in the late 1980s to help resuscitate the teaching of geography in the nation's schools.

NGS stopped providing direct financial support a few years ago, but many states still have active groups funded locally. The Alliances sponsor and conduct workshops and teacher-training institutes, prepare and disseminate lesson plans and other education materials, and publish newsletters — all focusing on geography as it pertains to specific curricula in the individual states.

You can check on the status of your state alliance through the National Council for Geographic Education at `https://ncge.org/connecting-educators/alliances`.

Index

A

AAG (American Association of Geographers), 345–346
abrasion, 105
absolute location, 32
accessibility, as a location factor, 275–278
acid precipitation, 335–339
acquiring geographic information, 20
Acropolis sites, 308
"act of God," 250
adaptation, 222–223
aerial photography, 75
Afghanistan, 194
Africa, 194
African lion case study, 21–26
Age of Discovery, 206
agriculture, mechanization of, 309
agronomist, 225
AGS (American Geographical Society), 346
air quality, 298–299
Al-Azizia, Libya, 169
Albany (New York), 308
Alliance of Small Island States (AOSIS), 129
altitude, 145–148
altitudinal life zones, 176
Amazon Basin, 228, 249
Amazonia, 249
American Association of Geographers (AAG), 345–346
American Geographical Society (AGS), 346
American High Plains, 117

amount, of flowing water, 110
analyzing geographic information, 20
Andes, 145
The Annals of the American Association of Geographers, 345
answering geographic questions, 21
Antarctic Circle, 142, 144
Antarctica, 116, 122, 128, 178
anthracite, 291
Anthropocene, 325–326
anticlines, 89
AOSIS (Alliance of Small Island States), 129
Appalachian Mountains, 90, 104
aquatic food chains, 329
aquatic impacts, 336–337
aquifers, 132
Arabian Peninsula, 249
Aral Sea, 123
Arctic Circle, 142, 144
The Arctic, 142
area, flat maps and, 48
Area, The Geographical Journal, 349
area source, 326
area specialist, 351–352
area symbols, 70–71
Argentina, 128, 249
Arkansas, 250
asking geographic questions, 20
Atacama Desert, 155
Athens, Greece, 308
Atlantic City, 271

Atlantic Coast boundary, 243
atlas, 54
atmosphere, 145, 325
atmospheric pressure, 154–157
Augusta, GA, 308
Austin, Texas, 281
Australia, 67, 194
Austria, 209
Author, in TALDOGS acronym, 60
automobile ownership, 311
average life expectancy, 190
azimuthal (planar) family, 52

B

barrier effects
  about, 226–227
  physical barriers, 227–229
  religion and, 232
  social barriers, 229–230
basin, 124
bathymetry, 76
Bay of Bengal, 122
bays, 122–123
beaches, eroding, 113–114
Bermuda Triangle, 357–358
Bhopal, 316
biodiversity loss, 333
biogas, 294
biomass conversion, 294–295
biosphere, 325
birth rate, 187
births, 187–189
Bitterroot Mountains, 243
bituminous, 291
BMW, 269–270

Bogota, 176
Bolivia, 253
boundaries
  about, 239–240
  boundary lines, 241–244
  consequences of, 245–256
  drawing/re-drawing, 240–241
  electoral district, 256–260
  ethnic, 242–243, 245–248
  functional disputes, 250–251
  geometric, 244
  land-locked states, 252–254
  multi-nation states, 245–246
  multi-state nations, 246–247
  natural (physical), 243–244
  positional disputes, 248–250
  resource disputes, 251–252
  shape of, 254–256
  size of, 254–256
  state-less nations, 247–248
boundary lines, 241–244
boutiques, 273
brain drain, 211
brain gain, 211
Brazil, 294
Buddhism, 231, 232
building codes, 95
Bundles, 139–140

C

California, 169
camanchaca, 135
Canada, migration and, 209
Canaries Current, 152
Canton Glarus, Switzerland, 225

carbonation, 106
cardinal directions, 60
career paths, 351–355
carrying capacity, 198
Cartagena, Colombia, 176
cartogram, 67
cartographer, 42
cartography, 42
Casablanca, Monaco, 152, 154
Cascade Range, 147
case studies, 21–28
Caspian Sea, 123
categorizing
  economic activities, 264–267
  tectonic processes, 102
CBD (central business district), 313–314
census-taking, 201
central business district (CBD), 313–314
channelized migration, 210
Charleston, SC, 152
Cheat Sheet (website), 4
cheese production, 272
chemical weathering, 106–108
Cherrapunji, India, 158
Chesapeake Bay, 323–324, 326, 327
Chile, 154, 249, 253, 255
China, 128, 183, 209, 274
Chinese Exclusion Act (1882), 208
Choropleth maps, 70, 71
Christianity, 231, 233
Churchill, Manitoba, 140, 152
cities
  about, 312–313
  central business district (CBD), 313–314
  downtown, 317–320

finding sites for, 306–308
  residential areas, 314–317
Classic Revival, 237
classifying climates, 160–161
clear-cutting, 326
cliffs, eroding, 113
climate change, 339–341
climate makers, mountains as, 88
climates
  about, 137–138
  altitude, 145–148
  atmospheric pressure, 154–157
  Bundles, 139–140
  classifying, 160–161
  in CLORPT acronym, 107
  coastal deserts, 154
  continentality, 148–150
  defined, 138
  desert, 168
  dry, 167–171
  factors causing, 138
  humid, 162–167
  humid mesothermal, 172–174
  humid microthermal, 174–176
  as a location factor, 280–281
  ocean currents, 150–153
  polar, 176–178
  rain, 140, 158
  seasons, 141–144
  snow, 140
  solar energy, 139
  tropical, 162–167
  vegetation and, 159–178
  wind, 150
climatology, 160

CLORPT acronym, 107–108
cloud catching, 134–135
clustering, dispersion vs., 184
coal, 290–291
Coast Range, 147
coastal deposition, 114
coastal deserts, 154
coastal plain setting, 111
coastlines, as natural (physical) boundaries, 243
Cold Canadian air, 358
Colombia, 176
colonies, 206
color, on maps, 66–67
Columbia, SC, 308
commercial economies, 268–269
commuter railways, 310
compact states, 254
competitive bidding, 314
complementary companies, 269
concrete production, 274
condensation, 131
confluences, 307
conformality, 56
conical family, 52
contagious expansion diffusion, 224–225, 235
containerization, 277–278
contamination, 132
Contiguous Zone, 127
continental drift, 85
continental glaciers, 116–117
continental shelves, 124–125
continentality, 148–150
continents, 84–86, 359
contour lines, 66

Cook, James (Captain), 205–206
corn belt, 175
cornucopians, 199–200
cost
  flat maps and, 51
  globes and, 50
  of labor, as a location factor, 274–275
  of land/rent, as a location factor, 279
  of trade, 254
Costa Rica, 176
country makers, mountains as, 88
creating maps, 77
crust
  about, 81–82
  categorizing tectonic processes, 102
  continents, 84–86
  earthquakes, 92–98
  Earth's interior, 82–84
  folding, 88–89
  Gondwanaland, 85
  magma, 86
  mountains, 88–91
  subducting plates, 98–102
  Theory of Plate Tectonics, 86–87
  Wegener, Alfred, 85
Cuba, 209
cultivation, shifting, 164–165
cultural convergence, 238
cultural diffusion
  about, 223–224
  contagious expansion diffusion, 224–225
  hierarchical diffusion, 226
  religion and, 231
  relocation diffusion, 224
cultural divergence, 238

cultural diversity, 221
cultural geography
  about, 219–220
  adaptation, 222–223
  barrier effects, 226–230
  contagious expansion diffusion, 224–225
  creating cultures, 237–238
  cultural diffusion, 223–226
  cultural diversity, 221
  hierarchical diffusion, 226
  isolation, 222
  language, 234–237
  number of cultures, 220–223
  physical barriers, 227–229
  religion, 230–234
  relocation diffusion, 224
  social barriers, 229–230
cultural landscape, 221
culture
  central role of, 284–285
  changing, 285–286
  earthquakes and, 94–95
culture makers, mountains as, 88
currents
  about, 113–114
  pollution and, 327–328
cylindrical family, 52

D

DailyGeo, 346
data storage
  flat maps and, 51
  globes and, 50
Date, in TALDOGS acronym, 60
Dead Sea, 123–124
death rate, 187
Death Valley, California, 169
deaths, 187–189
decentralization, 317
decomposers, 330
Deepwater Horizon oil rig, 328, 332
defensive sites, 308
deforestation, 333, 341
degradation, 326
democratic republics, 360
demographic transition model, 189–193
demography, 181
dependency ratio, 195
dependents, population and, 195
depletion, 133, 326
deposition, 109
desert climate, 168
desertification, 170
deserts, as a physical barrier, 228
detail
  flat maps and, 51
  globes and, 50
diffusing languages, 235
direction, flat maps and, 47
dispersion, clustering *vs.*, 184
dissolution, 106
distance
  flat maps and, 47
  scale and, 61–62
domestic migration, 210
Dominican Republic, 209
dots, 68
downtown, 317–320
drawing boundaries, 240–241
drip irrigation, 134–135
drought mitigation, 171
drought-maker, 155

dry climates, 167–171
Dust Bowl, 117
dust domes, 321

E

Early Expanding stage, of demographic transition model, 191–192
Earth
  comparing at different scales, 62–65
  interior of, 82–84
  rotation of, 42
  shape of, 42
earthquakes
  about, 92
  dangers of, 94–96
  humans and, 93–94
  magnitude of, 96–98
  New Madrid, 93
  San Andreas fault, 92–93
  tsunamis, 95–96, 97
Eastern Hemisphere, 38
ecology, 325
economic activities, categorizing, 264–267
economic development, 303–304
economic geography
  about, 263–264
  categorizing economic activity, 264–267
  economic systems, 267–269
  location factors, 269–281
economic systems
  about, 267
  commercial economies, 268–269
  subsistence economies, 267–268
economy, changing, 309–310
Ecuador, 145
edge cities, 318
educator, 352
EEZ (Exclusive Economic Zone), 127
El Niño, 152–153
El Salvador, 209
electoral district boundaries, 256–260
electric trolleys, 310
elongated states, 255
empowerment zones, 280
endangered rainforests, 163–164
endogenous forces, 88
enterprise zones, 280
environment
  about, 17
  defined, 325
environmental geography
  about, 323–325
  basics of, 325–326
  future challenges, 342
  global issues, 332–341
  pollution, 326–331
  role of attitudes toward nature, 332
  terminology for, 236
environmental issues, 320–321
environmental manager, 352
equator, 34–35
equinoxes, 142, 143
Eratosthenes (scholar), 10
eroding beaches, 113–114
eroding cliffs, 113
erosion, 108
"Essay on the Principle of Population" (Malthus), 197
ethnic boundaries, 242–243, 245–248
ethnic neighborhoods, 315–316

Ethnic Option, 241, 242
ethnicity, as a social barrier, 230
Eurasian Plate, 91
evapotranspiration, 131
Exclusive Economic Zones (EEZ), 127, 251–252

F

Falkland Island, 128
Fall Line, 308
fall season, 143
Fall Zone, 308
fallout, 331
fallow, 164
false color, 75
families, of projections, 52–53
farming, mechanization of, 309
fault lines, 90–91
faulting, 90
Fender, 275
fertile soil, 107
field of view
  flat maps and, 51
  globes and, 49
films, 24
fissures, 86
flash points, 233
Flat Earth Society, 360–361
flat maps, 44
floodplains, 111–112
flow lines, 69
flowing water, 110–114
folding crust, 88–89
food chains, 329–331
food surpluses, distribution of, 200
foodstuffs, forbidden and favored, 233–234
Foote, Eunice, 339
foothills setting, 111
foreign markets, inhibited access to, 252–253
foreign relations, 253–254
forests, as a physical barrier, 227–228
Former Yugoslavia, 233
fossil fuels, 289, 294, 298–299, 335
fragmented states, 255
Fray Jorge National Park, 135
freeway, 311
fresh water, 129–135
friction, religion and, 233
friction of resistance, 91
frigid zones, 159
frost action, 105
fuel, 125, 311–312
Fukushima disaster, 331, 332
functional disputes, 250–251
furlong, 63

G

Galicia, 361–362
geocaching, 74–75, 366
geographic alliance, 369–370
geographic information, 20
Geographic Information Systems (GISs)
  about, 72–74, 365–366
  global positioning systems (GPS), 74–75
  remote sensing, 75–77
Geographic Information Systems (GIS) technician, 353
Geographical Review, 346

geography
  about, 7–8, 19
  case studies, 21–28
  Geographic Advantage, 11–13
  history of, 8–9
  misconceptions of, 11
  as a modern discipline, 9–10
  online, 365–370
  Six Essential Elements, 13–18
  skills, 20–21
  uses of, 18
The Geography Teacher, 347
Geoguessr, 366–367
Geoinquiries, 367
geometric boundaries, 244
geomorphology, 81
Georges Bank dispute, 251–252
Georgia, 249, 279, 308
geothermal energy, 296–297
Germany, 209
Gerry, Elbridge, 257
gerrymandering, 257–259
ghettoes, 316–317
GIS (Geographic Information Systems) technicians, 353
GISs (Geographic Information Systems)
  about, 72–74, 365–366
  global positioning systems (GPS), 74–75
  remote sensing, 75–77
glacial till, 117
glacial troughs, 115
glaciers, 115–117
global grid, 32–35
global population, 182–183
global positioning systems (GPS), 74–75
global water supply, 120–124
globe, 49
GNI (Gross National Income) per capita, 193–194
Gobi Desert, 147
Gondwanaland, 85
Goode, J. Paul (doctor), 55
Goode's Interrupted Homolosine projection, 55
Google Earth, 367–368
Google Lit Trips, 368
gorged out, 116
graben, 90
gradational force
  about, 103–105
  changing the landscape, 109–118
  chemical weathering, 106–108
  defined, 82
  floodplain, 112
  flowing water, 110–114
  glaciers, 115–117
  gravity transfer, 109
  mass wasting, 108–109
  mechanical weathering, 105
  soil, 107–108
  wind, 117–118
grain storage, 200
grapes, 233–234
graticule, 39–40
Graticule Option, 242
gravity, 145
gravity transfer, 109
Gray's Atlas, 59
great circle, 38
Great Lakes, 124, 329
Great Lakes States, 243
Great Migration, 207

Great Salt Lake, 124
Great Victoria, 155
Greece, 308
green revolution, 199–200
greenhouse effect, 339–341
Greenland, 45, 46, 48, 116, 122, 128, 178
Grid, in TALDOGS acronym, 61
grid coordinates, 32, 33–34
grid pattern, 29–31
Gross National Income (GNI) per capita, 193–194
growing season, 172
growth
　of population, 185–186
　urban, 309–312
Guayaquil, 145
Gulf of Oman, 122
Gulf Stream, 151–152
gulfs, 122–123
Gullah, 229

H

Haiti, 209
halo effect, 269–270
hammada, 168
hanging valleys, 115
Hanson, Susan (geographer), 12
Harbor Hill Moraine, 117
Hawaiian "hot spot," 100
Hawaiian Islands, 205–206
head of navigation, 308
health services planner, 353
Hetch Hetchy Aqueduct, 132
hierarchical diffusion, 226, 235
High Stationary stage, of demographic transition model, 190–191
Highlands Climate, 176

highlands setting, 110–111
high-pressure system, 154–155
The High Seas (International Waters), 127
The Himalayan Mountains, 90–91, 147, 232
Hinduism, 232
Hipparchus (librarian), 32–35, 36, 40
historic events, as a social barrier, 230
hog operations, 272
home mortgage deductibility, 312
Homo sapiens, 240
Hong Kong, 182
hot pursuit, 127
hot-cold fluctuation, 105
hub-and-spoke networks, 276–277
human systems, 16–17
humans, earthquakes and, 93–94
Humboldt Current, 154
humid climates, 162–167
humid continental region, 174, 175
humid mesothermal climate, 172–174
humid microthermal climate, 174–176
humid subtropical climate, 172
humidity, 146
Hungary, 209
hunters and gatherers, 267
hydroelectric power (HEP), 297–298
hydrological cycle, 129–130
hydrosphere, 325

I

Ice Age, 128, 204–205
ice caps, 121–122, 178
icons
　explained, 3–4
　nominal, 68
　ordinal, 68

Ile de la Cite, 308
India, 157, 158, 183, 209
Indian Ocean, 122, 129, 147
Indian Plate, 91
Indonesia, 12–13, 255
Indus River Valley, 305–306
Industrial Revolution, 340
infant mortality, 190, 194
infiltration, 131, 132–133
infrared photography, 75
inner-city suburbs, 315
intensive subsistence agriculture, 268
International Meridian Conference, 35
Interstate Highway System, 311
inter-tropical convergence zone (ITCZ), 155, 156
intervening opportunity, 271
involuntary migration, 207
Iraq, 251
Ireland, 209
Irish language, 236
Islam, 231, 233
Islamic culture, 221
isolation, 222
isolines, 69
isotherm, 66
Israel, 233
Italy, 209
ITCZ (inter-tropical convergence zone), 155, 156

J

Jakarta, 12–13
Jamaica, 209
Jamestown, Virginia, 128
Japan, 194

Journal of Geography, 347
Judaism, 233
jurisdictional spillover, 318–319

K

Kalahari Desert, 155
Kashmir, 233, 248
Kentucky, 250, 308
Kidsdata, 348
Köppen, Vladimir (geographer), 160
Kremer, Gerhard, 53
Kurds, 248
Kuwait, 251

L

La Niña, 152–153
labor, cost of, 274–275
lahars, 101–102
lakes, 123–124
Lambert, Johann Heinrich (physicist), 56
Lambert Azimuthal Equal Area projection, 56
Lambert Conformal Conic projection, 56
land, cost of, as a location factor, 279
Land of the Midnight Sun, 361
Landform Option, 241
land-locked states, 252–254
language
  about, 234
  diffusing, 235
  physical geography and, 236
  as a social barrier, 230
  toponymy and, 236–237
lapse rate, 146
large-scale maps, 65

Late Expanding stage, of demographic transition model, 192
latitude
  about, 29–30, 36–38
  special lines of, 141–142, 144
latitude lines, 34
latosols, 164
law, letter of the, 259–260
"The Law of the Sea," 126–128
leaching, 164
League of Nations, 126
leeward slope, 147–148
Legend, in TALDOGS acronym, 60
letter of the law, 259–260
Libya, 169
LIDAR (Light Detection And Ranging), 76
Light Detection And Ranging (LIDAR), 76
lignite, 291
line symbols, 68–70
lines
  flow, 69
  isolines, 69
  nominal, 69
  ordinal, 69
lines of latitude, 141–142, 144
linguistic refuge, 236
liquefied natural gas, 291
lithosphere, 325
livelihood, opportunity of, 183–184
location analyst, 353–354
location factors
  about, 269–271
  accessibility, 275–278
  climate, 280–281
  cost of labor, 274–275
  cost of land/rent, 279
  proximity to markets, 272–274
  proximity to raw materials, 272
  taxes, 279–280
locations
  about, 29, 32
  absolute, 32
  case study on, 26–28
  equator, 34–35
  global grid, 32–35
  graticule, 39–40
  grid pattern, 29–31
  latitude, 29–30, 36–38
  longitude, 29–30, 38
  minutes, 40
  prime meridian, 35
  relative, 31
  seconds, 40
loess, 117–118
logarithmic numbers, 96
longitude, 29–30, 38
longitude lines, 34
Louisiana, 276–277
Louisville, Kentucky, 308
Lourdes, 232
Low Stationary stage, of demographic transition model, 193
low-cost fuel, 311–312
The Lower Huang Ho (Yellow River) Region, 305–306
low-pressure system, 154–155

M

magma, 86
magnitude, of earthquakes, 96–98
Maine, 150
malaria, 166

The Maldives, 129
Mali, 244
Malthus, Thomas R. (economist)
  "Essay on the Principle of Population," 197
Manaus, 140
Manitoba, 140, 152
mapping
  about, 41–42
  advantages of flat maps, 50–51
  applied geography, 46
  area, 48
  atlas, 54
  direction, 47
  disadvantages of globes, 49–50
  distance, 47
  earth's shape, 42
  families, 52–53
  flat maps, 44–48
  globe, 49
  Goode's Interrupted Homolosine projection, 55
  Lambert Azimuthal Equal Area projection, 56
  Lambert Conformal Conic projection, 56
  Mercator projection, 53–54
  Peters projection, 57–58
  projections, 42–44, 51–56
  Robinson projection, 55–56
  shape, 48
  Singapore, 45–47
maps
  Choropleth, 70, 71
  creating, 77
  defined, 42
maquiladoras, 275
marginal land, 170
marine west coast climate, 173–174
market analyst, 354
markets, proximity to, 272–274
mass wasting, 108–109
Massachusetts, 257
material impacts, 337
meander loops, 308
Mecca, 232
mechanical weathering, 105
mechanization, of agriculture and farming, 309
Mediterranean climate, 172–173
Memphis, Tennessee, 276–277
mental maps, 214–216
Mercalli, Giuseppe, 97
Mercalli Scale, 97–98
Mercator, Gerardus, 53
Mercator Projection, 47, 53–54
Mesoamerica, 305–306
Mesopotamia, 42, 305–306
metamorphic rock, 90
Mexico, 209, 275
Mexico City, 176
Mid-Atlantic Ridge, 86, 87
migration
  about, 203
  brain drain, 211
  channelized, 210
  choosing, 207–214
  domestic, 210
  history of, 204–207
  image problems, 216
  involuntary, 207
  mental maps, 214–216
  relocating within United States, 211–214

tourism, 216–217
trade and, 328–329
to United States, 208–210
voluntary, 207
migration fields, 206
Milankovitch Cycles, 144
mile, 63
milia passum, 63
mineral resources, 293–294
Minnesota, 308
minutes, 40
Mississippi River, 227, 250, 307
Missouri River, 307
mitigating effect, 149
modifying effect, 149
Monaco, 152, 154, 254
monsoons, 155–158, 165
moraine, 115
Mount Kilimanjaro, 176
Mount Mitchell, 104
mountain glaciers, 115
mountains
  about, 88–91
  as a physical barrier, 228
movies, 24
Mt. Everest, 91
Mt. St. Helens, 101
multi-nation states, 245–246
multi-state nations, 246–247
Mumbai, India, 157
Myanmar, 256

N

Namibia, 228, 256
National Aeronautics and Space Administration (NASA), 346–347
National Council for Geographic Education (NCGE), 347, 369–370
National Geographic Society (NGS), 347
National Oceanic and Atmospheric Administration (NOAA), 348
national security, oceans and, 126
natural boundaries, 243–244
natural gas, 125, 291–293
natural habitat, case study on, 21–26
natural hazards, 102
natural increase, in population, 187
natural resources
  about, 283–284
  central role of culture, 284–285
  coal, 290–291
  consequences of using, 298–299
  culture change, 285–286
  geothermal energy, 296–297
  hydroelectric power, 297–298
  life spans, 288–298
  mineral resources, 293–294
  natural gas, 291–293
  non-renewable, 289–294
  perennial, 295–298
  petroleum, 289–290
  power and, 286–287
  renewable, 294–295
  solar energy, 295
  wealth and, 287–288
  wind energy, 295–296
natural vegetation, 161
nautical mile, 127
NCGE (National Council for Geographic Education), 347, 369–370
neo-Malthusians, 197–199
Nevada, 211

Nevada del Ruiz volcano, 102
New Glarus, Wisconsin, 225
New Guinea, 228
New Jersey, 308
New Jersey-Pennsylvania border, 243
New Madrid, 93, 250
New Orleans, Louisiana, 276–277
New York City, 279
New Zealand, 193–194
The New Cartography (Peters), 57
NGS (National Geographic Society), 347
Nigeria, 245
Nile Valley, 305–306
nitrogen dioxide, 335–337
NOAA (National Oceanic and Atmospheric Administration), 348
nomadic herders, 268
nominal icons, 68
nominal lines, 69
nominal symbols, 70, 71
non-dependents, population and, 195
non-photographic imagery, 76–77
non-point source, 326
non-renewable resources
  about, 289
  coal, 290–291
  mineral resources, 293–294
  natural gas, 291–293
  petroleum, 289–290
North America, 116
North American Plate, 92
North Atlantic Current, 151–152
North Carolina, 308
North Pole, 35, 144
Northern Hemisphere, 36, 141, 142, 176
Northern Ireland, 233
The Northern Polar Region, 142

Norway, 255
nuclear power, 292

O

Obama, Barack (president), 252
ocean acidification, 334
oceans
  about, 122–123, 124–129
  currents in, 150–153
  ice ages, 204–205
  as a physical barrier, 227
office parks, 318
official language, 235
Ogallala Aquifer, 133
Oman, 249
"one person, one vote," 256–257
online geography, 365–370
opportunity zones, 280
ordinal icons, 68
ordinal lines, 69
Organism, in CLORPT acronym, 107
organizations, 345–350
organizing geographic information, 20
Orientation, in TALDOGS acronym, 60
orographic, 147
Ottelius, Abraham, 84
overfishing, 335
overpopulation
  about, 196–197
  carrying capacity and, 198
  neo-Malthusians, 197–199

P

Pacific Plate, 87, 92, 100
paleoclimatology, 348
Parent material, in CLORPT acronym, 108

Paris, 277
particulate matter, 108
peat, 291
Pennsylvania, 224, 256, 275, 289, 307
percent of population, 194
perennial resources
  geothermal energy, 296–297
  hydroelectric power (HEP), 297–298
  solar energy, 295
  wind energy, 295–296
perforated states, 256
permafrost, 178
permeable barrier, 227
The Perry-Castañeda Library Map Collection, 368
Persian Gulf, 287
Peru, 154
Peru Current, 154
Peters, Arno
  The New Cartography, 57
Peters projection, 57–58
petroleum, 125, 286–287, 289–290
Philippines, 209
photovoltaic cells, 295
physical barriers, 227–229
physical boundaries, 243–244
physical geography, language and, 236
physical systems, 16
physiography, 243
Pierre, South Dakota, 150
Pinnacles National Park, 92
Pittsburgh, Pennsylvania, 307
pizza, 273
places, 15
places of worship, 232
plant life, 125

plate tectonics, 91
point symbols, 68
polar climate, 176–178
police power, oceans and, 126
pollution, 326–331
population geography
  about, 181–183
  births, 187–189
  carrying capacity, 198
  census-taking, 201
  changes in, 187–196
  cornucopians, 199–200
  deaths, 187–189
  demographic transition model, 189–193
  dispersion *versus* clustering, 184
  growth of, 185–186
  infant mortality, 194–196
  neo-Malthusians, 197–199
  opportunity of livelihood, 183–184
  overpopulation, 196–200, 198
  percent of, 194
  pyramids, 195
  urban growth, 185
  wealth, 193–194
Population Reference Bureau (PRB), 348
Portland, Maine, 150
positional disputes, 248–250
power, resources and, 286–287
PRB (Population Reference Bureau), 348
precipitation, 131
pre-freeway, 311
pre-mechanization, 310
pressure belts, 155
primary activities, 264–265, 266–267
primary consumers, 330
primary producers, 329–330

primate cities, 304
prime meridian, 35
The Professional Geographer, 345
projections
  about, 43–44, 51–52
  Mercator, 53–54
proportional symbols, 68
prorupted states, 256
proselytization, 231
protected harbors, 307
Protestant, 233
Puerto Rico, 217
push factor, 207

Q

quaternary activities, 266–267
Quebec City, 308
Quito, 145

R

rabbits, 329
race, as a social barrier, 230
radar imaging, 76
rain, 140, 158, 361–362
rain shadows, 168
Rainforest, Tropical, 163–165
rainmaker, 155
rainshadow, 147
Raleigh, NC, 308
ranching, 268
raw materials, proximity to, 272
red lining, 317
re-drawing boundaries, 240–241
refuse, 321
regions, 15–16
relative location, 31

Relief, in CLORPT acronym, 107
religion
  about, 230
  barrier effects and, 232
  characteristics, 232–234
  cultural diffusion and, 231
  as a social barrier, 230
relocation diffusion, 224
Remember icon, 3
remote sensing, 75–77, 171
remote sensing analyst, 354–355
renewable resources, 294–295
rent, cost of, as a location factor, 279
rent gradient, 314
representative fraction (RF), 62
residential areas, 314–317
resource disputes, 251–252
resource ownership, oceans and, 126
RF (representative fraction), 62
RGS (Royal Geographical Society), 349
Rhode Island, 186
Richmond, VA, 308
Richter, Charles F., 96
Richter Scale, 96–97
ridgelines, as natural (physical) boundaries, 243
right of innocent passage, 127, 253
"The Ring of Fire," 98–99
river flow, 125
rivers
  about, 110–113
  as natural (physical) boundaries, 243
Robinson, Arthur H. (cartographer), 55
Robinson projection, 55–56
Roman Catholic, 233
root action, 105

rotation, of Earth, 42
Royal Geographical Society (RGS), 349
Royal Greenwich Observatory, 35
Rumaila Oil Field dispute, 251
run-off, 131, 132
rural-to-urban migration, 309–310
Russia, 183, 209, 254
Russian Arctic, 177
rusting away, 106

S

sacred sites, 232–233
Sahara Desert, 171
Salton Sea, 123
San Andreas fault, 91, 92–93
San Francisco, 132
San Gabriel Mountains, 307
San Jose, Costa Rica, 176
Saudi Arabia, 227, 249
savanna, 165–167
scale
  comparing Earth at, 62–65
  distance and, 61–62
  in TALDOGS acronym, 61
scale bar, 61
Sea Islands, 229
sea levels, 128–129
seas, 122–123
seasons, 141–144
secondary activities, 265, 266–267
secondary consumers, 330
seconds, 40
seismology, 92. *See also* earthquakes
semi-desert (steppe) climate, 169–171
Seven Seas, 362
sewage, 321

Seward, William H., 361
shading, on maps, 66–67
shape
  of Earth, 42
  flat maps and, 48
shifting cultivation, 164–165, 268
shopping malls, suburban, 317–318
Sierra Nevada, 132
Sierra Nevada Mountains, 147
Singapore, 45–47, 182, 194, 277
Six Essential Elements, 13–18
skills, geography, 20–21
small-scale maps, 65
smart phones, 274
smog, 321
Snider-Pellegrini, Antonio, 85
snow, 140
social barriers, 229–230
soil
  about, 107–108
  degradation of, 333–334
  infertility of, 164
  poor, 164
soil creep, 109
soil loss, 299
solar energy, 131, 139, 140, 154–155, 295
Somalis, 246–247
South Africa, 256
South Carolina, 152, 308
South China Sea, 128
South Dakota, 150
South Pole, 35
Southern Andes, 249
Southern Hemisphere, 36, 141, 143
Southern Philippines, 233
The Southern Polar Region, 142

Spain, 361–362
spatial systems
  about, 14–15, 59–60
  distortion, 67
  geographic information systems (GISs), 72–77
  making maps, 77
  scale, 61–65
  symbols, 67–71
  TALDOGS acronym, 60–61
  technology, 72
  topography, 65–67
"sponge effect," 112
spot height, 65
spring season, 143
St. Paul., MN, 308
Starbucks, 277
state-less nations, 247–248
statement of scale, 62
steel production, 272
steppe climate, 169–171
straits, oceans and, 126
streams, 110–113
strip mining, 299
subarctic climactic region, 174
subarctic climate, 175
subducting plates
  about, 98
  "The Ring of Fire," 98–99
  subduction, 99–102
subduction, 87, 99–102
subsistence economies, 267–268
suburban shopping malls, 317–318
suburbs, inner-city, 315
Suess, Eduard, 85
sulfur dioxide, 335–337
summer season, 143, 149–150
sun angles, 140
surface cover, of flowing water, 110
Switzerland, 225
symbols
  about, 67
  area, 70–71
  line, 68–70
  nominal, 70, 71
  point, 68
  proportional, 68
synclines, 89

T

Tahiti, 221
Taiwan, 274
Takla Makan Desert, 147
TALDOGS acronym, 60–61
Tampa Bay, 122
tarn, 115
taxes, as a location factor, 279–280
Technical Stuff icon, 3
technology, maps and, 72
tectonic processes, categorizing, 102
temperature zones, 159
temporal spacing, 99
Tennessee, 249, 276–277
terrestrial food chains, 329
terrestrial impacts, 336
territorial sea, 127
tertiary activities, 265–267
tertiary consumers, 330
Texas, 211, 250, 281
textile manufacturing, climate and, 281
Thailand, 221, 256
Theory of Plate Tectonics, 86–87

thermal imaging, 76
"thin air," 228
Three Gorges Dam, 298
tidal wave, 96
Tierra caliente, 176
Tierra helada, 176
Tierra templada, 176
Time, in CLORPT acronym, 108
Tip icon, 4
Title, in TALDOGS acronym, 60
topography, 65–67
toponymy, 236–237
torrid zone, 159
tourism, 216–217
trade
  cost of, 254
  migration and, 328–329
  oceans and, 126
Trans Alaska Pipeline, 178
Transactions of The Institute of British Geographers, 349
trans-boundary pollution, 338
transform fault, 91, 92
transportation
  about, 200
  changing means of, 310–311
transportation planner, 355
Trenton, NJ, 308
trilateration, 74
Tropic of Cancer, 141–142, 144, 151, 162
Tropic of Capricorn, 141–142, 144, 151, 162
tropical climates, 162–167
tropical monsoon, 165
tropical paradise, 362–363
Tropical Rainforest, 163–165
The Tropics, 141

tsunamis, 95–96, 97
tundra climate
  about, 176–177
  as a physical barrier, 229
turmeric, 233

U

UNCLOS *(United Nations Convention on the Law of the Sea),* 126
United Kingdom, 128, 209
United Nations, 126
United Nations Convention on the Law of the Sea (UNCLOS), 126
United States
  as a fragmented state, 255
  migrating to, 208–210
  relocating within, 211–214
United States Census Bureau, 349
United States Geological Survey (USGS), 350
United States-Canada dispute, 250–251
United States-Mexico dispute, 250
uranium, 292
urban area (UA), 303
urban geography
  about, 301–303
  cities, 312–317
  downtown, 317–320
  environmental issues, 320–321
  finding sites for cities, 306–308
  global perspective, 303–305
  urban growth, 309–312
  urban hearth areas, 305–306
urban growth, 185, 309–312
urban hearth areas, 305–306
urban heat island, 320
urban planner, 355

Index   389

Uruguay, 254
USGS (United States Geological Survey), 350

V
Varanasi, 232
vegetation, climates and, 159–178
velocity, of flowing water, 110
verbal scale, 62
vernacular language, 235
vertical mixing, 125
vertical rays, 139, 141, 142
vertical zonation, 176
Via Dolorosa, 232
Vietnam, 209
Virginia, 128, 211, 308
volcanos, 98–102
voluntary migration, 207
voting blocs, 258–259

W
Wailing Wall, 232
War of the Pacific, 253
Washington (state), 211
Washington, D.C., 308
water
  about, 110, 119
  currents, 113–114
  fresh, 129–135
  global water supply, 120–124
  oceans, 124–129
  rivers, 110–113
  streams, 110–113
  waves, 113–114
water cycle
  about, 129–130
  pollution and, 327
watershed management, 337
waves, 113–114
wealth
  about, 193–194
  earthquakes and, 94–95
  resources and, 287–288
weathering, 105–108, 138
websites
  American Association of Geographers (AAG), 346
  American Geographical Society (AGS), 346
  Cheat Sheet, 4
  Geoguessr, 367
  Geoinquiries, 367
  Google Earth, 368
  Google Lit Trips, 368
  National Aeronautics and Space Administration (NASA), 347
  National Council for Geographic Education (NCGE), 347, 370
  National Geographic Society (NGS), 347
  National Oceanic and Atmospheric Administration (NOAA), 348
  The Perry-Castañeda Library Map Collection, 368
  Population Reference Bureau (PRB), 348
  Royal Geographical Society (RGS), 349
  United States Census Bureau, 349
  United States Geological Survey (USGS), 350
  World Bank Open Data, 369
  The World Factbook, 369

Wegener, Alfred (geographer), 85
Welsh language, 236
West Africa, 232
West Virginia, 211
Western Hemisphere, 38, 177
wind
  about, 117–118, 150
  pollution and, 327
wind energy, 295–296
windward slope, 147–148
winter season, 143, 149–150
Wisconsin, 225

World Bank Open Data, 369
World Population Data Sheet, 348
The World Factbook, 369

X

xerophytes, 168

Y

Yangtze River, 298
Yemen, 249
Yi-Fu Tuan (geographer), 12
Yosemite, 132

About the Author

Jerry T. Mitchell is professor and chair of the department of Geography at the University of South Carolina. He holds a BS in History and MA in Geography from Towson University, and a PhD in Geography from the University of South Carolina where he returned to the faculty in 2004 after teaching for several years in Pennsylvania. His excitement for geography stems from a childhood spent leafing through the World Book Encyclopedia and frequent family relocations as a result of his father's work.

Jerry's research has focused on environmental hazards and geography education. He was the coordinator of the South Carolina Geographic Alliance for 17 years, providing geography learning opportunities for more than 40,000 teachers and students. Additionally, he served as the editor of the *Journal of Geography* from 2010 to 2019, was president of the National Council for Geographic Education in 2020, and was awarded the 2022 Gilbert Grosvenor Honors in Geographic Education from the American Association of Geographers.

Jerry lives in Columbia, South Carolina with his wife, Heather. His two grown children are currently in graduate and undergraduate school, respectively, with the youngest majoring in geography. He enjoys international travel and counts taking students on study abroad courses to Chile among his favorite professional activities. He spends his free time training for triathlons and is a two-time Ironman finisher.

Dedication

For Felix.

(I was encouraged to pick my best.)

Author's Acknowledgments

How about we do this chronologically?

Thanks go to my parents for instilling a bit of curiosity in their kid and encouraging me to share my excitement learning about this world with others. To my father, especially: Thank you for the international travel opportunities early. Twain was right.

And then there are the teachers. Miss Jung is first, a whirlwind of sixth-grade geographic fun. Who knows where I would be if not crowned Best Cartographer in her class at Bowditch Middle School? Holly Myers-Jones bats second. This Towson University geographer told me to go get a PhD. Don't ever doubt where you might send someone with a bit of encouragement. Finally, there's Susan Cutter — doctoral advisor, counselor, colleague, and friend. If I am a dummy, well then what are you?

As to this project, Charlie Regan, Executive Director of the National Council for Geographic Education, fielded a phone call about this effort and said he knew the right guy to write this thing. I appreciate the nod, Charlie. Greg Tubach, Vicki Adang, and Michelle Hacker at Wiley kept me on task, as did Daniel Mersey at Word Mountain Creative Content. Grayson Morgan cleaned up after me as our technical editor. As you can read here, there's a whole team necessary to make a book like this happen. My last thanks here go to Charles Heatwole, the author of the first version of this book. He provided me with a rich base of material that was indispensable.

The stories interwoven with this book's content were made possible by the many students I have had the pleasure of teaching at Bloomsburg University and the University of South Carolina. The outstanding K-12 teachers that I've worked with are the source of many ideas, too. Thank you to all of you.

Finally, and most of all, thanks to my wife, Heather, for supporting everything I've tried to do for the past 30 years. Let's keep exploring together.

Publisher's Acknowledgments

Acquisitions Editor: Greg Tubach
Development Editor: Daniel Mersey
Technical Editor: Grayson Morgan
Managing Editor: Michelle Hacker

Production Editor: Saikarthick Kumarasamy
Cover Image: © rangizzz/Adobe Stock Photos

Leverage the power

Dummies is the global leader in the reference category and one of the most trusted and highly regarded brands in the world. No longer just focused on books, customers now have access to the dummies content they need in the format they want. Together we'll craft a solution that engages your customers, stands out from the competition, and helps you meet your goals.

Advertising & Sponsorships

Connect with an engaged audience on a powerful multimedia site, and position your message alongside expert how-to content. Dummies.com is a one-stop shop for free, online information and know-how curated by a team of experts.

- Targeted ads
- Video
- Email Marketing
- Microsites
- Sweepstakes sponsorship

20 MILLION PAGE VIEWS EVERY SINGLE MONTH

15 MILLION UNIQUE VISITORS PER MONTH

43% OF ALL VISITORS ACCESS THE SITE VIA THEIR MOBILE DEVICES

700,000 NEWSLETTER SUBSCRIPTIONS TO THE INBOXES OF **300,000** UNIQUE INDIVIDUALS EVERY WEEK

# of dummies

Custom Publishing

Reach a global audience in any language by creating a solution that will differentiate you from competitors, amplify your message, and encourage customers to make a buying decision.

- Apps
- Books
- eBooks
- Video
- Audio
- Webinars

Brand Licensing & Content

Leverage the strength of the world's most popular reference brand to reach new audiences and channels of distribution.

For more information, visit dummies.com/biz

PERSONAL ENRICHMENT

9781119187790
USA $26.00
CAN $31.99
UK £19.99

9781119179030
USA $21.99
CAN $25.99
UK £16.99

9781119293354
USA $24.99
CAN $29.99
UK £17.99

9781119293347
USA $22.99
CAN $27.99
UK £16.99

9781119310068
USA $22.99
CAN $27.99
UK £16.99

9781119235606
USA $24.99
CAN $29.99
UK £17.99

9781119251163
USA $24.99
CAN $29.99
UK £17.99

9781119235491
USA $26.99
CAN $31.99
UK £19.99

9781119279952
USA $24.99
CAN $29.99
UK £17.99

9781119283133
USA $24.99
CAN $29.99
UK £17.99

9781119287117
USA $24.99
CAN $29.99
UK £16.99

9781119130246
USA $22.99
CAN $27.99
UK £16.99

PROFESSIONAL DEVELOPMENT

9781119311041
USA $24.99
CAN $29.99
UK £17.99

9781119255796
USA $39.99
CAN $47.99
UK £27.99

9781119293439
USA $26.99
CAN $31.99
UK £19.99

9781119281467
USA $26.99
CAN $31.99
UK £19.99

9781119280651
USA $29.99
CAN $35.99
UK £21.99

9781119251132
USA $24.99
CAN $29.99
UK £17.99

9781119310563
USA $34.00
CAN $41.99
UK £24.99

9781119181705
USA $29.99
CAN $35.99
UK £21.99

9781119263593
USA $26.99
CAN $31.99
UK £19.99

9781119257769
USA $29.99
CAN $35.99
UK £21.99

9781119293477
USA $26.99
CAN $31.99
UK £19.99

9781119265313
USA $24.99
CAN $29.99
UK £17.99

9781119239314
USA $29.99
CAN $35.99
UK £21.99

9781119293323
USA $29.99
CAN $35.99
UK £21.99

dummies.com

dummies
A Wiley Brand